Spirit of Place

A Shearwater Book

BOOKS BY FREDERICK TURNER

Geronimo: His Own Story (Ed.)

The Viking Portable North American Indian Reader (Ed.)

Beyond Geography: The Western Spirit Against the Wilderness

*Remembering Song: Encounters with the
New Orleans Jazz Tradition*

Rediscovering America: John Muir in His Time and Ours

Of Chiles, Cacti & Fighting Cocks: Notes on the American West

Frederick Turner

SPIRIT

OF

PLACE

The Making of an
American Literary Landscape

ISLAND PRESS

Washington, D.C. □ *Covelo, California*

Originally published in hardcover in 1989 by Sierra Club Books, San Francisco.

Library of Congress Cataloging-in-Publication Data

Turner, Frederick, 1943–
Spirit of Place: the making of an American literary landscape / Frederick Turner.
p.   cm.
Originally published: San Francisco : Sierra Club Books, c1989.
"A Shearwater book."
Includes bibliographical references and index.
ISBN 1-55963-180-5 (pbk.)
1. American literature—History and criticism.   2. Authors, American—Homes
and haunts—United States.   3. Literary landmarks—United States.   4. United States
in literature.   5. Regionalism in literature.   6. Local color in literature.
7. Landscape in literature.   I. Title.
[PS141.T8   1992]

810.9'22–dc20

92-14384
CIP

Book design by Abigail Johnston

Printed on recycled, acid-free paper

Manufactured in the United States of America

10 9 8 7 6 5 4 3 2 1

For Thomas Berger

Contents

Preface

As its title announces, this is a book about the making of an American literary landscape. Though its essential subject matter is literature, in a larger way it has to do with the processes of invention and accommodation by means of which the newcomers at last took imaginative possession of the land long since theirs in name. Its heroes are the men and women who learned in loneliness and silence and deprivation how truly to see where on the American earth they were: to see their specific places in such full and luminous detail that a radically native art, a literature, arose from a Massachusetts pond, the upper Nebraska panhandle, a New Jersey mill town, a New Mexico Indian pueblo.

To tell this story I have used the techniques of the cultural historian and the biographer. Except for the opening chapter, which establishes the context, I have also used some of the techniques of the travel writer, for I have traveled to the places the writers made famous in their works and described them as they are now. My hope in making pilgrimages that took me from Maine to Louisiana to California was to suggest in another way the variety and richness of our writers' achievements and the specificity of their struggles to come to creative terms with the unsung spots in which history had put them. I have hoped, too, that my experiences in these special places would cast a reflective light on the writers' struggles there. It may now be a bit easier in the wake of Thoreau and Faulkner to see the literary potential in any place, however apparently unpromising. But it might be just as difficult now to *be* Thoreau and Faulkner in Concord and Oxford as it was then for them.

Before Thoreau there wasn't any such thing as an American literary landscape. And before his contemporary, Thomas Cole, there wasn't a tradition of American landscape painting. Evident as such facts may now seem to us, they are nonetheless curious when we begin to muse on them just a bit, for they tell us that confronted with a magnificent, virginal landscape the newcomers could make no significant artistic use of it for better than two hundred years. Instead, American energies were expended in obliterating as much as possible of that original landscape and transforming the more accessible portions of it into a recognizable facsimile of the Old World. By the time of the Romantic movement here, those like Thoreau and Cole who wished to celebrate American places were inheritors of a littered, occasionally even blasted landscape bearing scant resemblance to that encountered by the first-coming whites. Before American places had ever been memorialized in art many of them had been so altered that no one could remember what they had once been like, nor could any recall what had happened in them. Americans believed they lived in an ahistorical landscape, one without spirit and without life except that with which they would presently endow it, and so instead of poems and paintings celebrating their lands they would have for their monuments toll roads, treeless villages with the names of antiquity slapped incongruously across them, and stump fields.

We have many more toll roads now and far too many denuded places that were once the sites of our groves. But thanks to our writers we also have a national literary landscape, places made special, if not sacred, because they have been the inspiration of literature: Concord and Hannibal, George W. Cable's French Quarter in New Orleans, Faulkner's Oxford, Mari Sandoz's sandhills in the Nebraska panhandle. The list I have made for the purpose of this book is only suggestive, and the process continues.

It is not a process of interest only to writers and their readers. The existence of this literary landscape—the places and their monuments—fulfills a vital function in our culture. The function of art in any landscape, says J.B. Jackson, our most distinguished landscape historian, is to allow people to participate emotionally in their place of living. That surely was one of the functions of the place-rooted myths of old. A part of that emotional identification came through the remembering that the myths inspired, for the places of the myths were the visible reminders of that ancient past that lives on in the present, of the enduring and the transitory, of eternity and mortality, of stewardship. In a culture with so little regard for its lands and still less for the histories enacted in those lands—all the silent sacrifices sunk like blood into any place—these works in their places remind us to remember and to care. And the end of that remembering is to know again that we do not live in some moral and biological vacuum— and never have.

Acknowledgments

Special thanks to the following individuals who volunteered their time and expertise in reading all or portions of the manuscript: Elise R. Turner, Christopher Merrill, Caroline Sandoz Pifer, John Hay, Pauline Pearson, John H. Hicks, Paul Mariani, Lucy Hilgendorf, Leslie Meredith, and Lucile Adler. The mention of their names here, while it is prompted by a generous motive, also serves a more practical one in exempting them from complicity in whatever errors the author may have committed.

Thanks also to my many guides to the literary places spoken of here: Walter Harding, Glenn Starbird, John Lyng, Winston Lill, Dorothy Lill, Christopher Botsford, Jane Botsford, the late Allan Jaffe, John Scott, Mrs. Edmond J. Lebreton, Dr. Wilbur E. Menery, W. Kenneth Holditch, Richard Thomas, the late Margaret Pond Church, Father Angelico Chavez, Orlando Romero, Mark Sorensen, Jack Donald, Flora Sandoz, the late James Webb, Richard Howarth, Lisa Howarth, William Ferris and the staff of the Center for the Study of Southern Folklore at University of Mississippi in Oxford, Merle Pearson, Gene Rowlands, Roy Diaz, Lorraine Diaz, Joyce Conrow, Terry Sala, Tessa Ward, Tiffany Nunes, Dr. William E. Williams, Lee Marmon, and Leslie Marmon Silko. In two instances I have changed the names of helpful guides, respecting thus their privacy, as they requested.

For uncommon kindnesses my thanks to Leslie Meredith, Lynne Moor, and Jim Cohee.

ONE

HOW DIVINE THESE STUDIES: THE CHALLENGE OF THE UNSUNG LAND

ON MAY DAY, 1893—once a day of agricultural rites among European peoples—a huge throng of those transplanted Europeans who called themselves Americans gathered at Chicago to celebrate Columbus's discovery of the New World. There President Grover Cleveland pressed the ivory button of a solid gold telegrapher's key to set in whirring motion all the shafting, drive belts, turbines, and machines of the Columbian Exposition. Out on the lake the warship *Andrew Johnson* saluted with all its guns, and a band launched into the "Hallelujah Chorus."

The opening of this monstrous event had missed by a good half-year the anniversary of the Admiral's landfall, but even so it was generally agreed that Frederick Law Olmsted and the other planners and architects had done an outstanding job in so swiftly transforming the vacant swamp that was Jackson Park into the White City that greeted the hundreds of thousands who poured through the gates on opening day. Who cared, really, that the anniversary had been missed? Here was without doubt the most spectacular gathering of art, science, and technology within the memory of man. The sculptor Augustus St. Gaudens had surely been right in remarking that the Exposition represented the greatest gathering of artists since the fifteenth century. In Machinery Hall, the Transportation Building, and the Electricity Building were a stunning array of new gadgets and devices, all of them calculated to make daily living easier and better for a nation in which the eight-hour, five-day work week was still a rumor.

No one seemed to care, either, that the architecture of the White City was a stew of styles from Classical Greece, Imperial Rome, Renaissance

3

Italy, and Bourbon Paris; or that one of its contributors, a reluctant Louis
H. Sullivan, had declared it a disaster of sufficient magnitude to set the
cause of contemporary architecture back half a century. On opening night
when they turned on the miles of tiny electric lights strung throughout
the grounds, it looked like a fairy land, and in that glow the jarring congeries of styles was harmonized into a spell-binding grace. Here was a spectacle, wrote one visitor to his wife, enough to have made the "Queen of
Sheba and poor Solomon in all their glory feel sick with helpless envy."

Among the throng of that first day were four members of a Midwestern farm family, two sons in the vigor of manhood and their parents. They
made a touching group, these four, as they toured the Midway with its
villages of Chinese, Irish, and Egyptians, hurriedly visited the exhibition
halls, then sat out on the shore of one of the lagoons, listening to the voices
of the singers floating past in gondolas under those fairy lights. The sons
took turns pushing the wheelchair their mother rode in. The father, bent,
work-blunted, observed the spectacle from within a moody silence. To
his sons he was sadly altered from that hard-handed, martial man who'd
survived the Civil War and then had relentlessly driven himself and his
family through nearly three decades of pioneer farming in Wisconsin, Iowa,
and South Dakota. In this setting of a gaudy urban civilization he seemed
a species of peasant, suddenly old and uncertain.

As for their mother, the sons were forced to acknowledge that she
was now a semi-invalid, broken by her pioneering years. When she heard
those sweet singers on the darkened lagoon waters their voices had not
come to her like a reward, a belated benediction, but rather as a final blow.
It was too much for a woman whose life had been spent in the kitchens and
yards of farmhouses set out like crates beyond even the most modest of
Midwestern hamlets. She leaned her head against the arm of her stocky son
and murmured brokenly, "Take me home. I can't stand any more of it."

The son, Hamlin Garland, had been determined for some years now
that his mother would not die on a desolate South Dakota wheat farm,
would not go the way of so many pioneering women whose names were
unrecorded in any annals save those of country churches, whose silent
sacrifices could at best be inferred from blurred dates on the slanting stones
of those churchyards. Grandmother McClintock, his mother's mother, had
gone that way: after a lifetime of toil she had appeared suddenly one afternoon at the doorway of her prison/kitchen, her face frozen in a stroke, and
then had fallen dead on the threshold. So, after the Garland family had
retreated from the spectacle at Chicago, Hamlin Garland and his brother

had taken their parents up to New Salem, Wisconsin, to scout out a retirement home for them, one with shade trees, a flower bed, and within walking distance of the amenities of town living. But neither of the brothers harbored illusions about what money could buy their mother at this hour of her life. She had suffered too greatly to take real pleasure in anything again, and the comforts they could now supply would never restore her health and spirits.

Hamlin Garland saw New Salem for what it was, a halfway house, "a pleasant hospital," as he was to write, "into which many of the crippled, worn-out and white-haired farmers and their wives had come to rest for a little while on their way to the grave." Watching its citizens doze on their neat little lawns in the shade of maples, or totter on canes along the boardwalks, he was not deceived into imagining that here was some small, serene place removed from the monotonous privations of life beyond, out there on the great sweeping prairies. He knew those privations too well and knew, also, how dearly these townsfolk had purchased their much-belated ease.

Garland's own apprenticeship to pioneer farming life had begun at an early age in what he was to call the "horse-killing hills" of western Wisconsin. Though he was then still so short he had to lift his hands above his shoulders to reach the plow handles, at ten it was his turn to bear a hand in the routine twelve-hour workdays of the family's farm. Eight or nine miles he trudged on those spring mornings of planting, the mud clinging to his oversized boots, the close-matted sod repeatedly upsetting the share so that he was obliged over and again to yank the heavy plow backward and make a new start. Then, after the nooning, another eight or nine miles, and then the evening chores.

Harvesting wheat in the heat of late summer proved an even greater test of the boy's endurance, a punishing dawn-to-dusk race against time and the weather. Nor was there any real rest after the wheat had been gathered, for there was corn to harvest and shuck. In the 1870s as the corn fields grew in extent the task itself stretched out, too, until it lasted almost into winter, the Garland boys husking the ice-glazed ears with their frozen mittens.

Like many of their neighbors the Garland family was tempted onward, following the northwestern migration of the wheat belt. Though the conditions of their daily lives fostered a tough realism, the family still believed in the pot of gold resting where the sun set, and the belief kept them moving. From the Wisconsin hills they came into northeastern Iowa, a land they found fertile and open, a place where a family might plant itself and prosper. But they didn't prosper there—at least not as much as Dick Garland had hoped—and so they went on. At last, on the scoured plains of South Dakota the Garlands came to a halt, stranded. The dream of greater prosperity to be found somewhere ever westward ended for them there, though some of Belle Garland's folks would eventually push on all the way to California before they, too, gave up the dream. Here in South Dakota Dick Garland had to look about him and admit defeat: he wasn't ever going to get rich following the wheat, the sun, the retreating tribes, and the buffalo. He had begun his sojourn in Maine, but he was now too much of a westerner to think of going back to New England. So when his sons summoned him to Chicago and then up to New Salem, he was a tired man and willing to accept their scaled-down vision of the future.

Years before Chicago, Hamlin Garland himself had determined that farming life was not for him. If he could not at that point save his mother, he could at least save himself. He had resisted the remove from Iowa in 1881, then had reluctantly staked a claim in South Dakota only to give it up and forever turn his back on farming life. He would try anything else, he thought, and for a few years he did: traveling bookseller, store clerk, itinerant laborer, teacher, lecturer. Maybe it was an undistinguished existence, even at times a hand-to-mouth one, but at least he was often in daily contact with books and ideas—things he had come to love— instead of sweating out his days in the lonesome deeps of the fields.

Eking out a living as a lecturer in Boston in 1887, the young man experienced a strange sort of epiphany one afternoon when he chanced to hear the scrape of a coal shovel in the alley beneath his garret. That single, random sound brought back suddenly, vividly the sound his own shovel used to make when he scooped corn from the wagon box out on the Iowa farm. In on that sound were borne a load of associations and feelings that set the man to dreaming—and then to writing.

"Up to this point," he recalled, "I had composed nothing except several more or less high-falutin' essays, a few poems and one or two short stories somewhat in imitation of Hawthorne . . . ." He had never attempted a delineation of the life he had known out in the western hills or on the prairies, those places that, so far as he knew, had scarcely been so much as noted in our literature. Now, driven to express something of what he had known there—the smell of rain on the grasses, winter's light, the flies in a summer barnyard—he felt himself without models or guides, a would-be breaker of the soil bereft of even the simplest of tools. Still, the grating sound of the shovel had inspired him to try, and subsequently he turned out an essay he called "The Western Corn Husking," which he placed in the *New American Magazine.*

Probably even so early Garland suspected he had begun on a new, uncharted career as a writer, for that same year he mapped out a return to the places of his youth, as if he knew he needed to go back there in search of details of a way of life he had passionately renounced. On his way back to Iowa he stopped in Chicago and visited with Joseph Kirkland whose novel *Zury* Garland had favorably reviewed earlier in the year. The older man listened to Garland's plans and encouraged him to try to get the life of the prairies and coulees into fiction, even as he had done in *Zury.* Before him, Kirkland's mother had written a brilliant, acerbic, semifictional account of pioneering life in Michigan, and Kirkland was convinced the life of these unsung places was a rich mine for the knowledgeable and the daring.

With Kirkland's advice and encouragement to bolster and objectify his own vaguer intentions, Garland went on into Iowa, a writer on the hunt now, absorbing atmosphere, as he said, like a sponge. For the first time, he looked upon the scenes of his boyhood and youth as potential materials for literature and not as the terrible scenes of trial. The discovery of this potential was as astonishing to him as Columbus's of the Antilles: here indeed lay an unsuspected new world. Why, he asked himself, had these places never been put into the national literature? Russian and English and French writers had known how to brilliantly utilize the provincial: that much he had learned in the long hours he devoted to self-improvement in Boston's public libraries. But the life he knew of America's Middle Border had not seemed available to American writers. Our writers, he had believed, wrote about other places, more accessible ones, more distinguished, places that had an established history—Boston, for instance. That was why he himself happened to be living there when he experienced his epiphany: it was where he felt most in touch with the great writers and their locales. Until he real Kirkland's *Zury,* about the only work he had come across that spoke to his background was Edward Eggleston's *The Hoosier Schoolmaster.* But as Garland contemplated his own prospects Eggleston and Kirkland hardly constituted a definable tradition of Midwestern literature within which he could hope to work. Artistically and emotionally speaking, it was as if the young man were standing alone and empty-handed on the prairie. It was his place, all right, but he would have to learn by himself how to work its tough sod. Nonetheless he felt compelled to try.

His efforts were rewarded by the achievement of a collection of stories, *Main-Travelled Roads* (1891), the book that remains Garland's major contribution to American letters. He dedicated it to his father and mother, "whose half-century pilgrimage on the main-travelled road of life has brought them only toil and deprivation." The stories in it aren't pretty, nor did their author intend they should be, and in some places there is an obvious attempt to create a bleakly realistic picture of farming life that would supersede the popular, sentimental image of it conveyed in "The Old Oaken Bucket." But the stories are intensely felt and their locales vividly realized, and it is astonishing that so inexperienced a writer could have written "Under the Lion's Paw," "The Return of the Private," and "Up the Coulee."

From the first it was more than an artistic achievement that Garland aimed for in his work. He had come to believe that the arts should serve a critical cultural function. They should be, he believed, the chief means by which a sprawling, heterogeneous nation might realize itself. As America neared the end of a century of unparalleled progress it seemed not to know what it was, what it contained. It had expanded so swiftly, so haphazardly that psychologically speaking it lagged far behind its own frontiers. America was more than Boston and New York, Philadelphia, and Virginia, seat of the old political aristocracy. Americans had mapped the continent, east to west, yet its interior realities remained in curious and significant ways blank spaces. It was, Garland thought, the job of the writer and the painter to deepen the national map, give it dimension. His own modest but real contribution had been to make Dutcher's Coulee, Cedarville, and La Crosse parts of the mental geography of the nation. If they were not so significant as Boston or Twain's Hannibal, still they were represented now, as before they had not been. His grandparents and parents had been pioneers. So now in his own way he was, too, making a way for countless others to reach imaginatively those coulees and prairies that his forebears had broken into a half-century ago.

By the time of the Columbian Exposition, Garland had earned a small but real reputation for what was then being spoken of as "regionalism." A year after the Exposition he came more boldly to the forefront with the publication of *Crumbling Idols*. This time he dedicated the book to

the "men and women of America who have the courage to be artists."
Inspired by his own success in turning the rough materials of his Middle
Border experience into art—and in part too by the bracing artistic climate
of Chicago in the nineties—Garland here proposed himself as an American
Hippolyte Taine. It was Taine who had written that the truly original artist would find the courage to work with the materials of his local environment, whatever these might be. Garland rang an American change on this
by making an artistic virtue of what had often been a term of opprobrium.
The trouble with American art, he wrote, was not that it was too provincial but that it wasn't provincial enough. It did not have its roots firmly
in the soils of specific places. It was not, as had popularly been supposed,
a question of literary nationalism. It was a question of correctly assessing
what the nature of American nationality was. Nor would he be misunderstood to be calling for more of the merely picturesque in our letters: there
had already been too much of this in the work of the local colorists.
Valuable as the precedent of these writers had been, their evocations of
American places had been more decorative than deep, he claimed, more
flash than substance. The kind of provincial literature he had in mind was
one that got at the texture, the spirt of a place, that made a statement about
its life that would strike readers as being "indigenous as plant-growth."
And he predicted that with the birth of this new, indigenous literature
there would come into being dozens of provincial art centers throughout
the land rendering the old east coast aesthetic axis a thing of the past. Now
the new songs, the true songs would arise out of the savannahs, the prairies,
the north woods, the magnificent Pacific Coast.

Granting (perhaps too easily) Garland the courage of his convictions,
there must appear to us something a bit shrill and historically shallow about
the claims in *Crumbling Idols,* as if the whole of literature, both foreign
and American, were merely prologue to what Garland and his regionally
oriented colleagues were presently to create. In the perspective of cultural
history as we now have it, it is obvious that what Garland was calling for
in *Crumbling Idols* had been in process at least since the Romantic movement of the American 1830s. And it might reasonably be argued that
American writers had been slowly learning to write evocatively about
American places ever since the first literary records of the New World were
shipped back to the Old. It simply takes time, we might say with Henry
James, for there to be enough of human history accumulated in a given
locality to make literature out of it. What was there, after all, in the Mid-

dle Border to write about before Garland's grandparents arrived there—
or, for that matter, until a couple of subsequent generations had fertilized
the new ground with their sweat and their bones?

But this is to ignore the great historic and indeed existential problem
American writers faced until about the time of the Civil War: how to learn
to trust yourself enough to take inspiration from the place where you were
rather than looking elsewhere, or backward in time, for models, guides,
landscape orientations. More: how to see in the unfeatured and often
dreary circumstances and details of your place of living the requisite
materials for literature. What was the spirit of a place, really, and how did
you get at it if the place had not been hallowed for you by long years of
habitation? The European Romantics—the Schlegels, Goethe, Madame de
Stael—had theorized that the wellsprings of any great national literature lay
in its folklore, its humble localisms. But didn't this really apply to old and
settled nations where there were old folk traditions, where every place bore
the marks of long and continuous human habitation, where the very place
names (such as those in England ending in "chester") were daily reminders
of the humus of human history? It was all well for Emerson to take his
cue from the European Romantics and call in the 1830s for an indigenous
literature. It was well for Whitman to echo Emerson in his *Democratic
Vistas* (1871). But how actually to do it was another matter, an individual
one, and here lay the burden of the American writer from the days of John
Smith through those of the Revolutionary War poet Joel Barlow, to those
of Hamlin Garland, who had at last to trust where he might be led by the
sound of a scraping shovel. Nor would Garland be the last to feel the
burden: William Carlos WIlliams would face it in Paterson, New Jersey
in the 1920s, just as Leslie Marmon Silko would on the Laguna reservation in New Mexico in the 1970s.

In his own time Emerson himself had actually been rephrasing—
though brilliantly so—what others had earlier called for: a literature that
based itself on a close scrutiny of the local and the commonplace and from
the first he had understood both the difficulty of the attempt and also its
cultural importance. Without American writers and artists to celebrate the
land, he thought, Americans' relationship to their lands was fated to remain what it was, shallow, meretricious, exploitative. But where were these
crucially needed celebrators? In the 1830s, as his own genius took wing,
more than two centuries had elapsed since Europeans first looked on the
wild splendors of the continent, and as yet there had appeared no writer
who had been able to evoke the spirits of these places. America as a whole,

America in its specific parts, remained an unrealized, unimagined country. Instead of singers and celebrators the country had to look to its politicians and speculators for an image of itself.

In the spring of 1834, just returned from Europe and thus especially alive to the differences between the landscapes of the Old World and the New, Emerson meditated on the need for an American mythology that would alter and shape Americans' relationship to their lands. On April 28, he wrote in his journal that the place where any man finds himself "is the true place, and superior in dignity to all other places." It was fruitless, he urged himself, to yearn for Rome or the high palaces of the world. "The place which I have not sought, but in which my duty places me, is a sort of royal palace. If I am faithful to it, I move in it with a pleasing awe at the immensity of the chain of which I hold the last link in my hand and am led by it. I perceive my commission to be coeval with the antiquity of the eldest causes."

On June 10, he noted that "Washington wanted a fit republic," then ticked off the names of heroes of the past—Aristides, Phocion, Regulus, Hampden—whose deeds had been celebrated in literature. "But," he concluded, "there is yet a dearth of American genius." The entry is slightly cryptic, a kind of personal shorthand signifying the need for American writers to celebrate America as the mythologists and historians of the past had done for their lands.

Then, on June 18, this remarkable entry:

> Webster's speeches seem to be the utmost that the unpoetic West has accomplished or can. We all lean on England; scarce a verse, a page, a newspaper, but is writ in imitation of English forms; our very manners and conversation are traditional, and sometimes the life seems dying out of all literature, and this enormous paper currency of Words is accepted instead. I suppose the evil may be cured by this rank rabble party, the Jacksonism of the country, heedless of English and of all literature—a stone cut out of the ground without hands;—they may root out the hollow dilettantism of our cultivation in the coarsest way, and the new-born may begin again to frame their own world with greater advantage.

The earthward direction of the entry, together with its strange and strangely compelling central metaphor—a stone cut out of the ground without hands—indicates that Emerson knew the problem of American literature was deeper than any had yet suggested. It was not merely a matter of na-

tionalism in letters. Emerson was far too skeptical to buy the bellicose rantings of those calling for a "national voice" in literature. It was not that
America still lacked traditions that could sponsor our writers. This was
a pseudoproblem (a century and more after Emerson it was still being said).
No, the root of the problem was a lack of attachment to what was already
ours, already possessed. It was a shallowness of tenure that expressed itself
in all aspects of the national life as well as in our derivative literature. And
Emerson, a committed Adams man, dared here to hope that the Jacksonian
Democrats would turn out to be the saviors of the culture. As they had
so upset the silly decorum of Jackson's inaugural ceremonies with their
muddy boots and worse behavior, so these rustics might tear up and rear-
range the cultural landscape and make it possible for subsequent generations to achieve a more primitive relationship with their lands: primitive
in the sense of original and fresh. Then from the new, raw beginning
perhaps an American literature might arise, one owing little or maybe
nothing at all to foreign models, beholden instead to the spirits of specific
places.

Emerson's metaphor of the stone is so compelling because it has
about it the untranslatable and mysterious feel of the mythological: the
stone as ancient, talismanic reminder of the sacredness of place. The year
previous while in England, Emerson had visited Westminster Abbey where
the English kept the sacred stone of Scotland, the Stone of Scone. In removing it in 1296, Edward I believed he was also taking with him the soul
of a rebellious people, for the Scots identified the stone with their origins.
The fact that the stone now resided in the national mausoleum was evidence that it belonged to the spiritual past and that the present age regarded
it as a curiosity of history. Nevertheless, scattered throughout the English
countryside were other ancient, talismanic stones, reminders that the belief
in sacred stones and sacred places had not been confined to the allegedly
barbarous neighbors to the north.

In fact, before the coming of Christianity all the peoples of the Old
World had lived in a numinous landscape spangled with sacred markers
and sacred places. The land itself was believed to be alive and under the
protection of numina, guardian spirits. In such a world one did not blithely
cut down a grove of trees, plow up virgin meadowland, dam a stream or
divert it. Any alteration of the landscape had to be carefully couched in
propitiatory rituals intended to appease the numina. Before plowing a field
a Roman farmer would make an offering of a pig and specially prepared

cakes; the same for thinning a grove, though here it was necessary to explain in prayer to the resident numen why the grove needed to be thinned. And what was true of the landscape in general was especially so of sacred places, whether these were groves, great boulders, caves, mountains, or springs: these were not be be tampered with, for they were charged reference points where the plane of this world was intersected by the plane of another. Hence, these were places of access.

The gradual conquest of Europe by the Judeo-Christian worldview changed all this. The one and only God banished the thousands of local numina just as he had earlier banished Zeus and Saturn. The sacred places were broken up, plowed under, cut down. Christian preachers carrying the messages of Paul, Jerome, and Augustine went through the countryside of the Old World, exorcizing it, desacralizing it, taming it. In their wake the only officially recognized sacred places were churches, cathedrals, and Christian shrines. And yet the triumph of Christianity in the land was never complete, and the numinous old landscape lived a buried but real existence, murmuring in an obscure tongue of the mythological past. In hilltop groves spared from foresting and cultivation, in dolmens and menhirs, in huge horses carved into chalk hillsides, bears incised on boulders, there lived a memory of the time when the land had been alive and charged.

Nor were these static reminders the only sources of such a memory, for in seasonal celebrations such as May Day, in games, masques, in children's rhymes, and in the iconography of the Middle Ages and the Renaissance there were countless aids to such reflection. Indeed, for Old World inhabitants—and still more shockingly and powerfully for those like Emerson who would, in a sense, re-encounter the Old World's landscape—it was impossible not to recall that all they looked on had once been invested with spiritual force. In the ancient, pre-Christian days the function of the myths themselves had been to encourage memory, to insist through narrative that humans remember who they were and where in the cosmos they stood; that they remember to remember (which a latter-day druid, Henry Miller, said is man's earthly mission). In the Christianized landscape of the Old World the varied physical forms of mythological survival did the same. They visibly reminded people of the long continuum of history enacted on the land, and in doing this they encouraged an emotional identification with it.

For the artists and writers of Europe all this was a rich, inexhaustible storehouse of narrative, incident, and psychological assurance. More, the myths and the visible memories of the once-numinous landscape were an extra-personal source of authority that could lend resonance to the work of even the most pedestrian of daubers and scribblers. And they knew this: in every age mythographers were careful to preserve and pass along the narratives and images, especially the Italians to whom the Renaissance was indebted for the fifteenth-century manuals of mythology compiled by Giraldi, Conti, and Cartari. Lope de Vega, Ben Jonson, Shakespeare, Dryden all drew on the Italian mythographers as well as on those daily physical reminders of the past in their places of living. It is hardly so simple a matter as the number or incidence of identifiable mythological images in this or that writer's work. It is instead the whole whelming context within which they worked, and it is safe to say that without this the work of even the greatest of the Renaissance and post-Renaissance artists and writers would be appreciably thinner.

Those who came to the New World in the aftermath of the Renaissance had hardly known, until they arrived, how vast an advantage they had enjoyed at home. But those who tried here to write or paint quickly felt the existential void of the unknown land like a draft upon their souls. For them, imaginative entry into the country was perhaps even more difficult than it was for the trailblazers, Indian fighters, fort builders, missionaries. The land in which they now found themselves, though it had from the first been invested with the dreams of the Old World, appeared to be utterly lacking in mythic associations or any visible aids to reflection. Here there were no ancient temple sites; no mountains within which were said to sleep warrior kings and their hosts; no storied pairs of trees—an oak and a linden—site of the transformation of Baucis and Philemon and told of by Ovid. Trees there were in the unnumbered millions of the climax forests of the east and west coasts and snarling mountain ranges, too, that barred access to the broad interior. But about none of these hung the least whisper of any understandable human past, and their silent presence was mighty and oppressive. America, as it was proving to the would-be exploiters and builders, was the enemy of progress and civilization. For the early artists and writers it was the enemy of the arts.

Here the writers had no recognizable cultural context within which they might work. They had lost that mental map that had been theirs in the Old World, that had allowed them to appreciate without precise formulation—and without the need for it—the rude megalith in the midst of the plowed and furrowed field. Where in this wild world were any such reference points, the places of imaginative access? Faced with such a mute landscape and the homely little settlements wherein they found themselves, our early writers felt themselves in a predicament Hamlin Garland well understood centuries thereafter.

Nor could they invest the native tribes with any alleged participation in some golden age. That was something in which stay-at-home *philosophes* might safely indulge. Here the aboriginal realities were too hand-to-hand for such paper flights. To the writers on the spot the Indians appeared quite as useless as they did to those mallet-handed pioneers and explorers who dodged the tribes when they could, fought them when they couldn't. Debarred by tongue and culture from appreciating the tribes' profound identifications with their lands, writers could but wonder at the significance of the tribal headmen who spoke so solemnly in treaty negotiations of what true ownership of the land meant. The whites "owned" more and more of the land, and yet, for the writer, what did this putative ownership amount to? Daniel Webster probably spoke for most of them when he observed that there was nothing "in the languages of the tribes as in their laws, manners, and customs, worth studying or knowing." Walter Channing thought otherwise, though it was after all a desperate otherwise that he thought. Surveying in 1815 the very modest beginnings of an American literature, Channing thought American writers might have to turn red if they hoped to achieve distinction. For in the speeches of the "Noble Red Man" Channing detected greater imaginative possibilities than those inherent in the use of a language (English) "enfeebled by excessive cultivation." Alas, it is clear from Channing's remarks on the alleged oratorical genius of the Indian that he had in mind the Indian as invented by the white man, a paper-thin simulacrum.

Early American literature, the literature of settlement and exploration, is thus a literature of resistance and bewilderment; or else it is an inventory of economic possibilities. In its pages the modern reader searches vainly for some authentic *assent* to the land, some significant effort to divine the spirits of its places. The chonicles of resistance and bewilderment often make interesting reading on one plane, so fierce are they and so black with their tales of loss, disaster, violent death. But in them there is not even the beginning of a coming to terms with America. And still less so in those dry, humorless inventories whose approach to American realities is a measured calculation of board feet and mineral tons. Here and there, to be sure, there is some wonderfully lively stuff, as in the work of the renegade Thomas Morton of Merrymount, or the Maryland writers George Alsop and Ebenezer Cook. But you can count on your fingers the writers prior to about 1820 who achieved some true imaginative break-through: William Byrd, William Wood, John Josselyn, especially William Bartram. The rest looked asquint at their unhallowed surroundings and tried to bring the Old World muses to bear on their predicament. For years some produced hopeless imitations of the current fashions in English letters. Or else, frankly despairing of America, they set themselves to translations of the ancients. Thus George Sandys, friend of the poet Michael Drayton, leaving for Virginia in 1621 carried with him Drayton's blessing to continue in the New World his translation of Ovid's *Metamorphoses*:

> And, worthy George, by industry and use,
> Let's see what lines Virginia will produce.
> Go on with Ovid, as you have begun
> With the first five books; let your numbers run
> Glib as the former; so shall it live long
> And do much honor to the English tongue.
> Entice the Muses thither to repair;
> Entreat them gently; train them to that air;
> For they from thence may thither hap to fly.

Sandys survived the Indian uprising of 1622 as well as other insults and incidents of settlement life to complete his translation of the last ten books, after which he returned to England. The myths had been translated in the New World—but not transported. And as far as an indigenous literature was concerned, this would not do. Nor would it do to dress the land in borrowed robes and rhymes as Joel Barlow had done with his New World epic poem, *The Columbiad* (1807), the reading of which is now in itself an epic chore reserved for graduate students of American literature.

By 1820, it was clear even to the most rabid of nationalists that, for whatever reasons, America had not developed a representative literature, that its writers lagged far behind its statesmen, orators, entrepreneurs, wilderness busters. The great land had not produced a great literature. The noble national ideals were not somehow ready and manageable themes for imaginative literature. Neither was the conquest of the wilderness (an irony deeper than almost any then suspected). Somehow the heroic contest of men against the forests and their savage inhabitants did not generate that usable drama that some had thought it should. The theme was there, the figures of the white hero and his red antagonist, too. But in the hands of our writers and artists the thing was a corpse.

There were traditions of the frontier as they sprang into being, but they were so bloody, violent, and uncouth that no writer with pretensions to acceptance would think of touching them: the foul jests, the cruel sports of backwoods clearings where men gouged out each other's eyes with blackened fingers; or bit off noses and ears in rough-and-tumble bouts. Also the legends of Indian haters, rank but authentic growths out of a darkened soil: Tom Quick cut his teeth on arrowheads, kept them sharp chewing lead slugs, and lived only to kill red men. Outlaws, too: the Harpe brothers, those terrible men, assassinated scores of unsuspecting victims, including women and children, and were themselves brutally murdered, their skulls left to splinter in the forks of a tree along the Natchez Trace. Not until Twain in the 1870s would a writer dare to utilize some of this rich and pungent material, and even he was intimidated by it.

Still, by the time of the 1830s, amid the hectic bluster of nationalistic sentiment and under the strengthening influence of European Romanticism, there were signs that American writers and artists were coming to understand that they would never learn what in their American world was worth writing about until they learned to look more carefully at where in that world they happened to be (the distinction is Wendell Berry's). In the paintings of Thomas Cole the American landscape emerged from its status as the stylized background of portraiture. William Sidney Mount's paintings captured the details of life in specific places. In letters, James Kirke Paulding, Washington Irving, and Cooper began to dig into the land. Writing in his popular *Salmagundi Papers* in 1820 on the much-belabored question of a national literature, Paulding made the crucial implicit distinction: it wasn't a question of how to sound like an American author, he thought, but rather of using what you already had on hand. His suggestions here are tentative, mere probes as they might appear to us now. Yet they helped mark out a faint trail. By utilizing, he writes, "those little peculiarities of thought, feeling, and expression," the American writer might eventually triumph over all his historical and cultural deficiencies. Paulding's own imaginative efforts—the narrative poem "The Backwoodsman" (1818) and the drama *The Lion of the West* (1831)—do not read very well today. Nevertheless, they show an attempt to come to terms with the spirit of life on the edge of an advancing civilization.

Farther along the trail was John Neal's novel of the Salem witch hysteria, *Rachel Dyer* (1828), which not only effectively utilized a bit of American history but also showed a feel for local details and for the spirit of a place and time. In the preface to the novel there occurs a remarkable passage on the use of the homely and the local. Of those who doubt the richness of the imaginative soil of America, Neal writes that they fail to understand that

> there are abundant and hidden sources of fertility in their own beautiful brave earth, waiting only to be broken up; and barren places to all outward appearance . . . where the plough-share that is driven through them with a strong arm, will come out laden with rich mineral and followed by running water.

This was to anticipate Emerson in the 1830s, especially in his "American Scholar" address in 1837, now popularly called the American Intellectual Declaration of Independence. Standing before the Phi Beta Kappa graduates at Harvard at the end of August, Emerson echoed Neal—and Benjamin Rush before him—as well as the theories of the European Romantics, just as one day Hamlin Garland would echo Emerson. We have, said Emerson, squandered our imaginative resources in this foolish lust for the approbation of the past, when all the while under our boots lies abundant material for the arts. "That," he pointed out, "which had been negligently trodden under foot by those harnessing and provisioning themselves for long journeys into far countries, is suddenly found to be richer than all foreign parts. The literature of the poor, the feelings of the child, the philosophy of the street, the meaning of household life, are the topics of the time." I ask not, he went on, his periods beginning to roll as he saw the full dimensions of his subject, "for the great, the remote, the romantic; what is doing in Italy or Arabia; what is Greek art, or Provençal minstrelsy; I embrace the common, I explore and sit at the feet of the familiar." What, he concluded, would we really know the meaning of? "The meal in the firkin; the milk in the pan; the ballad in the street; the news of the boat; the glance of the eye; the form and gait of the body. . . . "

But Emerson himself never knew how to do it. In his journal jottings he chided himself on his inability to make poetry or indeed any literature at all out of his farmer neighbors in Concord, preferring ever the abstract or the giant figures of history and myth. Then in the 1850s and 1860s as he ventured out into the Midwestern frontier on his lecture tours, he came face to face with the question all the writers had in one way or another to face: how to write about *this*? In Wisconsin in the winter of 1854 (then home to the Garland family), he was daunted by the sheer scale of things: the size of the lots, the girth of the trees, the gusto of the pioneers. He tried, momentarily, to make these people into mythic figures of a heroic age, living on wild meat. But it was no good. Here was no Virgilian idyll nor even the heroic savagery of the dawn of humankind. "The first men," he wrote, deep in Wisconsin's winter, "saw heavens and earths . . . we see railroads, banks, and mills. And we pity their poverty. There was as much creative force then as now, but it made globes instead of waterclosets."

But the evident contrast between the ages of myths and mills wasn't the essential issue here, and Emerson knew it. In the same entry he wrote that in order to really know something you have to do it. The man who would write of these prairies, the mud, the frozen rivers, broken roads, land speculators, landlords, half-breeds, and farmers would have to plunge himself into this life, "live in its forms—sink in order to rise."

He was not that man, and he knew it. His ornate periods, professorial habits, his Concord study-induced preference for the abstract debarred him from that imaginative plunge into frontier life or even the life he observed around him in Massachusetts. He noted with a self-deprecatory amusement that in Rock Island, Illinois, in 1855, he was advertised as "the Celebrated Metaphysician." But however divorced he was fated to remain from these rural realities, Emerson saw the Midwestern prairies aright, as symbolic of the challenge American writers faced and must successfully meet. He had known several who had the talent to imaginatively take on the circumstances of their lands, but thus far they had remained merely promising. One of them had been his neighbor, disciple, and friend for more than twenty years. Sometimes Emerson almost despaired of him but never quite. In a journal entry of January 1858, he wrote of him again:

> I found Henry yesterday in my woods. He thought nothing to be hoped from you, if this bit of mould under your feet was not sweeter to you to eat than any other in this world, or in any world. We talked of willows. He says 'tis impossible to tell when they push the bud (which so marks the arrival of spring) out of its dark scales. It is done and doing all winter. It is begun the previous autumn. It seems one steady push from autumn to spring.
>
> I say, how divine these studies!

TWO

NEARER HERE: HENRY DAVID THOREAU'S WALDEN

and

THE MAINE WOODS

I

  BY JANUARY 1858, EMERSON had already privately admitted his disappointment in Thoreau as a writer, though in his journal he could still admire the younger man's patient, persistent investigations of the natural world. But he thought Thoreau wanted ambition, and instead of aspiring to be what Emerson called the "head of American engineers," Thoreau was apparently content to be "captain of a huckleberry party," Emerson's disparaging reference to Thoreau's fondness for huckleberries and for leading troops of Concord children on harvesting expeditions.

Thoreau was, after all, then forty-one and presumably past that point where you can still style someone young and promising. And what, after all the promise, the heavy ruminations, had he produced thus far? Some scattered journalistic shots, some lectures, but only two books. One, a quirky, inconsecutive memorial to a dead brother, had been so dismal a commercial and critical flop that it was already forgotten by all but the author who had a garretful of remaindered copies. The other, while it had been better received, had been so intensely—so perversely, as Emerson must have felt—local that it could hardly speak to Americans about their condition, here in their lands of living. *Walden* was not, it seemed, the work of an engineer for America but instead a brilliant, idiosyncratic description of a two years' residence on Emerson's wood lot south of town. Thus, outside the tight little Concord-Boston axis Thoreau was almost wholly unknown, and, oddly, he seemed content that he should remain so. With a sort of churlish consistency he refused opportunities for travel, entrees to the wider literary scene, lecture dates that would take him from Concord. "I am so wedded to my way of spending a day," Thoreau wrote a New Bedford admirer, "require such broad margins of leisure, and such a complete wardrobe of old clothes, that I am ill-fitted for going abroad. . . . The old coat that I wear is Concord; it is my morning-robe and my study-gown, my working dress and suit of ceremony, and my night-gown after all. . . . *Cars* sound like *cares* to me."

Even within Corcord he by no means enjoyed an authentic repute. Some years before he and a companion had foolishly set fire to a sizeable stand of woodland, and townspeople still said after his retreating back, "Rascal" and "Burnt Woods." So far, he had tried school teaching and tutoring, manual labor and hermitry, carpentry, pencil-making, survey-ing, and authorship. Folks thought it queer that at any hour or condition of the day or night—sunup. sleet, moonlight—he might be seen, tramp-ing steadily across lots in his shapeless, old-fashioned clothes, bound out on errands wholly unimaginable to them, bound in bearing his treasures of plants, barks, arrowheads, the battered sheaf of notes he seemed so highly to prize. His Aunt Maria said she wished Henry would "find some-thing better to do than walking off every now and then." But he did not, and so when he'd gone out in 1845 to live on Emerson's Walden wood lot, the townspeople thought that a good enough fit for him: the pond was the habitat of shanty Irish, ex-slaves, and freedmen, of bums and loafers. If in 1845 Emerson had approved of this experiment, now, eleven years after after it had ended with Thoreau's return to the Emerson house-hold, it looked like another of Henry's vagaries.

Of which there were many. He had offended the town by resigning his church membership, and if he did not then set up as the village atheist, he did make it clear he had no interest in Sabbath services or any of the other orthodox observances he was pleased to call superstitions. Born David Henry Thoreau, he had reversed the order of his first and middle names and thereafter insisted Concordians address him that way. A col-lege graduate when this was a distinction not the rule, for the most part he lived no differently than the most illiterate of laborers and seemed to find a sort of honor in this, preferring to converse with farmers, railroad workers, ice-cutters, and wood-choppers than with his ostensible peers.

Nor was it a matter of exterior oddness only: those who knew Henry more intimately found him as cross-grained and full of knots as an old maple. While they testified to his talents and certain admirable qualities of mind and heart, they did also to traits they found less attractive. Emer-son, with whom he formed perhaps his closest relationship, eventually wearied of Thoreau's persistent negativism, his counter-puncher's habit of returning a "no" to your least assertion. Hawthorne, with whom he spent some pleasant hours skating on the Concord River and rowing along it in summer, found Thoreau the "most unmalleable fellow alive—the most tedious, tiresome, and intolerable—the narrowest and most notional," despite his "great qualities of intellect and character." And Ellery Channing, one of those "great promisers" who had also disappointed Emerson and who spent as much time with Thoreau as any, often found his friend simply impossible. He was an eremite, Channing thought, a Spartan saint mortifying himself, crucifying all the human virtues. And for what? "What is his compensation [?] Eternal solitude, and endless blundering, blunder after blunder." After Thoreau's early death, Channing, bereaved and still baffled by his wayward friend, observed that he'd never been "able to understand what he meant by his life."

The evidence of the journals Thoreau kept with astonishing steadfastness for almost a quarter century (1837-61) indicates that for a long while his own behavior was as mysterious and occasionally even upsetting to him as it was to neighbors and friends. Beneath the granitic resolve Thoreau showed the world and by which he is popularly known today, there ran in him a subterranean current of self-doubt and mistrust. He was rescued from it by the slowly revealed vision of a vocation, as the roiling clouds of a New England winter's day eventually break apart, revealing the presence of the abiding blue.

In 1837, graduated from Harvard, he returned to his village to cast about for a career and immediately found one open to him: that of the solitary scholar. It was open because it was a thankless, unpaid, self-created position of the sort society is perfectly willing a young person assume for as long as he likes. Emerson had made much of the advantages of solitude, had trumpeted the heroism of the solitary scholar, and now his young neighbor would try out this path the great man had talked about so confidently. In the flyleaf of the journal he began to keep that fall he copied out extracts from the work of three English Metaphysical writers. From Herbert he took the lines "By all means use sometimes to be alone, / Salute thyself. See what the soul doth wear." From the random treasure house of Burton's *The Anatomy of Melancholy,* Thoreau took this:

> Friends and companions, get you gone!
> 'Tis my desire to be alone;

> N'er well, but when my thoughts and I
> *Do domineer in privacy.* [Thoreau's italics]

And this from Marvell's "Garden":

> Two Paradises are in one,
> To live in Paradise alone.

But, as he was to discover, it is one thing to take on the career of the solitary scholar and still another to be able truly to live in it with assurance. While teaching school, doing odd jobs, and engaging in the appropriate activities of his self-styled career, Thoreau was beset by an insistent drumbeat of doubt. He was, in a sense, doing all the "right" things: he was studious, meditative, abstemious, solitary, out much in nature, keeping a journal. But what was the real point of it all and where could it lead? March 5, 1838: "What to do. But what does all this scribbling amount to?—What is now scribbled in the heat of the moment one can contemplate with somewhat of satisfaction, but alas! to-morrow—aye to-night—it is stale, flat—and unprofitable—in fine, is not, only its shell remains. . . . What may a man do and not be ashamed of it? He may not do nothing surely, for straightway he is dubbed Dolittle—aye! christen himself first. . . . But let him do something, is he less a Dolittle? Is it actually something done—or not rather something undone—Or if done, is it not badly done—or at most well done comparatively?"

March (8?) 1840, a lament in verse beginning with these lines:

> Two years and twenty now have flown—
> Their meanness time away has flung,
> These limbs to man's estate have grown,
> But cannot claim a manly tongue.
>
> Amidst such boundless wealth without
> I only still am poor within;
> The birds have sung their summer out,
> But still my spring does not begin.

Later in that same month in a mood of savage sarcasm he listed the vocations open to him—mail carrier in Peru, South African planter, Siberian exile, South Sea explorer—and then noted, "Thank fortune we are not rooted to the soil and here is not all the world. The buckeye does not grow in New England—and the mocking bird is rarely heard here."

Interspersed with such melancholy assessments of his condition are numerous efforts at consolation in which the young man repeatedly reminds himself that patience is the sovereign virtue of one who would be a solitary scholar and that neglect and the misapprehension of his fellows is the lot of such a one. But even deeper than this stream of doubt and running more surely was another stream, that of an obscure ambition that Thoreau had yet to admit either to himself or his journal: he wanted to be a writer. This was not the same thing as being a scholar, a term he had, after all, taken on from Emerson whose address, "The American Scholar," Thoreau had probably heard at his commencement exercises in 1837. The scholar as Emerson had defined it there was primarily a thinker, a philosophic observer of life, though in his view it was vital that in some form— poetry, teaching, architecture, political discourse—his thoughts be translated to the world. But a writer, while he could still be a philosopher, might be a man with a more active engagement with his immediate world. In his journals he'd certainly tried thinking in the high, heroic Emersonian mold, composing lengthy, ethereal entries on Truth, Friendship, and the like, but the fact was that when he looked at these they seemed more airless and commonplace than anything else. Was there not a way of bringing himself, Henry Thoreau, into a more direct confrontation with the facts of his experience, his world?

As far back as his undergraduate days Thoreau had been interested in the idea of an American literature, a subject much in the intellectual air of the 1820s and 1830s as the eagle of nationalism began to stretch and flap its wings. Emerson, Noah Webster, James Kirke Paulding, William Cullen Bryant, and a host of now-forgotten voices had all called for an indigenous literature. But there was something oddly vacuous and even seemingly imported about much of the debate. It wasn't addressed to the essential question: how to make your own place come alive on the page, how truly to write about what *you* knew. It was all well to rewrite the opinions of the Schlegels, Herder, and Madame de Stael, to say, we too in America must write our Bibles. But even the great Emerson (who said this) in his own attempts at creating an American literature remained on the level of the general and the abstract, and for all the local in his work, he might as well have written from London. At the end of his highly successful novel, *The Prairie* (1827), James Fenimore Cooper has his pioneering hero, Natty Bumppo, take his last and defiant stand against a flaming western sunset and pronounce the single prophetic word, "Here!" But where was "here" and how to write about it?

Little in Thoreau's background and almost nothing in American culture prepared him to make an answer. Like all undergraduates of his time he had been trained up into an appreciation of the classics and of the current pantheon of British writers. In answer to a class assignment on the idea of a national literature, he'd produced an essay indicating that he had read the European Romantics and knew their theoretic formulas for an indigenous literature. In that essay, however, he had broken a bit of new ground in writing that a truly American literature would have to be made up of robins and rail fences, not nightingales and formal gardens. Well, now he had the robins and rail fences, all right, had studiously noted them in his journals. But at this point they still seemed merely notes toward a something, his journals a steadily lengthening prolegomenon toward an entity he couldn't visualize.

Fall 1841 came to Concord, and as always at this turn of the season the solitary woods-walker's spirits lifted with the ending of the dog days, the crisping of the air and leaves. On September 4, he jotted down a bristling quiver of thoughts:

> I think I could write a poem to be called Concord—For argument I should have the River—the Woods—the Ponds—the Hills—the Fields—the Swamps and Meadows—the Streets and Buildings—and the Villagers. Then Morning—Noon—and Evening—Spring Summer—Autumn and Winter—Night—Indian Summer—and the Mountains in the Horizon.

There seems a kind of breathlessness here as if the writer had been ambushed by his own audacity in speaking the heretofore unsayable. Of course, it would be easy to make too much of this single entry and say here is the beginning of a writer's career, the beginning also of the making of an authentically American literary landscape. Still, we have nothing like this in the surviving jottings of Thoreau before this entry. And certainly before him in our literature there is nothing like Thoreau's evocation of the land and its spirits. Years later, looking back at Thoreau's work, the prominent nineteenth-century critic Thomas Wentworth Higginson remarked with a kind of astonishment that before Thoreau nature had never been described in the national letters. It is not too extravagant a claim, though what Thoreau actually described was far more than nature. But this is to anticipate considerably, for not only did the Thoreau of 1841 have to endure yet more years of probation and neglect, but in his lifetime his distinctive achievement remained largely unacknowledged, both by those nearest him as by the wider world.

But the September 4 entry is significant. Before it, the journals are muddy with self-recriminations and the whistlings of a man who can't see his path's trend. After it, a vocation voiced, the recriminations and whistlings diminish in frequency (they were never to cease), and the entries begin to take on the force of intent. He would not be, he was beginning to discover, a scholar in the Emersonian mold after all. He would be instead a writer in a mold of his own special casting. He would be a writer who positively gloried in the local, the homely. The very next day, perhaps still in the heat of that September 4 entry, he wrote of his admiration for Homer, who in his profound grasp of his place made even modern-day readers, far removed in time and geography, feel as if they were "autochthones of the soil." To feel that yourself, to make your readers feel so— *that* would be an ideal worthy of a life, for, as he observed about this same time, none of his fellow townspeople appeared to really know where it was they were living. They talked more of the prospect of Ohio than they ever did of their home ground. "This country," he wrote, "is not settled or discovered yet[.]"

So, though outward appearances remained steady with respect to Thoreau's movements and habits, a fundamental change had begun within. He would serve as a handyman at the Emersons', as a tutor to Emerson's nephew on Staten Island, and as a partner in his family's pencil-making business. He would continue his apparently random and aimless excursions about the New England countryside, his accumulation of plant specimens, arrowheads, written observations of nature's minutiae. In short, on into the mid-1840s Thoreau would continue to live as he had since returning from Harvard, drifting out of the morning of young manhood and toward the middle years with as little apparent direction as when he would take his boat into the center of Walden Pond and, lying on its bottom, let it take what direction it would until at last it gently nudged the sand and the indolent mariner arose to find to what strange shore he had been brought. But all this while Thoreau was training himself to see what otherwise passed beneath notice, the small but so significant facts of his place, and he was teaching himself how to write about them in such a way that they would bloom into the flowers of truth he sensed they were.

In 1842, he announced himself as a local writer with the publication of "A Natural History of Massachusetts" in the July edition of Emerson's *Dial* magazine. While ostensibly reviewing the sober state reports on

Massachusetts's natural life, Thoreau used the occasion to try out some of his own observations of the flora, fauna, and history of his neighborhood: spring spearfishing, the life of muskrats, the behavior of snakes. A good number of his lines were lifted from his journals and so retrospectively conferred upon those steadily piling pages the honor of a distinct purpose. Here for the first time in print is evidence of what would become his characteristic literary trait of seeing so deeply into a natural fact that at last it reveals its hidden spiritual dimension. Treading a fox's tracks in the winter's snow, he describes himself as on such a tiptoe of expectation it was "as if I were on the trail of the Spirit itself which resides in the wood, and expected soon to catch it in its lair." Tracing the fox's course "in sunshine along the ridge of a hill," he says, "I give up to him sun and earth as to their true proprietor. He does not go into the sun, but it seems to follow him, and there is an invisible sympathy between him and it."

Strictly speaking, there isn't a theme to "A Natural History of Massachusetts." But hunting here and there through its pleasant ramble, you can discern what will be the grand theme of the writer's mature works. "We must look a long time before we can see," he writes. If we can train ourselves to this task, we'll then see how "much more than Federal are these states!" America is meant to be more to us than a political confederation, more even than the idea of democracy, exalting as that may be. We live, Thoreau would now begin to claim, in a spiritual landscape as well as a physical one, and it can be one as exalting and poetic as the fabled Arcadian landscape of the Golden Age. But first, we must learn to see that this is really so. We must begin to discover our own places. It was becoming progressively clearer to him that he was himself destined to be a medium of that learning.

In the spring of 1845, Thoreau made plans to go to Walden Pond to live. In 1837, he and a college roommate, Stearns Wheeler, had camped in a hut by the shores of Flint's Pond, and the idea of a strategic literary retreat from the routines of ordinary living had appealed to Thoreau ever since. If there had been more than a little of the romantic in the experiment, now there was a good deal of the realistic to it. In 1845, Thoreau was a published writer, if of very minor stature, and he now felt his talent gathering to a force that required his undivided attention to it. Later, he would write that most men lived lives of quiet desperation and that from the desperate city they went out into the desperate country. If there was not now desperation in Thoreau's retreat, there was an urgency about it, as if he had worked himself into a position where he had to "put up or shut up." On Independence Day he moved into the cabin he'd built and the day following wrote:

> I wish to meet the facts of life—the vital facts, which were the phenomena or actuality the Gods meant to show us,—face to face. And so I came down here. Life! who knows what it is—what it does? If I am not quite right here I am less wrong than before—and now let us see what they will have.

"And now let us see what they will have." You have to like that: it is so direct and courageous, so open to the undiscovered possibilities his move had exposed him to. Here was that heroism of the solitary man to which he had long aspired, and now he had to prove equal to whatever the challenges might be.

The literary project foremost in his mind was a book based on a river trip he'd made with his brother John in 1839. John was now dead of lockjaw, and there may have been something of what the psychoanalytic critic Richard Lebeaux has called "grief work" in Thoreau's need to write of that idyll on the Concord and Merrimack. But of longer standing than his grief and his desire to memorialize his brother was his desire to make some sort of beginning, here, now, on his poem of Concord, to make his argument out of the sights, sounds, and tastes of this place in which he had been favored by the gods to live. Concord citizens might snigger at the wayward college man's defeated retreat to the pond's woody slums, and in later years they would uncharitably recall how much in contact with the town Henry had actually remained during his two years at Walden— how often he went to dinner at the Emersons', how on Saturdays his mother and sisters would bring him baskets of pies and doughnuts. But Thoreau somehow knew that he had found his appointed spot at Walden, that this was the place where he would become the writer he wanted to be. If in 1845 he did not know precisely what it was he would write or what form it would take, he knew at least that this was the time and place to make a beginning. And he was right on all counts: the pond was the right place, and he was ready for its rightness. While out there he experienced an astonishing surge of creative activity, and he rode its long, combing roll for the rest of his life. At Walden, Thoreau whipped through most of what became *A Week on the Concord and Merrimack Rivers* and completed a first draft of *Walden.* He wrote a long essay on Carlyle and what would eventually form the opening chapter of *The Maine Woods.* He also wrote out separate versions of these latter works for use on the lecture platform. And, as always, he was filling the pages of his journals.

He wrote poetry, too, though he had already come to the conclusion that his poem of Concord would be a poem only in the more general sense of that term, that poetry per se was a less heroic task than the writing of sustained, serious prose. A true account of the actual, he had come to feel, was in itself the essence of poetry. In his journal he essayed a thought he later refined in *Week,* writing that prose "proves an actual conquest and retention of territory. The poet often only makes an irruption like the cossack, or the Parthian horse—and are off again firing while they retreat but—the Prose writer has conquered like a Roman and settld colonies." Maybe the imagery here is a bit bellicose for the kind of possession Thoreau wanted to take of his own territory, but there is no doubt he wanted to take a sure possession of it and leave behind his aqueducts, too.

Week is in many respects an apprentice work. By 1846 its author had read hundreds of travel narratives—of Africa, the West, the Poles, the steppes of central Asia—and perhaps he was seduced by these into imagining that the story of any trip, even so modest a one as his, could be made to carry any amount of baggage. Much of the baggage is culled from his journals, and Thoreau had little compunction about loading up his narrative with autobiography, swatches of poetry (his own and others), local history, disquisitions on Friendship, Literature, Art, and Nature; speculations on the fate of nations and races; discussions of writers of antiquity, as well as Chaucer, Ossian, and Raleigh; Christianity and/or paganism. This list is by no means inclusive. The book's shape is thus as shaggy and indistinct as that of a bear glimpsed through a dense stand of woods.

But again, as in "A Natural History of Massachusetts," the grand theme: if we will begin to look carefully at where we are, we'll see that this is Arcadia or Rome and that we need not travel, either in imagination or by rail or steam, to any other. Our willing surrender (to use Frost's paradox) to the facts of our own landscape will prove our salvation. Launching out into the stream of his narrative, Thoreau writes that there is nothing inherently more famous about the Xanthus River or the Scamander than the familiar, homely Concord; it is only that those rivers have had their great singers while this one still awaits its own. "I trust," he writes, "that

I may be allowed to associate our muddy but much abused Concord River with the most famous in history."

> Sure there are poets which did never dream
> Upon Parnassus, nor did taste the stream
> of Helicon; we therefore may suppose
> Those made not the poets, but the poets those.

And in the middle of his week's excursion (the chapter "Wednesday"), he tells his readers that their own country "furnishes antiquities as ancient and durable, and as useful, as any; rocks at least as well covered with moss, and a soil which, if it is virgin, is but virgin mould, and the very dust of nature. What if we cannot read Rome, or Greece, Etruria, or Carthage, or Egypt, or Babylon, on these; are our cliffs bare?" The homely stone walls that fence New England's fields, he continues, are built of ruins as ancient as those of Rome and the Parthenon, and in the roar of a New England river or the wind in the woods may be heard sounds "older than the summer of Athenian glory. . . . " To clinch his point so none might mistake him, he concludes:

> What though the traveler tell us of the ruins of Egypt, are we so sick and idle, that we must sacrifice our America and to-day to some man's ill-remembered and indolent story? Carnac and Luxor are but names. . . . Carnac! Carnac! here is Carnac for me. I behold the columns of a larger and purer temple.

But the path that leads to an understanding of Concord as Carnac, says Thoreau, necessarily takes you through a confrontation with and meditation on its very discrete facts—stone walls, those mossy monuments that to natives represent only some farmer's lonely, back-breaking labor; the muddy, shallow river that unremarked runs through the fields and in flood time keeps the cattle out; the woods whose silent speech is of diuturnity—or was until they were cut down to feed the railroad: those very facts the writer had been patiently recording in his journals. Maybe, he suggests in the spirit of serious play, a good way to confront the facts of a place, a swamp, say, would be to stand "up to one's chin in one" for the whole of a summer's day, "scenting the wild honeysuckle and bilberry blows, and lulled by the minstrelsy of gnats and mosquitoes!" A day in the society "of those Greek sages, such as described in the Banquet of Xenophon, would not be comparable with the dry wet of decayed cranberry vines, and the fresh Attic salt of the moss-beds.''

A couple of years previous, Emerson had peevishly complained in his journal that the rhetorical trick of Thoreau's writing was soon learned: he consistently tried to upset the reader's expectations, Emerson wrote, by claiming for something the precise opposite of what was commonly assumed. Here, for example, Thoreau claims for a day spent submerged in swamp waters a comfort and instructiveness surpassing those to be had at a feast of ancient sages. Emerson was right, but only partly so: Thoreau does use this device a good deal. But what the older man didn't appreciate was that the device was less a literary one than a fundamental manifestation of Thoreau's vision: if you could steadily confront the meanest facts of your place, you would find them far richer in significance than anyone had supposed. We live, Thoreau would presently claim, in a world encrusted with shams, delusions, and received truths which on close inspection turn out to be falsehoods. Reality lies beneath and is fabulous. For him, the swamp day really was potentially the greater experience if you could subject yourself to it and its demands. If you could, then you possessed the swamp. Its facts had become yours forever, its truths vital and sustaining. And these truths, so meanly valued by others—the mosquitoes, the muddy waters, the bittern—showed you what life really was. Compared with this the redactional account of a legendary banquet was indeed stale and insubstantial fare.

His first try at enunciating something of this was, however, a commercial and critical flop—partly for reasons suggested above. For all his outward stoicism and his studied indifference to fame, it is pretty clear that Thoreau was devastated by the utterness of that flop. He had wanted a bit of worldly success, and Emerson had wanted it for him, too. He consoled himself with a bitter private laugh about the load of unsold books he trundled up to his garret in the family house while the town had a public snicker at the failure of the beak-nosed iconoclast who would presume to speak to them about their condition in these lands. Who knew Concord and its lands, anyway: those who worked them or some cloudy philosopher who didn't know what it meant to have to make a crop?

The courage of the man was the constancy of his conviction. He kept resolutely on with his self-appointed tasks as ''inspector of snowstorms and rainstorms,'' surveyor of forest paths and all across-lot routes, herdsman to the wild stock of the town, and patient observer of the ''unfrequented nooks and corners of the farm.'' If none attended his call (announced fleetingly here and there through the sprawling pages of his book) to take up on any spot a real residence and so realize an unsuspected harvest, he himself was bound to do so. And if a living as a literary man—to say nothing of fame—was not to be his, well, he would still write, continuing to be, as he said, a "reporter to a journal, of no very wide circulation, whose editor has never yet seen fit to print the bulk of my contributions. . . ."

From the beginning of his stay at Walden he'd been keeping notes on his daily life even as he went on with other literary projects. Then in the winter of 1846, when he lectured on Carlyle at the Concord Lyceum, he found his audience more curious about his life out at the pond than about his view of the great Scots writer. This was surprising, and it may have come to him at this point that he had another book on his hands, that the move to the pond had an additional significance now disclosed. At any rate, it was worth a lecture at least, and the next winter he gave "A History of Myself" at the Lyceum. It was so enthusiastically received he was obliged to repeat it a week later. A journal entry from this time preserves something of his introductory remarks to this lecture, the seed of what would become one of the most famous books in American literature. "I have heard," he drily joked,

> an Owl lecture with a perverse show of learning upon the solar microscope—and chanticlere upon nebulous stars[.] When both ought to have been sound asleep in a hollow tree—or upon a hen roost. When I lectured here before this winter I heard some of my towns men had expected of me some account of my life at the pond—this I will endeavor to give tonight.

By September 1847 when he left the pond for the Emersons' he had completed the first draft of *Walden*. Ahead lay nearly seven years of revisions and burnishings. There were complete drafts done in 1848, 1849, and 1852. In 1853 he completed the fifth and sixth versions and the year following a seventh version. As late as April of that spring he was still revising. Ticknor & Fields published the book that August.

At the outset it had been the spur of local curiosity that had set him to formal composition. To a very limited extent Thoreau was willing to gratify so common an impulse in his audience, though he had grave reservations about lecturing for mere entertainment and still less for what he called "profitless jest and amusement." But, as he indicated in his journal, he was in now for much bigger stakes than a single evening's stand in the public light. He was after the possession of the territory, for himself and for others, and so an autobiographical disclosure would not in itself be enough. Somehow autobiography would have to be extended so that one life might speak for others.

From his undergraduate days Thoreau had been intrigued by mythology. Now at the pond he resumed his study of it with a prescient intensity. For him at this critical point it had two sovereign qualities: it was antecedent to written literature and so the most ancient substratum of the literary impulse, the Adam of literature; and it was wonderfully suggestive—the ancient myths could be made to express almost anything you liked, could appeal with equal force to a child or a meditative adult. In a passage in *Week* adapted from a journal entry, Thoreau defined for himself the inexhaustible charm of myth when he observed that when we read that

> Bacchus made the Tyrrhenian mariners mad, so that they leapt into the sea, mistaking it for a meadow full of flowers, and so became dolphins, we are not concerned about the historical truth of this, but rather a higher poetical truth. We seem to hear the music of a thought, and care not if the understanding be not gratified.

But even before moving out to the pond, he had come to know with a certainty that it was no good attempting imitation of any aspect of the ancients, that a writer had to take his inspiration from where he was, not where others had been. So, three days after moving to Walden, he had written in his journal that he was glad to remember, sitting at his cabin door at evening, that he was "at least a remote descendent of that heroic race of men of whom there is tradition." Gazing out through the intervening trees on the pond's darkening waters, he thought himself on the "shores of my Ithaca, a fellow wanderer and survivor of Ulysses." But that wasn't the end of it: the survivor had himself to prove heroic here in nineteenth-century Massachusetts amid the blare of the new, raw mass media, the persistent rumble of the railroad not three hundred yards west of where he now sat. *Here* was where the engagement had to be made, and "now where is the generation of heroes whose lives are to pass amid these our northern pines? Whose exploits shall appear to posterity pictured amid these strong and shaggy forms?" Myth, then, as an importation was out.

But by the logic of his thought, Thoreau was led to reason that if he was successor to the Greek heroes and that this had to prove his Greece, then he must attempt the myth of his own time and place. Why couldn't he dare, heroically, to sing of Walden, of the Concord area, and make that trodden, misunderstood landscape come alive for his contemporaries (and even for future generations) as the ancient bards had their own landscapes? If the mythic impulse was true once, it remained so: anything truly seen once remained true forever. To see his own life as today's myth, this was a way at once of extending the autobiographical and making it permanent, while to spend his life in daily contact with nature was a way of attaching the myth to ancient green roots. He would, he wrote in his journal, start with the facts, the natural facts, those that would "tell who I am, and where I have been. . . ." But his way of stating these facts would make of them the materials of myth: "I would so state facts that they shall be significant, shall be myths or mythologic."

In those first days at the pond in the summer of 1845, he was again occasionally beset by the feeling that it was foolish and trivial for him, a college man, to be "wading in retired meadows in sloughs & bog holes in forlorn and savage places," but then following hard on this feeling came the more compelling one bidding him to see himself as a kind of Adam, released into the freedom of a beautiful and bountiful landscape which it was his incredible privilege to describe for the first time. "Enjoy thy dominion," he wrote on August 24, both god and hero at once, "—and name anew the fowl and the quadruped and all creeping things[.] Seek without toil thy daily food—thy sustenance—is it not in nature?"

From this perspective, reading through Thoreau's journals from the pond years to the publication of *Walden* is like walking the fox's footprints through the winter's snow, the writer's remarks about mythology and about his own myth-in-the-making leading you closer and closer to the lair of his intent. By 1851, three drafts into the book, he was writing that the peculiar opportunity offered the New World writer was to recreate in this virgin soil a new mythology that would match that of the Old World—but not imitate it. That nature, he said, "which inspired mythology still flourishes. Mythology is the crop which the Old World bore before its soil was exhausted. The West is preparing to add its fables to those of the East."

Later that same year he wished for the kind of grand anonymity of the old myths that had no authors but seemed to have been composed by the very land itself. A "man writing," he said, "is the scribe of all nature; he is the corn and the grass and the atmosphere writing." By 1853, as he worked through the fifth and sixth drafts, he could voice clearly the myth he was writing of his life. "Some incidents in my life," he now saw, "have seemed far more allegorical than actual; they were so significant that they plainly served no other use. That is, I have been more impressed by their allegorical significance and fitness; they have been like myths or passages in a myth, rather than mere incidents or history which have to wait to become significant." The landscape, too, had become mythological to him, lifted to another level of significance. So, he could now confidently write of angling not for fish in Walden but for the pond itself and with the "hook of hooks."

The year following, and on the home stretch now, he looked back at the work he'd made and in a remark later incorporated into the final version said that mythology provided the foundation for all literature and poetry. The imaginative efflorescence of the morning of the human race, it spoke always of beginnings. Perhaps he dared to hope that his book might provide a similar foundation for a new, autochthonous American literature.

On August 2, he noted that "Fields to-day sends me a specimen copy of my 'Walden.' It will be published on the 12th *inst.*" Publication was early. On the 9th: "To Boston, 'Walden' published. Elder-berries. Wax-work yellowing." But beneath this laconism there must surely have been a sense of excitement; a sense also of astonishment much akin to that of the artist of the city of Kouroo. In the magnificent myth with which he ended his book, Thoreau tells how that artist wished to make a perfect staff. "Having considered," Thoreau writes, "that in an imperfect work time is an ingredient, but into a perfect work time does not enter, he said to himself, It shall be perfect in all respects though I should do nothing else in my life." The artist's singleness of purpose endowed him unawares with perennial youth. "As he made no compromise with Time, Time kept out of his way, and only sighed at a distance because he could not overcome him." Through the rise and fall of dynasties, the origins and crumbling ruins of great cities,he patiently worked at his task. At last, when he had put the finishing touches to his staff,

> it suddenly expanded before the eyes of the astonished artist into the fairest of all the creations of Brahma. He had made a new system in making a new staff, a world with full and fair proportions; in which, though the old cities and dynasties had passed away, fairer and more glorious ones had taken their places.

With a kindred patience and making himself no compromise with time, Thoreau had made a world of full and fair proportions. He had wanted to describe fully an American place in which so many heedlessly lived, and now, like the artist of Kouroo, he looked with amazement at the pile of shavings at his feet. He ended *Walden* with the admission that "John or Jonathan" might not realize what this book revealed about their places. But it was there nonetheless. *Walden* in its narrative of one man's confrontation with the discrete facts of a particular place—hoot owls, the colors of pond ice, the hum of summer insects, golden pickerel beneath green waters—tells all who will listen that America, the New World, is first and last a grand spiritual opportunity. And wherever we confront the facts of our place so steadily that we see them glimmer on both their sides— physical and spiritual—*there* America is the New World still. A few years after the book's publication he was still inveighing against the "folly of attempting to go away from *here*! When the constant endeavor should be to get nearer and nearer *here*."

He had wanted to make a myth for his own time, one that would be available at some level for all readers. Whatever success we may judge him to have achieved in terms of availability, there can be little debate that he did make a myth and that in it a small pond in an American woods had become as historic and poetic as any landscape in Greece. In 1838, just embarked on his career as a solitary scholar, he had written in his journal that he would like to read Walden's first page, just as it had come fresh from the press of Creation. No one could quite do that. But Thoreau had done the next best thing: he had written a book that read like that first bright page and that still does so.

At the time few saw this. *Walden; Or, Life in the Woods* was accorded a modest reception. Reviewers called it charming. But it seemed so intensely local that it did not recommend itself to a wide audience. Even so perceptive a critic as Henry James, Jr., missed the larger point, and from the cultural safety of Europe James complained that Thoreau was so local he was positively parochial. Probably Thoreau would have accepted this as a tribute, since he had become accustomed to having his most hard-won virtue misunderstood.

At the same time there is good evidence that Thoreau himself knew well enough that Walden and Concord were not the whole world even as in his book he had made them appear so. There was in him a deep and radically American drive for the wild, and though he had learned that the wild was to be found under your very doorsill, yet there was another sort of wild that was out there beyond the pond and town, on the west side of the mountain, from which as yet no true report had come. There was in him also a powerful telluric drive that impelled him to dig into things, to search out the roots of trees, words, local history. And it was the combination of these drives that ultimately led him farther into the woods, searching now not the the fox's lair but for the spirit of the New World. It, too, he believed, had a kind of lair, and he would track it there if he could. At the pond he felt he'd found America. But there also he thought often about pre-white America, about what lay beneath the soil on which he walked, about what was beneath the furrowed fields of his neighborhood. Had he, as he claimed in *Walden,* finally reached that "hard bottom and rocks in place" or was there not a lower layer yet to be explored, a stratum that if probed might yield up the quintessential spirit of the land?

At least since his teenage years he'd been interested in Indians. He had a peculiar facility for discovering aboriginal artifacts wherever his walks took him in the Concord area, being able to detect their presence with what appeared to others an uncanny accuracy. Hawthorne remarked that Thoreau "seldom walks over a ploughed field without picking up an arrow-point, a spear-head, or other relic of the red men—as if their spirits willed him to be the inheritor of their simple wealth." But it was far more than the artifacts themselves that interested him, though these in their factuality, their thingness, were surely of interest. Rather, it was what lay behind them, what life, what spirit. They were native, radically so, and he wanted to get at that native quality.

In *Week* are to be found his first public comments on Indians. Most of these are pretty standard for the time, though in a few you sense an edge, as if the writer were beginning to think beyond the familiar Romantic stereotype of the Indian-as-at-one-with-Nature, et cetera. And by the time Thoreau had finished that book he was thinking beyond that easy image. Now he was interested in the Indian as the Cecrops of the New World, one who sprang up out of native soil, whose crafts and traditions were the natural expressions of the continent. In the Indian fact books he began compiling at Walden he noted reading in *Jesuit Relations* of the Hurons' belief in place spirits and perhaps also of their emergence myth, which tells how that tribe came up out of the soil and rock. In his bean field at the pond his hoe struck down through the sandy soil and rang against the buried implements of the vanished first inhabitants, and he turned up to the light of the present these silty, unexpected gifts of his husbandry.

The buried artifacts he unearthed were also symbolic of his predicament in his search for the aboriginal New World in that they were about all there was left of the Indians of his neighborhood. Occasionally he would encounter some old Indian woman moping along a country lane; or a man selling baskets door-to-door in the village. Such encounters hardly constituted the sort of authentic contact Thoreau more and more consciously sought, but he knew where it might be had. His father had once traded with the Indians up in Maine, and as a younger man Thoreau had himself made a brief excursion there in 1838. In his fragmentary notes on the experience there is a single sentence that stands out. He had been talking with an old Indian who sat at the end of a dock at Oldtown, idly slapping his moccasined foot against a piling. Then, pointing up the Penobscot River, the old man exclaimed, "Two or three miles up that river one beautiful country." In 1846, he interrupted his stay at the pond to search for the beautiful country that lay hidden in the dense forests upriver of the last of the white man's settlements.

This was the first of what were to be three entries into the wild world in which the Indian was still much at home, that aboriginal America that, as he wrote his friend H.G.O. Blake, began where white America left off. He didn't get much of the Indian on this first one, however: the Penobscot guides they thought they had engaged to meet them upriver never showed up until Thoreau's party was coming back down to civilization, and Thoreau was disgusted to learn that the reason for their nonfeasance had been an extended drunk from which they were still recovering. He did, however, make significant contact with the Indian's world. Here was no woodlot on the edge of a prim village, but Nature "stark & grim," a huge, primitive force that required a set of adaptive responses—physical, mental, spiritual—such as the tribes themselves had had to make. Thoreau was powerfully, ineradically impressed, almost overwhelmed.

The terminus of the trip, both geographically and psychologically, was Mt. Katahdin, which Thoreau's party climbed and camped out on. Thoreau himself on the second day went on alone and achieved the summit. On his return to Walden in mid-September he was still very much in the grip of that mountaineering experience, and from his notes he composed a draft of a long piece that was to become the first chapter of *The Maine Woods*. In it he described his sudden, mysterious feelings of awe and terror as he came down from Katahdin's summit. We haven't really seen nature, he writes, "unless we have seen her thus vast and drear and inhuman." Nature on the mountain's summit was not a kindly mother, dandling her children on an ample knee.

> Nature was here something savage and awful, though beautiful. I looked with awe at the ground I trod on, to see what the Powers had made there. . . . This was that Earth of which we have heard, made out of Chaos and Old Night. Here was no man's garden, but the unhandselled globe.

In such a setting Thoreau had no trouble imagining the Indian's spiritual world: he was there in its midst. This was a "place for heathenism and superstitious rites—to be inhabited by men nearer of kin to the rocks and to wild animals than we." If he did not there make an offering to the old Powers, he surely felt their presence. He had long spoken of the gods, plural. On Katahdin he came to know them, those unseen but real guardian spirits. Here for certain was contact with the soul of the country. "What a place to live, what a place to die and be buried in," he exclaimed. And to think that all this lay unsuspected beyond the fringes of contemporary civilization! "You have only to travel," he continues,

> for a few days into the interior and back parts even of many of the old States, to come to that very America which the Northmen, and Cabot, and Gosnold, and Smith, and Raleigh visited. If Columbus was the first to discover the islands, Americus Vespucius and Cabot, and the Puritans, and we their descendants, have discovered only the shores of America.

He wanted more contact with that world, and in 1853 he got it, this time going up from Bangor to Monson, across Moosehead Lake in a steamer, then by canoe up the West Branch of the Penobscot to Lake Chesuncook. On this trip he also made contact with his Indian guide Joseph Aitteon, descendant of a long line of Penobscot chiefs and successor to that station a few years hence.

The contact with Aitteon was both disappointing and bracing. Aitteon was hardly a savage—or "salvage," the archaic spelling Thoreau liked because its root was in "silva," woods. Instead he was in many respects as much a man of civilization as he was of the wilderness. He whistled pop tunes, used the white man's slang, was cheerfully dependent on civilization's provisions during his wilderness excursions. The tribes of New England, Thoreau was learning, had all been more than merely touched by the white man's civilization. To find the kind of unalloyed tribal life he was on the trail of, he would have had to travel far beyond the Mississippi.

Still, Thoreau found in Aitteon and his fellows some of those qualities he had learned to admire. Aitteon was a superb boatman and accomplished in woodscraft. Thoreau was excited to hear him call to the moose from a birch horn—though the actual shooting of a moose cow and her butchering was a process that revolted him. During their days together Thoreau found what he most had hoped to find by his exploring: that this Indian really was an autocthon of so grim and shaggy a place, was thoroughly at home in it however much he now lived below in a village. He could find his way about in it and was familiar with its most recondite phenomena. One night, sitting about the fire with Aitteon listening intently for the sounds of moose, Thoreau and a white companion

> heard, come faintly echoing, or creeping from far, through the moss-clad aisles, a dull, dry, rushing sound, with a solid core to it, yet as if half smothered under the grasp of the luxuriant and fungus-like forest, like the shutting of a door in some distant entry of the damp and shaggy wilderness. If we had not been there, no mortal had heard it. When we asked Joe in a whisper what it was, he answered,—"Tree fall."

Back in Concord once more, he set to work on an essay he called "Chesuncook," which he eventually sent to *Putnam's Monthly Magazine.* Maybe it was the dreariness of that endless wall of woods and the bloody ending to the moose hunt that caused him now to confess he felt the wilderness was best taken in occasional doses, that it would not do as a steady diet. "It was a relief," he writes, "to get back to our smooth, but still varied landscape." The poet must, he says at the end of the essay, "from time to time, travel the logger's path and the Indian's trail, to drink at some new and more bracing fountain of the Muses, far in the recesses of the wilderness." It is a somewhat lame ending to what had really begun as a vision quest, as if the actual experience of the Penobscots in their native place had been too much for the Concord traveler.

But it wasn't an ending to the quest. Thoreau went back yet again in 1857, on this excursion going beyond Lake Chesuncook, along the Allegash, then turning south on the East Branch of the Penobscot and so down to the river's main course. And as on this third and final trip he went farther into the wilderness than ever before, so in spirit did he also, drawing nearer to his guide, the Penobscot Joe Polis, than he had to Aitteon. On his first trip he had been introduced to the debauching effects of white civilization on the Indian. The second trip had showed him the blood-and-guts realities of that hunter's life so often sung in Romantic literature. But also on that 1853 trip he had spent a night around a campfire listening to a group of Penobscots converse in their native tongue, and he had felt then "as near to the primitive man of America" as any of the earlier explorers had gotten. Now on this third trip he was to get closer yet to his ultimate goal of seeing the Indian, native on his grounds. Passing over the lakes and up the streams, he heard Polis retell the legends of the various spots they looked out on. He took notes on the man's methods of woods orientation and found them unfathomable. He watched Polis make thread from the roots of the black spruce, then failed successfully to follow the Indian teacher's example. And he heard him talk to a muskrat. Nature, he concluded, "must have made a thousand revelations to [Indians] which are still secrets to us."

But the best and most lasting revelation was made to the reverent white traveler, too. In the weird glow of moosewood in a campfire's embers, Thoreau saw the presence of the Indian's place spirits. These woods, he writes in "The Allegash and East Branch," "were not tenantless, but chokefull of honest spirits as good as myself any day . . . and for a few moments I enjoyed fellowship with them. Your so-called wise man goes [to the woods] trying to convince himself that there is no entity there but himself and his traps, but it is a great deal easier to believe the truth." Here indeed was that discovery he had gone forth to find: the place and its resident spirits and the man at home with both. What he had found for himself in the tamer landscape of Concord was true even more evidently in wildest Maine. Maine, shaggy and unimproved, was America, too, however removed it was from the messy march of Progress, and he had now taken possession of these farther shores in a way unthought of by the more prosaic explorers of old.

Thoreau never finished his Maine woods book. In his last months, shot through with tuberculosis, he had worked hard at it, spending grueling days trying to untie a knot he found in its concluding section, but it wouldn't come undone for him. The manuscript's subjects, he said sparely, were "the Moose, the Pine Tree & the Indian," and when he died on a May morning in 1862, he was still musing on this work, muttering in his final moments, "Moose . . . Indian. . . . "

II

❧ WRITING ON THOREAU AND AUTOBIOGRAPHY in America, the critic James M. Cox says that when we visit Walden today "we inevitably recognize that we did not really have to go there." The distinguished cultural historian Leo Marx says the significant location of Walden is not outside the village of Concord, realistic as the book's setting may at first appear. "On inspection," Marx writes, the setting "proves to be another embodiment of the American moral geography—a native blend of myth and reality."

From their points of view they're both right, of course. In considering the significance of *Walden* as text or cultural artifact, the actual pond and surrounding woods recede into the remoteness of metaphor. From the standpoint of literature viewed as art there may be something inherently silly about a literary pilgrimage—maybe about any pilgrimage, for that matter. There is a kind of oafish literalism lurking about the edges of such acts, as if by going there (birthplace, gravesite, shrine, pond) you could surprise magic, catch it in its lair, as Thoreau playfully suggested he might in tracking the fox also track the Spirit of the Wood. The appreciation of art, and especially great art, lies, as we know, beyond pilgrimage and any kindred denotative effort, and I suppose that in following Joyce's traces through Dublin, or seeking out Hardy's Wessex settings around Dorset, you may only be adding to the already substantial heap of the Errors of the Vulgar.

On the other hand, there's a lot to be said for the exploration of places authors have vividly made us see since clearly they were important, and in some cases crucial, sources of literary inspiration. Whatever these places become in the works themselves—metaphors, moral geography, or merely rich atmosphere, as, say, in the works of W. H. Hudson—the places themselves are still significant, and a purposeful tour of Dublin's winding streets and along the Liffey can give you insights into Joyce's achievement available nowhere else. There isn't much unimproved heathland left in Dorset, but in the few patches there are, you can come closer to Hardy's intent when you see sunset on them or the steel of a morning's lowering light. The Wessex landscape is a character in his novels, complete with spirit, force, pedigree, past, and an actual encounter with it enhances understanding in a way quite the opposite of any literalism.

As far as Thoreau and *Walden* are concerned, I confess I always had trouble with him and his most famous book until I saw the pond itself. When as a college sophomore I was introduced to him in an American literature survey course, I found him by turns boring ("Economy") and insufferable ("Higher Laws"). Occasionally these chapters still strike me so, the former putting me into a nod, the latter rasping across my spirit like a bow played roughly aslant. I admired Thoreau; who but a thug could not? And I assented, sort of, to the quiet heroism of his life. But both admiration and heroism hadn't much blood to them. It wasn't until I had moved to Massachusetts and been taken by a friend to see the pond and its woods that I was at last able to come to loving terms with Thoreau and *Walden*.

It was a bright, late October day. The inevitable end-of-fall weekend of rain had come, and driven by the wind they used to call the "Montreal Express" it had slashed the glory from the trees so that now the wooded sides of Walden's bowl wreathed the still waters like smoke. Fortunately for me, my friend was one who had had his own meditative hours at the pond, and he left me much to myself, understanding my mingled awe, my instant and consuming hunger. I wanted everything there all at once, suddenly face to face with all I had previously failed to imagine. On impulse I stripped in some pondside bushes downslope from the cabin site and plunged into the burning waters, emerging immediately with a talismanic stone clutched in my numbed fist. I wish I had then cried, "Contact! Contact!" as Thoreau had on his way down from Katahdin, but I didn't have the breath.

Thrilling as that first visit was, it was not finally an exuberant experience. Seeing the placid pond that day as it lay waiting for winter to seal it for a season, browsing the woods with the tatters of the still bright leaves underfoot, hearing the wide-spaced bird calls—all of this seemed somehow somber despite my almost desperate excitement. Here, I felt for the first time, was a life lived in earnest. Here was where the lone man had fought past his long misgivings, his profound fears and puzzlements, to wedge his feet downward "through the mud and slush of opinion, and prejudice, and tradition," to a hard bottom and rocks in place. The pond seemed still and forever signed with the integrity of his intention, proof against the designs and attitudes of all who should come to it in his wake.

While I lived in Massachusetts I saw the pond in all seasons. In winter, alone or in company, I built bonfires on its thick ice at the point where I judged Thoreau had found its greatest depth intersected by its greatest length. Warmed by their winter-edged glow, I would read relevant passages from "The Pond in Winter," glancing up now and again to have the prospect of the northern shore and the cove above which the cabin had stood. I liked then to imagine the curl of smoke from that unseen chimney, rising with a cheerful fortitude above the naked clack of the trees. Or to imagine myself one among those crews of ice-cutters whose jolly company Thoreau found so heartening.

As spring came grudgingly to New England, I would wade in toward the pond through the slushy snows of the first half-reasonable day in March, the clouds scudding through the blue, the sun flitting down along the dank trunks of the trees. Then the best place was the northeastern shore that Thoreau identified as the pond's spring sunspot. There was a bench there, and one afternoon a friend and I sat on it, wrapped in over-coats and mufflers, drinking wine, reading favorite passages from *Walden* and toasting the author's achievement *in situ*. At that season the skaters had been warned off by the small winding rivulets you could trace along the blinding surface of the ice, as if these were the courses of underground streams.

In summer I swam naked in the shadowed green waters of the south-western shore not far from the railroad embankment. Once, paddling about there, I disconcerted a family in a fishing boat that had drifted near enough to espy my innocent indiscretion. The man stood up hastily in the stern, pulled his cord, the doughty Seahorse pulsed an oil-rich smoke and stood out for respectability. Then I had occasion again to remember that Thoreau himself was an inveterate nude bather, here and along the streams and rivers where he commonly walked with only a hat or a shirt on to shield his head and shoulders from the sun.

The cabin site was always slightly disappointing, and over the years I learned to avoid it. There was, of course, no cabin there, though replicas were to be seen in a number of other Concord sites. There were only markers outlining the cabin and a boulder bearing a plaque. And after a while the plaque was gone, too, doubtless pried off to confer some tem-porary honor on a college dorm room. There was also a granite shaft that told you the site had been discovered November 11, 1945 by one Roland Wells Robbins. For some reason the site always struck me as the pond's one empty spot.

When I visited Walden again it was after a lapse of years. I had moved from Massachusetts into that West Thoreau thought more and more about in his late years and of which he had but a fleeting vision when, a dying man, he saw Minnesota and the Sioux, then poised on the edge of their fatal 1862 uprising. My guide this time was Walter Harding, author of the standard Thoreau biography and almost a dozen other books on Thoreau and his circle. It was summer and Harding was in Concord to lead a National Endowment for the Humanities seminar for high-school teachers. "It's supposed to be on Thoreau, Emerson, and Hawthorne," he said as we talked by phone, "but with a proper sense of proportion we spend four weeks on Henry and a week each on Emerson and Hawthorne." He couldn't, as it happened, take me around the pond that day but could the next. In the meantime, he suggested, why did't I have a chat with Roland Robbins, who had discovered the cabin site and excavated it.

I found Robbins at his house outside town on the old Cambridge Turnpike. He was mounted on his tractor-mower in New England's humid haze, his machine adding to the local miasma with its blue exhaust and the green mist of minutely mulched lawn. In his late years Robbins had come to regard himself as a work of art, an attitude, as Constance Rourke once observed, that is the origin of the comic spirit. When we were settled in one of the two replica cabins he'd built on his property, he wanted me to understand that though he had excavated the cabin site, sure enough, he had done many more important excavations and reconstructions elsewhere: Crown Point, Fort Ticonderoga, Sleepy Hollow, and Salem Village, where he'd found in the detritus of the West Indian slave Tituba's house some scattered gold beads—bright relics of a rumored black magic.

The cabin thing, though, he was telling me through the steady, bubbling bounce of his talk, was more or less a whim. He had known, of course, the significance of July 4, 1945, and had gone out to the supposed site of the cabin on that centennial date to "find the thing, and, by God, I found it!" Not that day, though. He showed me his brown-backed log books of the search, including the entries for November 11 and 12 of that year.

"I was all ginned up by that point," he said with a twinkle. "I knew I was close. I found the chimney foundation on the eleventh, and on the twelfth—*one hundred years to the day Thoreau finished it* [the cabin]!—I excavated it. Do I think he *wanted* me to find it? You're damned right I do! I think he wanted me to stay right there until I found the damned thing! Oh, I'm supposed to be psychic and all that, but I don't bother with it. It was a hell of a lot of hard work, I'll tell you."

He wanted me to understand that this was no hut we were talking about, no playhouse, either. "It was a sure-enough house," he said emphatically. "Have you read *Walden*? Yes, well you go back and read it again. You'll see there it was a beautifully finished house, too, and I have 500 pounds of brick and plaster and window glass up in my attic to prove it." He pulled out a box of rusted, twisted nails from a place on a shelf next to a set of Thoreau's journals and copies of his own *Discovery at Walden.* Then he showed me another box within which were a few fragments of brick with clots of mortar still clinging to them. The relics were oddly, deeply impressive.

But soon enough Roland Robbins had said all he wanted to on the subject of Thoreau and his cabin. There were a good number of other aspects to his career he wanted me to know about, and so I stayed on to hear all about the excavations at Fort Ticonderoga and Salem. And it was worth it, too, as much for the intrinsic interest of these matters as for Robbin's narrative style. Here, he would have a visitor know, was no One-Shot Willie, despite the local fame of his discovery at the pond, but a sure-enough archeologist. At the end of a stunning afternoon of his talk I at last bid him goodbye, though it was evident he hadn't begun to get to the bottom of his anecdotal bag. But even so, he had ventured freely so much of himself that as I stood at the threshold of the model cabin I found myself saying, "You know, Mr. Robbins, you're a character."

"Oh, I know I'm a character," he shot back with a high laugh, "and I love it!"

After his NEH seminar had concluded its business the day following, Walter Harding and I drove out of the leaf-shaded town with its neat lawns and handsome old homes to a parking lot just above Brister's Spring, from which Thoreau used to haul water when that of the pond was too warm in late summer's dog days. Harding was a sturdy, ruddy man in his late sixties who on this day wore a blue t-shirt with the legend, "Simplify, Simplify.—Thoreau." I was certain, I told him, that in taking me around the pond, he would make me see things that I would never notice. He smiled, accepting the tribute, and we went up the hill and across Route 2 that runs from Boston out into the western hills of the state. "We'll go in here," he said at the northeastern edge of the pond's woods. "I want to show you something." We walked through a stand of dwarf oak with an understory of briars and years of accumulated leaf mold. Harding was obviously searching in here for bearings and at last marched straight ahead until we came to the edge of a faint declivity.

"This is where Zilpha's cabin stood in Thoreau's time," he said. referring to the ex-slave woman whom Thoreau mentions in *Walden* and in his journals: a local fortuneteller, a weaver of yarn and fates, whose singing was once heard in these woods. "I don't think this has been that much remarked on," Harding was saying now as we stood there in the filtered light with the ground under us occasionally shaken by the passing of the heavy diesels on Route 2, "but when Thoreau came out here, this was sort of the slum area of town. This was where people came to live if they couldn't make it in town. And there's something appropriate in that: I mean, in terms of local opinion about Thoreau. You know about the woods-burning incident, and so forth. I don't think they ever quite forgave him for that."

This was the burden of our talk (mostly his) as we passed from Zilpha's place westward along a fire road, past the old Lincoln Road that branched away in a gentle curve, showing us its age by the depth of its grassy bed. Here was where Thoreau had his bean field though now the site was grown up to oaks, some of substantial proportions. Returning to his theme at Zilpha's house site, Harding said that there remained a faint but still definite residue of mistrust and dislike of Thoreau in the Concord area, a kind of substratum of local contempt that would be surprising to those thousands of pilgrims who come here each year to walk the pond's circumference and meditate at the cabin site. "They're always surprised if they encounter it," he said, "but it's real enough. The folks over there at Walden Breezes (nodding in the direction of the trailer court across the road from the pond's eastern shore), they'll tell you—if they admit to knowing who he was at all—that Thoreau was a loafer and a drunk. You could tell the routes of his walks, they'll say, by the trail of empty rum bottles he left behind him." Part of this he attributed to bitterness. Since the pond became a shrine, a series of state rulings has forced the removal of a small concession stand at the court and decreed that though its residents may remain the durations of their lifetimes, no new residents may come in. In consequence, the value of the court property had shrunk, and the residents felt themselves isolated on the edge of a phenomenon they neither shared in nor understood.

But it wasn't, Harding went on, just the folks at the trailer court: it was more established, genteel opinion in town as well. When the first Thoreau Society meetings were held in Concord, Harding said they occasioned a good deal of surprise, even mystification. "The attitude was," he recalled, " 'Why would you want to hold a meeting about *him*? Emerson's our great man.' "

Here we turned south toward the pond itself, and descending the gentle slope Harding pointed out in the leaves and brush a gray, gnarled stump, the jagged edges of which now barely protruded above the thrust and decay of succeeding generations of plant life. This was a remnant from the stand of white pine Thoreau had set out some years after his residence here and about which he had cared so much. In the winter of his last illness, his journal entries slowing, spotty, he noted his sadness at seeing one of his pines bent almost double by its weight of snow, and then later his joy at seeing the tree upright once more—as he must have surmised he could not hope to be again himself. Harding pointed out that the stumps, though hard to make out, were in a row, and pausing at one I broke off a sliver and caught in the instant's rupture the merest whiff of ancient fragrance. Most of the trees had been consumed in a railroad fire of the 1890s and the rest in the great hurricane of 1938. Harding said there was yet one standing, and later, when we'd reached the opposite shore, he pointed it out to me, defying me to go back over there and find it. I tried the next day and couldn't.

At the cabin site we ate a light lunch while in the background a shirtless young man balanced atop the rock cairn there and turned in the four directions, clearly in the grip of a Thoreauvian ceremony. Harding said he'd seen marriages performed there and baptisms and all sorts of other, private rituals. Many a delightful day he had spent at the cabin site, pretending to read but actually listening to the talk of the pilgrims. He said he'd even slept out there one night, "though they won't let you do that any more. They patrol through here now several times each night. But years ago that wasn't the case, and one night I put my sleeping bag right down over there next to the cairn. In the middle of the night I felt something: it was a fox standing on my feet. It was a moonlit night, and I could see him clearly. When he saw what he was standing on, he jumped right over my head."

Then, walking the edge of the pond around to the south shore, following the old aboriginal trail deeply incised into the slope, he told me a clutch of anecdotes relating to Thoreau and Walden, including one about the old Concord judge, Prescott Keyes, "old enough back then in the 1930s to have remembered Thoreau himself. Anyway, the case brought before him was that of two young men arrested for swimming nude in Walden. The young men seemed in danger of at least a fine, but the old judge was so deaf the bailiff had to repeatedly shout the charge to him: 'Nude swimming, your honor! In Walden Pond!' Finally, the old boy got it: 'Goddammit! That's the *only* way to swim there! Case dismissed!' "

In our circuit Harding took some pleasure in the physical condition of the premises—very good these days, he said, except for the shore erosion due to the generations of walkers. Back in the late 1940s and early

1950s, he recalled, "this really was the slums. One morning Ed [Edwin Way] Teale and I made a circuit of the pond and jotted down a list of all the junk we found strewn about or floating on the water. At that time Ed had a syndicated column, and he printed the list. It got a good bit of play, and I think it was instrumental in getting the pond cleaned up and policed."

Indeed, the pond did look spruce, and at its eastern shore, below the frown of Walden Breezes, the famously forward-working gears of American culture had miraculously been turned back, for the noisome old bathhouses I'd remembered were gone and so was the long, unsightly concrete pier. Now the area was a grassy open slope on which this day sat a small group of children from the Boston Center for the Blind while their teacher explained to them the significance of this unseen place. Some of them ran extended palms lightly over the tops of the grass, their heads bent down in attention.

On our return to his rented summer home, Harding went back over his long career as a Thoreau scholar, one, as he could now see, that had its beginnings back in high school at Bridgewater, Massachusetts, where a brilliant, inspiring teacher first got him interested in American literature. "I started reading all around him," Harding said with a faint smile as we rolled into his driveway just off the Commons, "rather like the kid who eats all the rest of the cake and saves the icing for the last. Then I read *Walden,* and I said to myself, 'This is the boy for me.' And he has been."

I could see one of Harding's NEH seminar students waiting for him at his door, shifting politely from one foot to the other, and no doubt Harding would spend the next several hours with him, talking as enthusiastically of Thoreau as he had with me. But I had to detain him for a last question. After the half-century he'd spent thinking about Thoreau and with all he'd learned about the man, was it better to have had to know him solely through the books? Harding tilted his head back against the headrest and stared upward a moment. "I have no doubt he would have been a very difficult person to be around," he said slowly. "Especially if you were an affected or pompous sort of person. I'm sure he could be cantankerous and contentious." Then he turned toward me with a great, sunny smile. "But, gee! I wish I could have taken a walk with him!"

Thoreau had said the subjects of *The Maine Woods* were the moose, the pine tree, and the Indian. I thought of them as the moose, the mountain, and the Indian, and I wanted to experience something of that triumvirate for myself, to extend, after all my own years at Walden, my sense of Thoreau's whole career. For surely Maine was a significant extension of it. It used to be said that the Maine woods experiences had shown Thoreau that he could really love nature only when it was fairly well pruned and tamed, and it's true that he makes remarks tending in this direction at the end of the Chesuncook section and that he let them stand in his late revisions of the manuscript. But there are other remarks in the manuscript that indicate that at last Thoreau had fought past his first reservations about the wilderness up there and that the Maine experiences deepened and extended his love of nature in the raw, that he met the challenges of the woods, the moose hunt, and the awesome mountain. In doing that, he had extended the literary landscape beyond the reaches of the white settlements, those shores of America, and on into the green heart of the New World. So, like Thoreau, I left behind the assurances of Concord and Walden and on a late summer's day headed north for Maine.

By prearrangement I stopped at Bremen, Maine, to pick up naturalist and author John Hay, whom I'd met several years before while living near him on Cape Cod. That had been a savage spring on the Cape, full of dark, stunted days of rain and sudden, unseasonable snowstorms. I nursed a lingering cold and grievances against the cruelly exposed spit of land I'd chosen to live on, and Hay had tried to cheer me up by taking me for walks along the beaches and marshlands he knew so well. One day near Brewster we watched the annual up-river run of the herring, but I fear I wasn't feeling much of the quiet joy my companion expressed over this phenomenon.

Hay was just short of seventy, his handsome patrician face weathered by seasons spent out of doors watching the herring and the herring gulls, the migratory patterns of birds, peering into the underworld of mosses. It was a bright morning at his seaside cottage, the fog lifting off Broad Cove, and a few gulls skimming the woods and shoreline. He took me out in a wooden skiff to a rock where more than fifty seals sunned themselves in the new day. Across a small channel lay another large rock cluttered with cormorants, terns, and gulls. We swung tightly around them, the birds moving nervously, watchful, and most of the seals dropping off with plump splashes. Hay told a story or two of these creatures, of the amazing adaptability of the herring gull and the 18,000-mile flights of the arctic tern. "These ought to be matters," he said, "for general admiration, instead of which we regard such creatures as freaks. Whenever an unusual creature comes within our reach, we don't know how to react."

His jaw was set and he looked at me with something of a glare in his eye. "We don't seem to know where we live. On the Cape recently there was a bald eagle at the dump, and people gathered there, not to marvel at it, but to *stone* it! Then there was a dying whale on the beach, and it was still alive, rolling its eyes. People gathered around and threw sand at it, and some of them dropped lighted cigarettes into its blowhole." The putt-putt of his outboard took over for a few moments as Hay lapsed into a silence. Then he said, almost to himself as we turned back for his dock, "I don't know. . . . Maybe I'm just getting cranky with all my complaints about modern civilization." It seemed more a rhetorical remark than a genuine admission of self-doubt, and I could hardly imagine him giving up a jot of his admiration for any of the natural world—or any of his Thoreau-like amazement that more of his fellow humans didn't share it.

Our plan was to roughly retrace Thoreau's first extended visit to Maine in 1846, ending with a climb of Katahdin. Hay said he was primed for the ascent, and I had no doubt he was, seventy years or no. That made our first stop Old Town, where Thoreau had hoped to get Penobscot guides for that first trip and where on his second and third trips he'd had the fortune to engage Joseph Aitteon and Joe Polis. This last stand of the Penobscot nation sat just above Bangor (in Thoreau's day the jumping-off point) on the river named for the tribe. Indian Island was the tribal seat now, and a bridge took you across to its narrow streets and wooden houses built the century before. When we were there a huge new tribal center was under construction, its concrete walls decorated with traditional designs. Across from it was the office of Glenn Starbird, tribal genealogist.

Starbird told us his title was now, formally, Rights Protection Researcher, but under whatever designation he knew the tribal lines of descent thoroughly, and when I told him I'd like to meet any descendants of either Aitteon or Joe Polis, he said there weren't any direct ones. Aitteon, he said, had died heroically on the West Branch near Millinocket in 1870, trying to save the lives of three fellow boatmen wrecked on a log jam. He left no known direct descendants. Joe Polis had had no children. But Starbird said he could introduce us to Priscilla Attean (the current spelling), a descendant of both the Aitteon and Polis families, and in the outer office he did so.

She was a broad, handsome woman with square teeth and intelligent, inquiring eyes behind her glasses. Currently, she was serving her first term in the state legislature. Before that, she'd taken a degree from the state university just down the road at Orono, and it was there she'd encountered literary evidence of stories she'd heard growing up: how her relations had once guided a white man up into the woods and that he had subsequently written about his experiences. Her family, she told us, hadn't said much more than that about Thoreau, and so she knew only what was there in the books. But it was clear from her remarks that she knew *The Maine Woods,* all right, could rehearse its details if she wished. There was little point in that. I asked her if the Polis house was yet standing, and she obliged with easy directions to it.

When he'd seen the house in 1857, Thoreau had been somewhat surprised by its substantiality, remarking that it was a "two-story white one with blinds, the best looking that I noticed there, and as good as an average one on a New England village street." He had already expressed in his notes his particular desire to have an aboriginal guide on this third trip, and not just any one, either, having already rejected one man because he was "too dark-colored, as if with African blood . . . and too young for me." Perhaps, then, viewing the stout, four-square home of Joe Polis he might have wondered whether this man would prove aboriginal enough. In the event, Polis's house was indeed a reflection of his wealth and social standing but not at all of his Indianness: he had maintained all his essential connections to the wild. The house was still impressive in its stoutness and its spanking coat of white paint. From its roof protruded that lodge pole of modern civilization, the TV antenna.

Joe Polis was less than a hundred yards away in the old burial ground where, off to one side, were the graves of the man and his wife. Mrs. Polis's shaft lay face down in the rank grass, and Hay and I righted, for a moment anyway, this terminal wrong, then read the blurred inscriptions on both stones. Standing there you couldn't help but think of what went into Thoreau's Indian quest, his desire to have just the right guide. Complex motives here, for sure, and I told Hay I had had similar feelings myself from time to time. The white man's desire for an Indian guide, a teacher, is both pathetic and foolish in some ways, and when you happen to be on a reservation or at a tribal dance you sometimes see a startling reflection of this in the sardonic glance an Indian man will throw your way: they know the impulse, are not fooled by it. This said, there are certain circumstances under which the impulse can have genuine value, as in Thoreau's case it did. He wanted more than the dusky-skinned romance of the thing, the easiness of Fenimore Cooper's idyll. He wanted to get at the radical nativeness of the Indian, and in apprenticing himself to the man whose bones now moldered under our feet he got a measure of what he'd sought.

From Old Town we swung north again to Millinocket, a place name Thoreau learned was Penobscot for "Place of Islands," and from which his two white guides on his 1846 trip had come. It was a dreary mill town now, dominated by the spectral smokestacks of the Great Northern Paper Company, the current baron of the north woods. J. Parker Huber, retracing Thoreau's 1846 trip with scrupulous fidelity, discovered that the route now led smack through the Great Northern parking lot.

The company was now in big trouble, victim of that lethal cartel of bad international exchange rates, plant deterioration, and wage differentials that allow workers in Canada, the southern states, and Finland to produce more tons of paper per diem than can be done in Millinocket. It used to be said that high-school kids here partied till dawn at graduation, then lined up at Great Northern for their jobs. No more, and maybe never again. In what some residents say was a last-ditch effort to save their investment and the towns of Millinocket and East Millinocket as well, Great Northern several years ago proposed damming the Big Ambejackmockamus Falls, thus creating a source of power that in generating megawatts would have saved both megabucks and lots of jobs. Environmentalists fought the proposal from the start, though, since it would have flooded a stretch of internationally famous white water and adjacent scenery. In what developed into a classic and bitter grudge match that still smolders here, the environmentalists claimed the project wouldn't ultimately save the jobs anyway. Later, the company seemed to agree, admitting that its problems were of such magnitude and complexity that, dam or no, the jobs were gone. So, the "Big A," as the project was called, is dead. And so is Millinocket: in whatever business establishment you enter—supermarket, motel, bar—the gloom of the superseded pervades the premises like a film of psychological soot. The faces of the residents strike the visitor as sad and resigned. Since Thoreau had a good many trenchant things to say about the wasteful practices of the primitive lumber companies of his day as well as about the crucial national need for places of scenic beauty, it seems clear he wouldn't have shed any tears over the fall of this industrial giant, whatever the human, household costs. There was that sort of moral ruthlessness to him. His view remained fixed on other phenomena, like the great blue mountain that rises above the town like a master spirit.

In the dwindling afternoon, Hay and I drove out of town toward its solemn, bear-haunted slopes, as far as the road would take us. "This is the end of civilization here," he said at last as we curled into a sandy space at the road's ending. "Here's where you *have* to see that something else begins."

And it was true. The road ended in a tangle of low ferns behind which was a stand of deciduous trees, then pine, and at last, looming blue in the faded light, Katahdin. Even so far from it you felt dwarfed and very finite. Hay wandered off into the woods a bit and soon called to me. He wanted to show me the tracks of a moose in sand pocked by an early afternoon rain; they were deep, definite, and about as large as a domestic cow's.

On the way back to town and dinner the experience at the road's terminus hovered over us like Katahdin itself, and the talk turned to mortality. "Do you," I asked Hay, "ever find yourself thinking of yourself as a kid? I guess I was thinking back there of Thoreau up here in '46, and how he must have thought of himself then as pretty much a kid. He was always talking in his journals about his unrealized life, how slowly his seasons revolved, how much he had left to do. Then, in *Walden* he says he left the pond because he had other lives to live when, in fact, he'd already lived most of his."

"Sure," Hay answered, "all the time, Especially in my dreams. I'll be seventy at the end of August. How does it happen that suddenly I wake up and find I'm *seventy*?!" He looked over at me with a fierce astonishment.

"But," he went on, in a quieter tone, "I hope I don't live to be so old I lose interest in the natural world. That would be like death to me. I read somewhere that Thoreau said in his journal that he found himself losing interest in the natural world."

"Yes," I said, "he feared he was from time to time, especially when he was sick or fatigued. I really think that's attributable to the onset of t.b. I had it myself when I was a kid, and I can still remember how absolutely listless it made me feel: I didn't want to do anything. The only good that came out of it was that I missed most of sixth grade."

At dinner I urged Hay to talk. He is by no means a monologuist, but under my questionings and perhaps prompted too by the martinis I insisted upon, he did talk about his background and his beginnings as a naturalist. The grandson of the John Hay who was Lincoln's private secretary and later Teddy Roosevelt's Secretary of State, Hay himself had tried to enter the State Department after World War II but, as he was saying now while I made notes on my paper doily, "It wasn't any good. I found out that Grandpa wouldn't do." Earlier, before the war, he had been the

Washington correspondent for the Charleston (S.C.) *News and Courier,* and had waited—through Munich, the sinking of the *Athenia,* and the invasion of Poland—for his number to be called. During this edgy period he lived on the Cape at Brewster, where he still lives when he isn't in Maine or off, say, in Costa Rica studying damselflies. It was there he had a signal experience that perhaps his precarious time of life prepared him to understand.

Conrad Aiken was then living at Brewster, and Hay spent his days with him, going to movie matinees and drinking in the evenings. "Conrad would start promptly at five," he recalled. "He'd drink the cheapest gin you could get: Dixie Belle and orange juice. And then he'd get off on the subject of consciousness, about which he could go on for hours.

"One day during this period, I was out along a stream there at Brewster and saw the herring coming upstream against the current, and it suddenly came to me: this is the Life Force. I'd been missing the connections between all natural phenomena that are tied together by this thing. It had tremendous impact on me.

"Later, having a drink with Conrad in the backyard of his house, I observed that consciousness, about which he had so much to say, was a shared thing, that humans weren't the only things in the universe that were conscious. 'Nonsense!' he said. 'Sheer nonsense! You're barking up the wrong tree there. You've bought that Schweitzer stuff about "reverence for all life." *We are superior!'*

"Well, we aren't at all. That's what's so wonderful about a place like Katahdin: it shows us our limits. If we really look at nature, we see this every day. But we don't know how to look. If we did, we'd see those limits, and we'd see how much life there is that hasn't anything to do with us but is absolutely right just where it is."

That night in the motel room we shared I drifted off into an uneasy sleep, hearing the mutterings of thunder, as if what Thoreau had called "the Powers" were warning us off our quest. Maybe there were, as Thoreau suggests in "Ktaadn," places where it was rash and even irreverent to venture. And as I fuzzily mulled this matter over, there came a tremendous explosion of thunder and a flash of lightning that exposed the shabby elegance of the room. I foresaw us rained out of our climb, and maybe I even partly hoped we might be, the mountain having intimidated me somewhat and Hay's talk having also induced a sense of humility and proper proportion. Who was I to think of clambering up this grand phenomenon, thoughtlessly entering into realms I took to be mine in my human arrogance? Thoreau on the mountain felt that huge awe, felt as if some vital portion of the climber there had escaped through the "loose grating of his ribs." Through the thunderous hours I had ample opportunity to doubt the wisdom of my trip. Hay slept soundly.

As it turned out, the day dawned bright and brisk. Though Katahdin itself was wrapped in a heavy white and gray mantle, the thick woods at its base were spangled with early sun and last night's rain. The mountain sat in the midst of Baxter State Park, the magnificent bequest of former Maine governor Percival Proctor Baxter. As a young man Baxter had read and been deeply impressed by *The Maine Woods,* and in 1930 he began his purchases of land from Great Northern and other holders, eventually deeding to the state a wild domain of 201,018 acres. When you climb Katahdin you are thus, thanks to Thoreau and Baxter, climbing through a landscape largely untouched since the former's day.

From the Roaring Brook Campground we climbed steadily for two hours through golden birches, canoe birches, and the undergrowth of mountain wood sorrel, hobblebush, maple leaf, viburnum, bunchberry, and sheep laurel. Near noonday we made Chimney Pond and lunched there. Across the pond rose the astonishing, sheer soar of Katahdin's granite bowl, the faces of the uppermost sides still slick from the night rain. Looking up I was glad when Hay suggested we take the more moderate Saddle Trail to the top rather than Thoreau's conjectured route up what was now called the Dudley Trail.

Even so, the Saddle Trail was another two hours of climbing, the last portion of which was increasingly steep, and at last almost hand over hand. Finally, we came out onto a green, ground-hugging sward dotted with the climbers' cairns of many years. Beneath, to the south, was that Maine country which Thoreau had come to see, a vast broad stretch of green. It looked this day and from this height just as it had to Thoreau in 1846: "No clearing, no house. It did not look as if a solitary traveller had cut so much as a walking stick there."

Up here, Great Northern and the dying towns it had spawned were invisible. Lake Millinocket flashed in the afternoon light like a bit of glass in a lawn, and the black shadows of clouds chased each other across the dense slopes. It was delightful then to simply lie back on the down of purple crowberry and dream up at the clouds; then to sit up at your leisure and dream down at them. Hay busied himself with prying gently into the composition of our soft bedding, alert as ever for some new phenomena that would, to use Thoreau's words, "extend the limits of the actual."

The way down was if anything more arduous than the way up: you've already put yourself out some in the ascent and so have that much less energy. Your knees have little tremors in them and the scenery that cheered you with its rugged newness on the way up now appears in the guise of a woody, rocky barrier to your ease. Then, too, I was conscious of the fact that Hay was after all seventy, but he kept pace with me and seemed no more fatigued than I.

I had given up on sighting the moose that would have given me what I thought of as the magic triumvirate of the trip. Our way up the mountain was simply too populous with other hikers for me to expect to see one of these shy creatures that seemed to Thoreau like a species of frightened cow. Even our stop at a small sheltered lake halfway to the summit had been punctuated with the sights and sounds of other climbers. Now down at Chimney Pond we paused to rest and share a last can of soda while the fading sun highlighted the mountaintop and the sheer face of the opposite shore. Then came a young woman running along the shore toward us, crying breathlessly, "There's a moose over there!" We went to where she'd pointed, and there in a dense grove of pines lay a huge bull moose, contentedly munching his dinner of seasonal greens. It was so dim in his regal hall that at first it was hard to distinguish his details, but the longer we stared, the more of him began to emerge from that green gloom: the broad boss of the antlers, the grand brown expanse of the head and neck, the calm, almost somnolent glance of his liquid eyes. Ropy filaments of green hung from his steadily munching mouth. As long as we cared to look, he tolerated our curiosity, and at last, the light going more quickly now, we left him there and made our final descent.

That, clearly, was the fit ending to our excursion, and we said so to one another. Fit, not only because we had seen the moose as Thoreau had, but fit also because we'd seen the animal alive, vigorous, and at his ease beneath the pines. Some part of Thoreau had wanted to participate in the moose hunt of 1853: he'd wanted to witness the aboriginal hunter at his livelihood, pursuing his food through the primeval woods. But when Aitteon had actually killed the cow and Thoreau had to confront the reality behind the image, his soul shrank back. The "still warm and palpitating body" that the Indian pierced with his knife, exposing the "ghastly naked red carcass"—this had sickened the white man.

No one knew better than Thoreau how much death was a part of life. The myths he so loved told him this with a magisterial, poetic insistence, and so did his own careful investigations of the progress of the seasons where every year he noted the green of new life pulsing up through the slough of decay. Later, his friends bore witness to the depth of his conviction of interconnectedness when in his own last days death drew its bonds tighter around him with each breath. No man, they said, faced his end with more cheerfulness than Thoreau. Nevertheless, there was something about that dead moose that haunted Thoreau in those last days as he heroically tried to clean up the manuscript, something he seemed to wish to alter, as if he thought the animal ought not to have died at the hand of the red slayer but instead on its own terms. And maybe he was still trying to restore its life with his own last breath as he spoke the words, "Moose . . . Indian. . . . " Leaving the Maine woods, John Hay and I had the image of the bull moose by the pond to think about, and I was pleased to imagine that in having that we had in so small a way completed something for Thoreau that he had wished.

THREE

A PILOT'S MEMORY: MARK TWAIN'S <u>LIFE</u> <u>ON THE</u> <u>MISSISSIPPI</u>

and

<u>HUCKLEBERRY FINN</u>

I

BY 1874 HE WAS THE BEST-KNOWN, best-loved author in America and enjoyed possibly an even greater repute in Europe. He was prosperous beyond the dreams of almost any writer in the world and certainly beyond anything he himself would have dared to dream of earning by his pen. He was blissfully happy in his domestic circumstances, the joyful, mercurial master of a household that included his beloved wife Livy, their three daughters, and six servants, all of them sumptuously ensconced in a newly built steamboat palace in Hartford where he received the famous and near-famous of his time. He was a forceful director of the publishing company that issued his books and insisted that the books be sold only by the subscription method. This meant that his industrious agents hived out into America's towns, villages, and hamlets carrying the prospectus for Mark Twain's latest book, rapping on doors, then showing housewives and husbands what grand, fat, funny volumes, what telling graphics and imperishable bindings were theirs for a few hard dollars.

All of it—the whole gaudy edifice of his life—rested on Twain's reputation as a funny man. He had earned that reputation in the most recondite fields of the republic: in the rough mining camps of the Sierra where he heard the violently colored tales of his fellows and remembered every one of them; as a journalist in the instant city of San Francisco; on the boards of the lecture circuit that so drained a man with missed travel connections, poor rooms, the greasy victuals of nameless hamlet hostelries, and then the gut-busting, nerve-wrenching performances in which he expertly milked his audiences for laughs.

He now had four books behind him. He had begun in 1867 with a collection of far-Western sketches headed up by "The Celebrated Jumping Frog of Calaveras County," a tale he'd heard in California and had so artfully embellished that it instantly established an Eastern audience for him. *The Innoccents Abroad,* issued two years later, was a farcical, irreverent, and at last tedious account of an American guided tour of Europe and the Holy Land. "It sells right along just like the Bible," he gleefully told a friend. Then two years after that Twain looked back to his youthful days in California and capitalized on his mining adventures there, padding out the book to the specifications of the subscription trade with sixteen additional chapters on the Sandwich Islands. The result was *Roughing It.* In 1873 he made his debut in fiction with *The Gilded Age,* a novel of murder, legislative venality, and financial skullduggery written in collaboration with his Hartford neighbor, Charles Dudley Warner. Thus he was famous before the age of forty, and still he had not so much as begun to mine the great mother lode of his childhood experiences in the Mississippi Valley of thirty and forty years before. Had he never gotten around to doing so, Mark Twain would be remembered today as the funny man he seemed in 1874 instead of as one of the authentic giants of our literature.

For many an author an imaginative return to the experiences of childhood is postponed to the later years, as if only having satisfactorily explored new worlds does the retrospective impulse insist upon itself. Twain's two great contemporaries, for instance, William Dean Howells and Henry James, waited until their mid-fifties to write *A Boy's Town* (Howells, 1890) and *The American Scene* (James, 1907), and in James's case it was not until 1913 when he was sixty that he really revisited his personal past in *A Small Boy and Others.* At forty no writer seemed less given to intimate retrospection than the frenetic Twain, who appeared then as the archetypal American in the midst of a ceaseless journey taking him ever farther from his origins: the Sierra, Europe, and now a settled Victorian respectability in Hartford. True, he had utilized his mining and newspaper adventures in the far West in "Jumping Frog" and *Roughing It,* but you have to look hard in those pages for any evidence of the man himself. As for his earliest and formative days in the seedy little town of Hannibal on the banks of the Mississippi, these were buried beyond apparent recall.

To an extent this appearance was the reality. Twain was the archetypal American of his time, rootless, in love with movement, mechanization, gadgetry, all forms of novelty. In some of his abruptly changing moods he could be vicious to those who lived in the past or even occasionally thought back fondly on it. He almost killed off a life-long friendship with his childhood friend, Will Bowen, by telling Bowen he was too old a man to be indulging in the mental masturbation of thinking about the old Hannibal days. "Every day that is added to the past," he snarled, "is but an old boot added to a pile of rubbish." Writing to Frank Burrough, a newspaper colleague from St. Louis days, Twain enlarged on the theme, telling Burrough he'd stood Will Bowen's sentimentality as long as he could but couldn't stand it any longer. "There is one thing," he wrote, "which I can't stand and won't stand, from many people. That is sham sentimentality . . . , the rot that deals in the 'happy days of yore,' the 'sweet yet melancholy past,' with its 'blighted hopes' and its 'vanished dreams'— and all that sort of drivel." He had told Bowen so, he said. "And I said there was but one solitary thing about the past worth remembering, that was the fact that it *is* the past—can't be restored." A large portion of his measureless contempt for his older brother Orion derived from the fact that Orion was literally stuck back in the landscape of the past: out there in Keokuk, Iowa, just up-river from Hannibal, pottering about in a series of menial jobs. Well, he, Twain, had passed through his Keokuk phase long ago, in his meteoric rise having once lectured there to a larger house by far than the great Emerson had been able to draw in that same town.

With Howells, another village boy born in the same era in the big valley, Twain could spend hours raging against the past. Howells had his own reasons to fear and hate his past, and once after a joint session in retrospective vilification he had wondered aloud to Twain why the past was so odious to them both. Howells recalled that Twain had responded from the "depths of his consciousness" that the past was hateful because it was "so damned humiliating."

But as always with Twain there was more to his attitude toward the landscape of childhood than rage and contempt. In other moods he could be as sentimentally nostalgic as he found Will Bowen. In 1866, on the lip of national literary fame, he wrote his mother and sister from San Francisco saying, "I wish I was back there [home] piloting up and down the river again. Verily all is vanity and little worth—save piloting." He thought then of writing about his life on the river and in the little village in which he had grown up, and in fact two of the first things he did write after coming east for good at the end of 1866 were reminiscent sketches of Jimmy Finn, the town drunkard, and the Cadets of Temperance, a transitory village organization to which little Sam Clemens had briefly belonged.

Again, in February 1870, four days after his marriage and while his bride slept upstairs in the handsome house her father had bought for them, Twain waxed deeply nostalgic in a letter to Bowen, recalling once more Jimmy Finn, the landscape of Hannibal days, the time the tramp burned himself up in the village jail, and the various kid pranks of that old and unforgotten time. That same year he made a more formal cast in the direction of Hannibal when he began a sketch called simply "Boy's Manuscript." Its hero is a village boy named Billy Rogers who numbers among his friends one Bob Sawyer and who has a little sweetheart named Amy. Billy's chief ambition is to become a pirate like Kidd or Lafitte, but the ambition remains unfulfilled because after a few pages the author lost interest in the subject and dropped it.

There were more engrossing and indeed easier things to think about. Rolling now in unprecedented wealth (his 1870 royalties from *Innocents Abroad* were $190,000), delighted in marriage, he contemplated a book on Noah and his sons, became part-owner of a newspaper, accepted a handsome fee for a monthly contribution to a New York magazine, got fired up with the idea of a book on South African diamond mining and dispatched a friend there to research it for him, and instead of writing about the river and the village he wrote *Roughing It.* He went back on the lecture circuit, fiddled with ideas for inventions such as a steamboat paddle that could cut through winter ice, thought of a satirical book on England (where he'd been received as a celebrity), and eventually collaborated with Warner on *The Gilded Age.* "Boys Manuscript" seemed utterly forgotten and his Hannibal days useless for literary purposes—or any other. Only a few scattered hints in *The Gilded Age*—the faded Virginia aristocrats, a couple of backwoods villages, and a persistent undercurrent of talk about the East Tennessee Land Scheme—are there to show us that Twain was at some level still mulling over his Missouri days.

In the summer of 1874 while the monstrous house was being finished in Hartford and Livy was pregnant with their third daughter, Twain took the family to Livy's sister's farm outside Elmira, New York. He thought Livy needed the comparative quiet of the place, and he could work in the handsome octagonal study built for him there. In it, surrounded by windows through which he let the breezes blow as they would (holding down his manuscript sheets with brickbats), he began to let himself drift back along the current of memory into the past. Once more he took up "Boy's Manuscript," and through the rest of that summer he wrote happily along about that time and place he had told Howells he found "so damned humiliating." Now his hero was named Tom Sawyer, and in constructing his character Twain drew on his recollections of Will Bowen. Fully present for him now were his memories of Hannibal and also his memories of what he would later style the "almost invisible village of Florida, Monroe County, Missouri" to which his parents had come in the irreversible descent of their fortunes and in which he'd been born in 1835. For Twain the landscape of childhood was Florida as well as Hannibal, and it was this fusion that he now summoned up as the locale for his boy's fiction.

In Florida, Judge John Marshall Clemens had joined his wife's relations in what was then and forever remained a backwater settlement on the Salt River: two streets, log cabins, rail fences, corn fields, and beyond, the woods. There was a log church with the standard puncheon floor, and under it the hamlet's hogs rooted, scratched, and sometimes squealed, interrupting the Sabbath services. There were two general stores, one of them run by Judge Clemens and his brother-in-law, John Quarles. Both stores dispensed the same items: calico, salt mackerel, coffee, New Orleans sugar, hardware, gunpowder. "If a boy bought five cents' worth of anything," Twain was later to recall,

> he was entitled to half a handful of sugar from the barrel; if a woman bought a few yards of calico she was entitled to a spool of thread in addition to the usual gratis "trimmin's"; if a man bought a trifle he was at liberty to draw and swallow as big a drink of whiskey as he wanted.

What Twain remembered most about those earliest years was the Quarles farm outside Florida where he spent much time and to which he was periodically to return until about the age of twelve. It was, he said, "a heavenly place for a boy." He remembered the double log house with its roofed breezeway connecting the halves and on which the family ate in sultry weather; the rail-fenced yard with dozing hounds; the smokehouse, orchard, slave quarters. Then, stretching toward the encircling woods, were the tobacco fields. He recalled with special vividness his visits to the slave quarters and the old woman he and the other kids saw there, her scant white hairs knotted to ward off the witches and whom the kids believed to be a thousand years old. There he heard the slave songs and ghost stories told by the kindly Uncle Dan'l. He remembered all the sights and sounds of that country life: the mysterious aura of the woods, the sounds of woodpeckers and wood pheasants, the way a hawk looked hanging high above the treetops, the sweet scent of high-heaped maple fires, the sound a watermelon made when you split it on a hot summer's day.

The Florida years ended for him in 1839 when Judge and Jane Clemens took their family eastward to the river village of Hannibal where the Judge hoped for a change of fortunes. Hannibal in 1839 was a community of about a thousand engaged in the manufacture of cigars, the distillation of whiskey, and the processing of pork. Its few and unpaved streets were quagmires in spring, deep in dust in summer, and iron-rutted in winter. At regular intervals citizens had to quit the streets altogether when the hog drovers came through, marshaling their grunting herds toward the two slaughterhouses. When the wind came from that quarter and from the tanyard, the village lay under a heavy, pungent canopy. The local boys often played at the slaughterhouses, using castoff pig bladders as balloons.

But more significant than any of these features of village life was the fact that Hannibal was a port on the Mississippi, and its major streets ran down to the river, locus of the town's emotional as well as economic life. Twice daily steamboats called there, revving the village into a brief and furious activity, the boat whistle shrilling, merchants running toward the wharves, idlers taking their hands from their pockets and ambling down to see the show, and the juvenile populace crowding in for an awed look at the gilded boats and their swarthy, swearing crews. For Hannibal boys, the ambition to become rivermen was about the equivalent of later boys' dreams of becoming professional baseball players.

For Sam Clemens and his friends, the river was an inexhaustible source of instruction and entertainment. They fished it, boated on it, swam in it, hooked rides on the great rafts that glided past the village, hanging on to their sterns for a mile or more below Hannibal before dropping off to swim ashore. In winter they skated the black ice, thrilling to the constant sense of all that locked power just beneath the thin surface they skimmed across. When they weren't on the river they had other diversions nearby: nutting in the woods, berry picking on the winding country road that led past a section of the village called "Stringtown." They had sword fights with coopers' lathing, Indian fights along the brushy banks of Bear Creek. On the river a couple of miles below the village was a vast, labyrinthian limestone cave they timidly explored.

In short, in some respects here was a boy's paradise, equal in its possibilities for play to that left behind in Florida, and those possibilities were not at all abridged for the Clemens boy by the fact that his parents lived in genteel poverty—and after the death of Judge Clemens in 1847 in circumstances considerably worse. Nor were the hated routines of school and church serious hindrances: Sam Clemens managed to have his share of fun in the ample margins of regular Hannibal life. He continued to find his avenues for amusement when the workaday world closed in more tightly around him with the death of his father. Then he was apprenticed to a local printer to help Jane Clemens make ends meet, but as his autobiography makes clear, life as an apprentice had its moments of high hilarity and the village was still a place of interest and a certain sort of communal warmth.

Yet there were good reasons why Twain had been able to stay away from writing about this outwardly happy place until he once again in the summer of 1874 took up the story of the boy he now thought of as Tom Sawyer. And there were good reasons why, returning imaginatively to it, he should do so in a guarded, partial way that reveals the murky misgivings he held about it. For it would be too easy—and misleading—to simply conclude that the chief reason for his delay in writing about the village, the river, the backwoods farm was that it had merely taken him this long to regard his own life as a work of art.

There was, of course, the poverty, the poignantly mingled strands of his father's Virginia aristocratic background and subsequent failures (land speculator, lawyer, storekeeper, justice of the peace, bankrupt), the succession of ever shabbier dwellings leading at last to the thin little house on Hill Street where Jane Clemens was obliged to take boarders and young Sam to live with the printer, Ament. In his autobiography Twain remembered Hannibal as a "little democracy which was full of liberty, equality and the Fourth of July, and sincerely so, too; yet you perceived that the aristocratic taint was there." Judge Clemens surely perceived it: a proud man, he keenly felt his fallen status, as did the other members of the family.

Then too, for all its paradisiacal aspects for boys, Hannibal was just another little place on the big river, an undistinguished place, a whistle stop, to which in his later years of international celebrity Twain would apply such adjectives as "beggerly" and "shabby" and "poor." If now the world recalls the town's dominant color as white—the white of its buildings and whitewashed fences—to Twain its true color, symbol of its status in the wider world he had come to know, was its pervasive slate-hued mud. Visiting Hannibal at wide intervals in his maturity, he noted with some astonishment the changes that had occurred. But when he saw the old, inevitable mud he recognized home. In his late years, as if seeking a yet further and final distance from the mud and what it stood for, he would affect an all-white costume and wash his hair daily, preferring, so he said, to be clean in a dirty world.

So there was a certain sense of shame at such humble origins, a sense no doubt encouraged by the high society he preferred in the long years of his prodigious fame. There were also, and more significantly, feelings of terror and anger associated with the place, and these were even harder for the author to come to terms with. By the age of twenty-three Sam Clemens was on intimate terms with death, and his early and constant exposure to it left unannealed scars. In his last years, bereft of so many old friends, bowed down by the losses of Livy and his daughters Susy and Jean, he was still talking obsessively of those deaths he'd experienced in his Missouri days. A sister Margaret had died in 1839 just before the family moved to Hannibal, and then three years later his brother Benjamin died, an event Twain recalled with the vividness that marks a traumatic experience. Forty-seven years afterward he could recall his mother holding him by the hand as the two knelt at the deathbed while the tears flowed unchecked down his mother's cheeks. She was moaning, Twain said. "That dumb sign of anguish was perhaps new to me, since it made upon me a very strong impression—an impression which holds its place still with the [mental] picture which it helped to intensify and make memorable."

While still a lad he had one evening entered the office his father kept as justice of the peace and there had stumbled over a corpse: a man had been murdered and his body stored there pending an inquest. About this same period he either saw or heard from his father about the murder of Sam Smarr—shot by William Owsley at the corner of Hill and Main streets, just a few doors away from the Clemens home. If he did not see the actual shooting itself, Sam Clemens did see the dying man laid out in a drug store, gasping out his life—so Twain would later claim—under the burden of a heavy Bible some thoughtful soul had provided for the occasion.

Then when Sam was twelve his father died, that "silent, austere man," as his son remembered him. In notes Twain made later on the event he said he recalled the deathbed scene and his father's embrace of his daughter, "the first instance of affection" the boy had ever seen in him, after which John Clemens muttered, "Let me die" and did. That same day, aghast at the fall of his household's aloof monarch and at the same time profoundly curious, Sam secretly witnessed the post-mortem through the keyhole of the death room, then told Orion what he'd seen. When, many years after, Twain encouraged Orion to write about it, William Dean Howells was appalled at the blazing, naked realism of what Orion produced and urged Twain to suppress the reminiscence forever.

Other encounters with the great dark followed in close succession. There was the cholera epidemic of 1849 that so frightened and bewildered the village and especially the sensitive, impressionable Sam Clemens. He saw a slave struck down and killed in the street by a white man who hurled a chunk of slag at him for some trifling offense. He saw a man stabbed to death by a drunken companion who buried his knife to the hilt in his erstwhile friend's chest. Drink figured in another homicide the boy witnessed. A young California emigrant, crazed with whiskey, lurched up Holliday's Hill to menace a respectable widow who lived up there. Clemens and a friend named John Briggs followed at a safe distance and saw what happened. The widow, Twain recalled, had stood in the shadows of her porch holding a loaded musket and "warned the man that if he stayed where he was while she counted ten it would cost him his life. She began to count, slowly; he began to laugh. He stopped laughing at 'six'; then through the deep stillness, in a steady voice, followed the rest of the tale: 'Seven . . . eight . . . nine'—a long pause, we holding our breaths—'ten!' A red spout of flame gushed out into the night and the man dropped with his breast riddled to rags."

A less violent but no less shattering end came for a contemporary of Sam's, Lem Hackett, who fell out of a flatboat while fishing and drowned. Another boy called "Dutchy" drowned before the horrified eyes of his playmates when he became entangled in some hickory hoops the town's coopers had sunk in the creek to soak. It was left to Sam Clemens to dive down to discover what had happened to Dutchy and to feel the dead boy's hand trail limply across his own in the turbid water.

Living on the banks of the Mississippi, Sam Clemens and his fellows were given constant reminders of the river's immense power and of the hazards of work and travel on it: tales of steamboats killed on hidden snags, tales and local newspaper stories of steamboat boiler explosions with frightful loss of life and disfiguring burns. Indeed, these stories came home to Clemens with a fearsome reality when his own brother Henry was fatally burned in a boiler explosion aboard the steamboat *Pennsylvania*. Corpses, often badly mangled or decomposed, frequently washed ashore in the Hannibal area, and the far-ranging boys saw them. Once, poking about near Glasscock's Island on the Illinois side of the river, the boys' poles accidently dislodged the submerged body of a runaway slave: it rose, hideous, head first, showing Sam and his friends its sodden and mutilated face.

Of all these horrors, the one that stayed closest to Twain, a shadow tracking him through life, was the accidental death of the tramp who burned himself up in the Hannibal jail. Sam and his gang had been teasing the tramp until Sam felt contrite enough to give the man a pack of matches. Then in the night to hear the fire bell and to run toward the glare and find that stray gift turned into an instrument of death, the black figure of the trapped man etched against the white heat while outside the townsmen battered futilely at the jail's heavy door. At the end of his life, writing and finally dictating his disheveled autobiography, Twain was yet telling himself in italics that he was *not* responsible for the tramp's death.

Here is a lot of death and destruction for a kid to handle, and for Twain his spiritual armor rendered him peculiarly vulnerable. The construction of that armor began with little Sam's converse with the slaves on his uncle's Florida farm. The survival of genuine African beliefs among the Quarles slaves must be highly conjectural, but elsewhere in the Mississippi Valley such survivals were common enough at this time and even later. What Twain remembered of slave lore suggests the Quarles slaves did retain some of their African beliefs, for the general impression they gave him was of the power and omnipresence of the spirit world, the way it intersected and interacted with the visible world. In West African belief systems the spirits of the dead remain actively interested and involved in the affairs of the living, and these spirits must be constantly placated lest they do harm to the living.* To such surviving beliefs in the omnipresence of the spirit world and the potential malevolence of the dead was added the slaves' painfully acquired sense of the precariousness, the vast uncertainty, of existence in the alien world of their white masters. Hanging about Uncle Dan'l and the other slaves, hearing their songs of tribulation, listening at night to their ghost stories while the firelight glimmered along the log walls and clay chinking of the slave cabins, little Sam Clemens perhaps first came to feel that existential dread that was to become so conspicuous a feature of his world view and his work.

Not conjectural are the effects of white Protestantism on the boy. The Judge was a freethinker, though on his deathbed he acknowledged Christ as his savior. But the prevailing atmosphere of that time and place was frontier Protestantism, and through his mother and school and church

*West Africa was the area from which the large majority of slaves was taken.

Twain was thoroughly versed in Presbyterianism and Protestant folklore. Jane Clemens was in many ways a cheerful person, no mean psychological feat considering her economic situation and the fact that her marriage had not been an especially joyful one. At the same time, however, she was devoted to the darker, necromantic side of Christianity, much interested in discerning portents of mortality in dreams, omens, far-off disasters, the deaths of neighbors. So from this official quarter of his culture Twain had reinforced the insistent sense that the daylight world of action and events was only a pale shadowing of the brooding, portentous reality behind. A vengeful if not actively malign deity oversaw the most inconsequential happening in the most out-of-the-way place and weighed its inherent righteousness in the divine scales. "Accidents" were such in name alone. Actually, the drownings, incinerations, steamboat sinkings, et cetera, were warnings to surviving sinners to repent before they too should be summoned unshrived before the throne.

Periodically spasms of intense godliness convulsed the community. Camp meetings on the outskirts of town drew hundreds to their brush-and-sapling shelters where evangelists spraying the sweat of exhortation brought cringing sinners to the mourners' bench. Ephemeral societies were gotten up in town promoting some special aspect of holiness: Sam Clemens briefly joined the Cadets of Temperance before he lapsed back into cigar smoking. Occasionally on swings through his spread-out, unorganized parish some first-water ministerial talent would visit Hannibal and spark off another revival. Twain recalled the visit of Alexander Campbell of the Campbellites during Twain's apprentice years and how agog the town was in his august presence.

During these spasms of communal religiosity Sam Clemens repented his habitual sinning. And in the wake of those deaths of his young years he was more than merely repentant: he was actually terrified. He was terrified because, as he says in his autobiography, he was "educated" and "trained" to be so. "I was a Presbyterian," he writes, "and I knew how these things were done. I knew that in Biblical times if a man committed a sin the extermination of the whole surrounding nation—cattle and all— was likely to happen." The clustering disasters of his young life, Sam Clemens was convinced, "were inventions of Providence to beguile me to a better life." In the daytime the youth was able to carry on cheerfully enough, trusting that in his case anyway God's patience would continue everlasting. But nights were different. When a terrific thunderstorm lashed the town after the drowning of Lem Hackett, young Sam was certain this was a divine visitation. Nor was it only on nights following some terrible event that the boy suffered, for, he said, with the going down of the sun "clammy fears" gathered about his heart. "Those were awful nights," he remembered, "nights of despair, nights charged with the bitterness of death," nights when he "begged like a coward, begged like a dog." We don't have to descend to the dime-store variety of psychoanalysis to see that for a fiercely proud and famous man the memory of those wretched nights would have been profoundly humiliating. But more: the memory would have enraged and disgusted him since in his mature view his cringing terror was all so unnecessary, the consequence of a belief that was as foolish as it was unwarranted.

Mark Twain never outgrew that sense of life's dreadful uncertainties he got from the blacks and whites of his childhood and youth. In many ways he remained captive of the old beliefs, and the inescapable condition compounded the rage he felt toward this part of his past. "In my age," he tells us in his autobiography, "as in my youth, night brings me many a deep remorse. I realize that from the cradle up I have been like the rest of the race—never quite sane in the night." To go back to Hannibal then was to voluntarily re-encounter that fright-fraught, God-haunted world. He had not gone back physically since 1867 and was not to do so again until 1882. Nor until this Elmira summer of 1874 had he really allowed himself to revisit it in his imagination. In going back now it is not surprising that he should have chosen to accentuate the smiling, sunshiny, white-washed aspects of his Hannibal world.

The opening chapters of *The Adventures of Tom Sawyer* are just what he claimed them to be, a hymn to boyhood, and their author does not care to specify the ages of his young characters except to indicate by the kind of play they engage in that they are pre-adolescents—living, that is, in that provisional paradise before the guilts of sex and the torments of moral choice are visited upon them. Writing of such characters playing at life, Twain felt safe enough in his re-entry into Hannibal, keeping its darker aspects resolutely at bay, but at the end of the summer he stopped. The new house was almost finished, the baby was due, and there were a hundred obligations that required the family's presence in Hartford. Then too, as he said, his creative tanks had run dry. The boy's book had come to a halt and wouldn't go any further. This was Twain's way of saying that he no longer wished to write on a particular project: it had happened on a number of previous occasions and would again, the reasons ranging from loss of interest to his own inner sense that the project was wrong for him—or wrong for him at that time. Here a good surmise is that memory had borne the writer far enough back so that the old, disturbing recollections were beginning to surface like that death's-head slave who had so suddenly bobbed into view, destroying the boys' idyll of nutting along the Illinois shore.

But once raised into view the landscape of the past seemed stubbornly insistent. That fall while finishing touches were applied to the house, Howells wrote Twain asking him for a contribution to *The Atlantic Monthly,* of which Howells was now editor. The funny man from the West was hot copy now, welcome even in the hallowed, polished pages of the magazine of the Eastern cultural establishment. But Twain wrote Howells back in a peevish tone, saying that there was so much commotion attendant to the finishing touches on the house—the incessant demands of the architect, the groundskeeper, the billiard-table man—that any sort of literary concentration was impossible. Later that same day, though, Twain, in escaping the house for a walk with his close friend, Joe Twichell, fell into a revery about old times on the river when he'd been a pilot (and thus sovereign and free, unlike his burdened present condition). And the more he talked, the more it became evident that here was a subject to which he might instantly and without effort turn his whole attention. "Would you," he wrote Howells for the second time that day, "like a series of papers to run through 3 months or 6 or 9?—or about 4 months, say?" Howells was delighted by the prospect.

As always with any new project, Twain himself was delighted. The man Livy called "Youth" was ready as soon as he heard from Howells to hurl himself headlong into the writing, certain that he could give the *Atlantic*'s readers a view of old-time river life they could get nowhere else. He had his notebooks from his piloting years (1857–61), but more than these he had his recollections, these preserved in the alembic of his rigorously schooled pilot's memory. "I think," he was to write, "a pilot's memory is about the most wonderful thing in the world." Give a man, he said, a "tolerably fair memory to start with, and piloting will develop it into a very colossus of capability." Whether piloting was responsible for Twain's own phenomenal memory or whether his powers of recall were merely sharpened by his experiences on the river is a question. Whatever the case, the man could memorize entire lectures with brilliant accuracy and apparently remember any story he'd ever heard. But such a memory had also its significant drawbacks, as he already knew and as he was soon enough to be reminded in imaginatively returning to the

Mississippi, for like the river itself Twain's mental return to it was fraught with submerged snags.

As was his practice in any non-fictional work, he wanted to begin with the facts. Planning for *Roughing It,* he found he could remember few details of the overland trip to Nevada he had taken with Orion in November 1861, but he did remember that Orion had taken notes, and he was willing to pay Orion for their use. Whatever extravagant divagations Twain might make, there was always a core of factuality to them— authentic details of dress, speech, behavior, occupation—that gave those divagations their punch. Always he was contemptuous of those who presumed to talk or write on a subject they did not know in an intimate and factual way. In the matter of piloting he had the facts all right, but facing the final fact of his own potential composition it became clear to him that the living truth of that vanished pre-war life on the river was a style of living and especially a way of talking that would be extremely difficult— impossible maybe—to get down on the page. His notebooks gave notations of landmarks, soundings, crossings; and he knew the sources of background information on the steamboat trade—numbers of ships and their names, dates and times of famous races. But how to capture the perishable wind, spirit of a bygone era? And, especially, how to capture and tame into acceptable literature a style and spirit that were very far indeed from the custom of the *Atlantic's* pages? For as Twain well knew that style and its talk were violent, vulgar, and gaudy. To faithfully reproduce this in polite letters was unthinkable. It would all have to be somehow suggested and how to suggest the knock-down, powerhouse quality of the river's life without hopelessly bowdlerizing it was a problem of gigantic proportions. Indeed, Twain's struggles with the books that have established his enduring reputation—*Tom Sawyer, Life on the Mississippi,* and *Huckleberry Finn*—indicate the difficulty of the problem. He wrestled with the problem for a decade (1874–84), after which he retired from that field and retreated yet further into an impersonal historic past in *A Connecticut Yankee in King Arthur's Court, Joan of Arc,* and science-fiction efforts like "The Mysterious Stranger" and "Three Thousand Years Among the Microbes."

The life of the river and the huge valley it had carved were, depending on your point of view, either the "Body of the Nation" (Twain's view) or its sewer (the view of a number of foreign travelers and some members of the Eastern cultural establishment). But even if you held the former view, you still had to reckon with the facts of life out there. Here was in truth an unkempt, brawling, masculine subculture, turbulent and powerful as the river itself. The lives of the steamboatmen and the raftsmen who had preceded them were rough to the point of barbarousness, tremendously exacting and radically unpredictable. They lived daily with the vicissitudes of broiling sun, electrical storms, floods, snags, shifting channels and consequent groundings, sunken wrecks that could snatch a ship's hull out of her and sink her in minutes. In the lust for quick gain, steamboats were hastily and cheaply thrown together and rushed out into the trade. The results were those frequent fires, boiler explosions, and sinkings of Twain's youth, the legends of lost ships and lost lives that quickened for a moment the talk of villages on the river's banks.

Nor was all the river's commerce legitimate. A sizeable amount of the money changing hands did so in the activities of pimping, prostitution, gambling, a variety of con games, outright theft, and murder. For these reasons the arrival of a raft or steamboat at one of the river's port towns, carrying its full freight of roustabouts, gamblers, pimps, whores, con men, and bullies was a mixed blessing, however essential the contact was commercially. As for the towns themselves, the ones large enough to cater to the river's motley personnel earned reputations without precedent in American history: the St. Louis waterfront, the "Gut" in Memphis, Natchez-under-the-hill, the "Swamp" in New Orleans. In these places the rivermen went at play with a spectacular ferocity, drinking, dancing, whoring, gouging, gambling, shooting at targets and at each other. In the first half of the nineteenth century a rude pantheon of quasi-legendary figures arose from this life. Brawlers like Mike Fink and Bill Sedley; river pirates like Wilson of Cave-in-Rock and Colonel Plug; land pirates—the Harpe brothers, Mason of the Woods, and the robber chieftain, John A. Murrel—who preyed on solitary travelers following the Natchez Trace or on merchants returning on horseback from New Orleans.

Pervading all this and celebrating it was the talk of the rivermen, backwoodsmen, and hunters of the great valley: loud, boisterous, bottomlessly profane, often hilarious. On the perishable breath of oral tradition an informal collective history of the Mississippi Valley was spoken into being by thousands of unknown men telling broad jokes of physical cruelty and discomfort, smutty stories of sexual conquest, of shooting and drinking contests; telling tall tales and kindred lies, creating and passing along legends of Mike Fink, the Harpes, of steamboat races where they fed live slaves to the furnaces to put on extra speed and lighten their loads ("We're fallin' behind, Cap'n! Throw in another nigger."); stories of hunts for "creation b'ars," of gouging contests that left both champions blinded or missing ears, noses, even their tongues.

Travelers into the region from the settled East or from abroad were aghast at the rude vigor of the life they encountered, the coarse and colorful talk that flavored it. Here truly was the authentic yawp of barbarism. The European romantics had theorized about "folk-say," and about the wellsprings of a great national literature lying in such hinterlands as this. But out in the gloomy woods and canebrakes, in the muddy little river villages, on the decks and in the bars of steamboats, the folk-say of these Americans seemed so hopelessly vulgar that no imaginative bridge could ever be constructed from it to pages a self-respecting person would read.

Still, by the early 1830s written versions of Mississippi Valley folklore began to appear in the sub-literary channels of newspapers, sporting journals, almanacs, joke books, plays, and also in the travel accounts of foreigners. The local collectors, all men who themselves relished the pasttimes and humor of the folk, were generally men of some learning: country newspaper editors or doctors. A good number of them were lawyers who heard the stories on their rounds, then swapped them with their colleagues at taverns and hostelries when district court met. To be sure, these were toned-down versions, but nevertheless the literature of what was then known as the Southwest (southwest, that is, from northeast) preserved a good deal of the bite of the folkways of Tennessee, Georgia, Alabama, Louisiana, Arkansas, and Missouri.

Writing in the sporting journal *Spirit of the Times* in 1851, an anonymous contributor managed to cram into a paragraph the spirit of his time and place. Describing a day in the life of a little village somewhere in the old Southwest, he claimed they had

> two street fights, hung a man, rode three men out of town on a rail, got up a quarter race, a turkey shooting, a gander pulling, a match dog fight, had preaching by a circus rider, who afterwards ran a footrace for apple jack all round, and, as if this was not enough, the judge of the court, after losing his year's salary at single-handed poker, and licking a person who said he didn't understand the game, went out and helped lynch his grandfather for hog stealing.

A.B. Longstreet, a Georgia lawyer who collected his newspaper sketches of the early 1830s into *Georgia Scenes* (1835), described an epic rough-and-tumble match between two local champions, Billy and Bob.

Billy's backers swore their man needed but one lick at Bob to "knock his heart, liver, and lights out of him," while Bob's backers were convinced their man would throw Billy so quickly that by the time he hit the ground the "meat would fly off his face so quick that people would think it was shook off by the fall." Both groups proved right in the event, for shortly after the fight had begun the observer noted that "Bob had entirely lost his left ear and a large piece of his cheek" while "Bill presented a hideous spectacle. About a third of his nose . . . was bit off, and his face so swelled and bruised it was difficult to discover in it anything of the human visage."

In Tennessee, George Washington Harris collected tales he heard on his travels through the state and began writing these up in the 1840s. In one he published about a Tennessee hill man named Jo Spraggins, he characterized Spraggins as one who "hates a circuit rider, a nigger, and a shot gun—loves a woman, old sledge, and sin in eny shape." When Spraggins advertises a dance at his cabin, by sundown of that night the girls "come pourin out of the woods like pissants out of an old log when tother end's afire." The dance ended, naturally, in a free-for-all, and sixteen days later the narrator's girlfriend was still picking gravel the size of "squirrel shot" out of his knees. Harris also developed a vicious picaresque hero he called Sut Luvingood, modelled in part on a man he met in southeastern Tennessee whose escapades are a relentlessly repeated mixture of violence, cruelty, and sexual bravado.

Finally, hear Thomas Bangs Thorpe's example of the tall talk Sam Clemens heard as a boy on the wharves of Hannibal and later as a pilot on the river. Thorpe's character is aboard a Mississippi steamboat and has been asked about the wild game in his native state. He answers that game is so abundant there that he almost never bothers to shoot something as trifling as a bird. "I never did shoot at but one," he says, "and I'd never forgiven myself for that, had it weighed less than forty pounds. I wouldn't draw a rifle on any thing less than that; and when I meet with another wild turkey of the same weight I will drap him." The listeners in the steamboat cabin are incredulous: a turkey weighing forty pounds? Yes, says the man,

> and wasn't it a whopper? You see, the thing was so fat that it couldn't fly far; and when he fell out of the tree, after I shot him, on striking the ground he bust open behind, and the way the pound gobs of tallow rolled out of the opening was perfectly beautiful.

A cynical Hoosier asks where this event was supposed to have happened. "Happen!" the tale-teller exclaims,

> happened in Arkansaw: where else could it have happened but in the creation state, the finishing-up country—a state where the *sile* runs down to the centre of the 'arth, and the government gives you a title to every inch of it? Then its airs—just breathe them, and they will make you snort like a horse. It's a state without a fault, it is.

Twain admired such gaudy talk all his life, and in the book he was eventually to write about life on the river he evidently was thinking of Thorpe's braggart when he composed a hilarious variation on the murderous mosquitoes of Arkansas.

But from the outset Twain knew it was out of the question to reproduce this sub-literature of the old Southwest and still less the oral materials that underlay it. In the first place, he knew that much of that sub-literature was already outdated less than half a century after publication. It had traded far too heavily on regional peculiarities and orthographical japeries to be either nationally appealing or durable, and the audience for which it had been published no longer existed. Then, too, much of it was simply too crude for assimilation on the level for which Twain was now writing.*

No, he would have to turn pilot once again, only this time he would have to run this river of literature pretty much on instinct, the old markings and notations being of only the most general use. The true and essential genius of the man is never more clearly displayed than in the *Atlantic* articles that began to appear in January 1875. Here for certain was a "lightning pilot" who ran a daring course through the hazards of dark old childhood fears, the sordid aspects of river life, and the temptations of literary imitation to produce an account that more than a century afterward seems as fresh and vivid as it did then. You hardly notice the things left out of account altogether or only hinted at, and "Old Times on the Mississippi" appears to us complete and absolute, the definitive account of how it really was.

---

*Here and there in Twain's published work are examples of the undiluted stuff. He often finds humor in physical discomfort, deformity, even dismemberment, and in "The Invalid's Story" (1882) we get a fair sample of what the old Southwest found hilarious. It is a story in which two characters confuse the smell of runny Limburger cheese with that of a putrifying corpse.

Beginning with himself, his childhood memories of the river and its life, Twain soon enough went beyond the self and into the rich milieu in which that self had existed: not only Sam Clemens as he had been in Hannibal and then on the river in the years before the war, but the great river itself flowing through time and human history, carrying on its brown waters Indian canoes, black-robed Jesuits, the raftsmen and steamboats, and draining a huge valley—black mud, woods, woodyards, backcountry hamlets, riverbank villages, the life they harbored. So the man who had long delayed an imaginative return to his past now became the immortal scribe of his place, the one who gathered up into a gorgeously vivid narrative what otherwise might well have been lost. Twain is, to be sure, a persistent presence in his narrative. But at last it isn't Twain the reader carries away from these articles. It is the river, and in the pages of the *Atlantic* (significant name), the Mississippi—a geological phenomenon heretofore unclaimed imaginatively—became a part of the nation's literary landscape.

To set his scene at the outset, Twain wrote one of the most unforgettable single paragraphs in American literature when he described the daily arrivals of the steamboats at the sleepy little village of his youth: the bellow of the black drayman, the sudden spate of furious activity, the appearance of the boats themselves—white railings, gingerbread woodworkings, and flags snapping—the scrambling passengers and lordly captain. Above all, there is the river, "the great Mississippi, the majestic, the magnificent Mississippi, rolling its mile-wide tide along, shining in the sun; the dense forest away on the other side; the 'point' above the town, and the 'point' below, bounding the river-glimpse and turning it into a sort of sea, and withal a very still and brilliant and lonely one."

With this image to work from Twain launches into his narrative which, like the river itself, does not move in a straight and orderly line but is filled with meanderings. He tells us of the raftsmen and in one of the very few references to the seamy side of river life characterizes them as "heavy drinkers, coarse frolickers in moral sties like the Natchez-under-the-hill of that day . . . elephantinely jolly, foul-witted, profane. . . ." He introduces us to the rank and dignity of piloting and to the awful responsibilities of that profession, here and there slipping in bits of technical talk to give readers the tang of it. He describes the great days of steamboat racing and an epic race for which the boat was so stripped down that the captain "left off his kid gloves and had his head shaved." He tells us of famous pilots, of dangerous crossings successfully run, and of his own trying initiation into the mysteries and lore of the piloting profession.

But best of all was the talk, the inimitable linguistic inventiveness of the rivermen. There was the simple lie, for instance, a commonplace of everyday conversation, as when Twain remarks, passingly, that he once served on a boat so slow "we used to forget what year it was we left port in." This boat, he says, "was so slow that when she finally sunk in Madrid Bend, it was five years before the owners heard of it." There was the compound, elaborated lie, known as the tall tale, such as the one Twain spins of the somnambulistic pilot who one night took his boat through a hellishly dangerous crossing while asleep, then calmly retired to his berth below. The feat, says Twain, caused one who witnessed it to exclaim, "*I* never saw anything so gaudy before. And if he can do such gold-leaf, kid-glove, diamond-breast-pin piloting when he is sound asleep, what *couldn't* he do if he was dead!" And there was a variant of this last, the shaggy-dog story. In a chapter reminiscent of "The Story of the Old Ram" in *Roughing It,* Twain subjects his readers to a long-winded discussion of the crucial necessity of memory in the profession of piloting. But what we actually get is an absurd digression on the mindlessly encyclopedic memory of a pilot named Brown who couldn't forget anything. Such a memory, the narrator says innocently, "is a great misfortune."

> Mr. Brown would start out with the honest intention of telling you a vastly funny anecdote about a dog. He would be "so full of laugh" that he could hardly begin; then his memory would start with the dog's breed and personal appearance; drift into a history of his owner; of his owner's family, with descriptions of weddings and burials that had occurred in it, together with recitals of congratulatory verses and obituary poetry provoked by the same; then this memory would recollect that one of these events occurred during the celebrated "hard winter" of such and such a year, and a minute description of that winter would follow, along with the names of people who were frozen to death, and statistics showing the high figures which pork and hay went up to. Pork and hay would suggest corn and fodder; corn and fodder would suggest the circus and certain celebrated bare-back riders; the transition from the circus to the menagerie was easy and natural; from the elephant to equatorial Africa was but a step; then of course the heathen savages would suggest religion; and at the end of three or four hours' tedious jaw, the watch would change, and Brown would go out of the pilot-house muttering extracts from sermons he had heard years before about the efficacy of prayer as a means of grace.

Finally there was the profanity, brought to a mighty art by several generations of professionals in this line. The kid who paddled out from shore with his friends to hook rides on the passing rafts was enthralled by the "rude ways and tremendous talk" of those crewmen. Later, when he shipped himself, he was treated to the magisterial profanity of the hands, mates, and pilots. There was, for instance, a huge mate, much adorned with tattoos, whom Twain remembered as being "sublime" in his command of cussing. Telling his charges to start the gangplank forward, the mate would bellow, "Dash it to dash! are you going to *sleep* over it! '*Vast* heaving. Vast heaving, I tell you! Going to heave it clear astern? WHERE're you going with that barrel! *for'ard* with it 'fore I make you swallow it, you dash-dash-dash-*dashed* split between a tired mud turtle and a cripple hearse-horse!'" Then there was his mentor, Horace Bixby, whom the cub pilot angers with some stupidity. Bixby searches frantically for some object worthy of his wrath and spies a trading-scow. He threw open the window, Twain writes, "and such an irruption followed as I never heard before." When Bixby at last closed the window, "he was empty. You could have drawn a seine through his system and not caught curses enough to disturb your mother with." Still later the cub repeats his offense, and his "gunpowdery chief went off with a bang . . . and then went on loading and firing until he was out of adjectives." There are numerous other descriptions of this art and all of them so inventive the reader hardly notices that these are, after all, polite substitutions. Indeed, they are probably preferable to what must have been a fairly numbing daily barrage of oaths and epithets.

With the August installment of "Old Times" Twain brought his ship safely into harbor, closing with a tall tale of an impecunious pilot named Stephen who devised an impossibly dilatory method of repaying his creditors. In a sense Twain, too, had paid some debts—to his past—and it must have seemed to him that he had done so more or less on his own terms. He had been able to extract and utilize selected parts of the past while avoiding most of the snags that lurked in those waters. Perhaps it was for this reason that now in the summer of 1875 he felt easy about picking up the neglected boy's manuscript again. Under the happy illusion that the thing was writing itself and he merely the amanuensis, he whipped on through it. He was back in Hannibal once more, experiencing the deeply familiar—the layout of the streets, the old weathered buildings of his youth, the surrounding natural features, the quality of those lost days. It seemed easy enough now simply to remember. Years later, when this book, *Tom Sawyer,* and *Huckleberry Finn,* too, had become world famous, he would claim they were "simply autobiographies." To an extent, this was true, but to a further extent it was misleading since here as elsewhere Twain used the facts of his life imaginatively. Like Ovid Bolus, the legendary liar of the old Southwest of whom Joseph G. Baldwin had written, Twain had long since "torn down the partition wall between his imagination and his memory." So, working over the remembered facts of Hannibal, Twain now transmogrified Will Bowen into Tom Sawyer, John Briggs into Joe Harper, his mother into Aunt Polly, Tom Blankenship into Huck, and Laura Hawkins into Becky Thatcher.

Maybe it was too easy. Allowing himself to remember it all, the dark things of that place and time began to obtrude like shadows on a sunny field. The hymn to boyhood began to turn into a literary blues, its tone darkened with terror and tragedy. The sham battles with wooden swords, the playing at robbers or at Robin and his merry men fade before the ghastly reality of a midnight murder in a graveyard closely resembling that in which John Clemens moldered, and the boys, Tom and Huck, who had played pirate find themselves menaced by a real pirate, the vengeful and implacable Injun Joe. Twain concludes the book with the death-by-starvation of Injun Joe, lost and immured in the great limestone cave, and the fact that the boys subsequently become rich through the find of Injun Joe's treasure is an obvious sentimental sop that didn't fool its inventor. He said he did not know whether *Tom Sawyer,* as it had turned out, was a children's book or one for adults. It almost seemed as if the book had turned against him, betrayed him. The result was this hermaphrodite, neither one thing nor the other.

He continued to wrestle, Jacob-like, with his problem, hoping for a last blessing. The following summer, again at the Elmira farm and the octagonal study, he wrote at two pieces that indicate, in their differing ways, that continuing contest. One was a privately circulated item he called *1601; or, Conversation as it Was by the Social Fireside in the Time of the Tudors.* In it Twain employed all the profanity and gross sexuality he had had to exclude from "Old Times" and to carefully skirt in *Tom Sawyer.* In the case of the latter manuscript, Howells had gone through it with his unusually sensitive Victorian geiger counter and deleted most of the mildly indecorous words and situations, and what of these were left standing were taken out by Twain himself and/or Livy. Now in *1601* Twain had in a way restored them all as well as those he'd been compelled to leave out of his description of old-time river life. It was, as Twain's biographer Justin Kaplan so well observes, a "covert way of scribbling dirty words on Tom Sawyer's fence."

The other piece he worked at in these summer months was a boy's manuscript in much the same vein as that in which he had begun *Tom Sawyer*. The boys, Tom and Huck, are back. They play at pirates once more, fool Miss Watson's big slave, Jim, raid a Sunday-school picnic. But even earlier than in *Tom Sawyer* the tone darkens—in the very first chapter, in fact. As with the author himself in his Hannibal days, clammy fears gather about the heart of Huck Finn as the sun sets. Sitting in his room at the Widow Douglas's, Huck hears the mournful rustle of the leaves, an owl whooping over a dead person, a whippoorwill, a dog crying over someone about to die, and the wind whispering some indecipherable message to him. "Then away out in the woods I heard that kind of a sound that a ghost makes when it wants to tell about something that's on its mind and can't make itself understood, and so can't rest easy in its grave and has to go about that way every night grieving."

Whatever the writer's conscious intentions may have been as he began this ostensible sequel to *Tom Sawyer,* such a beginning indicates that Twain was here verging on a much fuller, more direct confrontation with the spirit of his old place: its haunts and attendant terrors, the brooding spectre of death, the ominous burden of the Afro-American lore he heard there, the cruelty and violence of river life. True, the sun shines here also in the early pages of *Huckleberry Finn*. But this is almost from the outset a dark narrative of violent death, feuds, child abuse, cruel chicanery, and the pervasive, unacknowledged wrong of slavery. The unseen world of the spirit is everywhere in it, manifesting itself in a congeries of beliefs meant to fend off malign visitations.

A drowned man floats on his stomach; a woman on her back. Overturning a saltcellar is bad luck; to ward it off, throw a pinch of salt over your shoulder. Nails in the shape of a cross on your boot heel keep the devil away. A bit of quicksilver in a floating loaf of bread will direct the loaf to the location of a drowned corpse. Birds skipping from bush to bush presage a storm; to catch one behaving so means death. Counting the things you're going to cook for dinner is bad luck. So is shaking a tablecloth after sundown. When the owner of a beehive dies his bees must be told before the next sun-up or they will sicken and die. Talking about dead persons invites them to haunt you. Handling a snake skin is bad luck. If you kill a snake, its mate will curl about it and strike you when you come near. Looking at a new moon over your left shoulder is bad luck. "It looked to me," Huck thinks, "like all the signs was about bad luck, and so I asked [Jim] if there warn't any good-luck signs." The slave replies that there are "mighty few—and *dey* ain't no use to a body. What you want to know when good luck's a-comin' for? Want to keep it off?" Huck doesn't, of course, but even so far as Twain took his story that summer it is plain that Huck's luck is basically bad. Thus his concern with ominous portents.

Evidently at the point in the story where Huck's and Jim's raft is rammed by a steamboat Twain wondered what sort of story it was he had embarked on, how deep into that dark, shunned netherworld he was letting himself be taken even as he was letting the river take his characters ever farther south, ever deeper into slave soil. Abruptly he shelved the manuscript, writing Howells that he might finish it sometime or might possibly burn it.

He did not burn it, but neither did he touch it again for two years and then only feinted at it before laying it aside again until 1882, six years after he had begun it. While its paper curled and gathered dust the manic man thrust himself into other affairs, other writings. He wrote a play with Bret Harte, gave a disastrous after-dinner speech at the Whittier birthday celebration, in which he savagely lampooned Emerson and the other New England greats, traveled abroad and wrote about it in *A Tramp Abroad,* wrote *The Prince and the Pauper,* and began to sink money into James W. Paige's new typesetting machine.

It is possible he might never have returned to it, that it would have remained one of those bleached, unfinished ships he always had on his literary ways. But in the spring of 1882, Twain made a physical return to the river, and that experience, profoundly if obliquely impressive, brought him back to the manuscript of what the world has come to regard as his masterwork. The occasion for the return was a new contract signed with the publisher James Osgood to deliver a properly fat book on river life as he had known it in his youth and as he found it now. He already had about half the book written in the form of the *Atlantic* articles of 1875. Now he needed a bridge into the present day, and for that he needed a trip on the river. Direct observation, he had come to feel, was less vital to him in writing than recollection, but the recollecting had all been done, and there was nothing for it but to go out and look. In April he left with Osgood and a stenographer for a five weeks' excursion down to New Orleans and back.

He had hoped to travel quietly as a private citizen and so gather his impressions in an unhectic way, but it was far too late for that, and quickly his flimsy incognito was exposed. He was also exposed, mercilessly, to the current of time that had rendered the steamboat and all that life utterly obsolete and which had even altered the very course of the river he had once had by heart. Some of the old hands, true, were still on the river— Lem Gray, his old gunpowdery chief, Horace Bixby—but most of the others were gone, dead, vanished, retired to the backcountry farms from which they had originally been lured. "Old Times on the Mississippi" really were more old times than Twain had known when he wrote about them for the *Atlantic,* more than he wanted to learn in this spring of 1882. By the time he had gotten to St. Paul at the end of his trip, it had become something of a nightmare voyage and he was tired and depressed. Ahead lay the tedious business of making a book, *Life on the Mississippi,* out of the old, once fresh articles and this recent, dispiriting trip, and by the time he had finished it he was in a rage at what he called "this wretched God-damned book."

Twain's weariness with the project and his understanding, brought home to him daily on the river, that the days of his youth were done obviously influence the latter portions of *Life on the Mississippi.* There are dead stretches in it, dry statistics, and digressions that are not really diverting. Yet the trip was far from a failure, however much it may have seemed so to Twain at the time, for on his way up-river from New Orleans he had stopped at Hannibal for a few days, and that experience turned him back to Huck Finn's story.

He had arrived early in the morning and passed, he says in *Life,* through the Sabbath-sleeping streets, wondering whether this was the reality and the rest of his life a long dream from which he was only now awakening. He climbed Holliday's Hill for a comprehensive view of the town and its changes, that same hill on which he and John Briggs had witnessed the killing of the drunken California emigrant. While musing up there Twain says he was interrupted by an elderly stranger to whom he put a series of questions of the "what-ever-happened-to" sort. Whether such an encounter ever took place or whether the writer simply invented it as a convenient method of retailing the information he subsequently gathered on his childhood friends is unknown. But of much more significance is the fact that Twain uses the conversation as a transition into a real engagement with the old fears so inextricably bound up with this place. He tells of the deaths of Lem Hackett and Dutchy. More, he reveals his abject, cringing terror in the wakes of these, and he fingers the religious source of that terror. He concludes the section on the Hannibal visit with an unsparing description of the fiery death of the jailed tramp and of his own subsequent nightly torments of conscience. This is precisely the sort of moral dilemma he was shortly to construct for his ragamuffin hero, Huck, a boy, Twain said, of sound heart and deformed conscience.

So, during those months when he was thrashing about in the thickets of *Life on the Mississippi,* Twain's genius was also at work on *Huckleberry Finn,* the one book on paper obliquely feeding the other one as yet much in the mind. When he had finished the first, he turned the fullness of his talent on the other and carried it speedily through to conclusion in the summer of 1883. He had meant, he had told Howells years earlier, to carry his boy hero through a cross-section of river life in the 1830s and '40s, and this is what he now did, drawing not only on his own memories but also on information, stories, impressions he'd picked up on the river in the spring of 1882. The feud, for example, that he heard about from the old mate aboard the steamer *Gold Dust* becomes the Grangerford-Shepherdson feud in which Huck witnesses the murder of a boy his own age. Steaming down-river to New Orleans, Twain had been reminded again and again of the decaying little towns he'd seen in his piloting days, and now, taking Huck and Jim through the same stretch of river, he wrote of them with the vividness of renewed acquaintance. The cruel and ignorant loafers, the baseless pride of the country aristocracy, the hysteria of camp meetings, the work of heartless con men, the abiding racial bigotry—this was the world Twain had known in youth, and traces of it were still in evidence in 1882—enough, anyway, to spark and quicken his memories.

There was also, and more significantly, the look, the feel, the smell of the great river itself as Twain re-encountered it, and even taking into account his tremendous powers of recall, it is hard to imagine he could have written so powerfully of it had he not taken this obligatory tour of inspection. Once again, as in his return to Hannibal, it was as if he were awakening to the reality of his life from a long dream. *This* was where he had lived, where he still lived in his mind, and he was able to impart this insistent sense to the novel and to its hero. Early in the novel Huck himself awakes from a sound sleep to find himself on the moonlit river. At first, he says, "I didn't know where I was for a minute.

> I set up and looked around, a little scared. Then I remembered. The
> river looked miles and miles across. The moon was so bright I could

'a' counted the drift logs that went a-slipping along, black and still, hundreds of yards out from shore. Everything was dead quiet, and it looked late and *smelt* late. You know what I mean—I don't know the words to put it in.

Clearly, these are the words.

It is not a perfect book—what is?—its chief flaw being the lamentably silly business of the last eleven chapters where Tom shows up at the Phelps farm (the old Quarles farm transported down-river to Arkansas). It could have been worse: at one point Twain thought of extending the boys' adventures down into the Louisiana swamps through which they would ride on a stray circus elephant. As it is, it is as though, having written so faithfully of the world he had known in youth, Twain wanted to be forgiven for having done so, wanted to be taken back into his readers' affections as the genial funny man who had written *The Innocents Abroad* and *The Prince and the Pauper*. No matter. The book easily triumphs over its conclusion, as perhaps its author knew it would. He had gotten a place and a period on the page as few in American letters had before him.

Perhaps in some recess of his mind that knowledge provided a clandestine kind of comfort in the darkening years that lay ahead of him; those who themselves try to write will hope it did. During these years of financial ruin, the deaths of his wife and daughters, the drying up of his creative energies, Twain made periodic gestures toward bringing Tom and Huck back, suggesting at once how deeply and autobiographically he was attached to them and that he knew the magnitude of his achievement in their books. There were *Tom Sawyer Abroad* and *Tom Sawyer, Detective* as well as fragments varying in length from the substantial "Huck Finn and Tom Sawyer Among the Indians" and "Tom's Conspiracy" to mere notebook jottings in which Twain thought of bringing the boys back to Hannibal as old men. In "Three Thousand Years Among the Microbes" (1905), surely Huck's final appearance, he is a cholera germ (the same disease that terrified Sam Clemens during the epidemic of '49) in the body of a tramp who tellingly resembles Pap Finn. His name, "Huck," it is explained, is only a rough translation of his real name, which consists of three clucks and a belch. And the Mississippi, the "majestic, the magnificent Mississippi, rolling its mile-wide tide along and shining in the sun," as he'd described it in *Life,* is now dwarfed into a ridiculous size in the microbe's eyes by the veins and arteries of the tramp's diseased body. "The Body of the Nation" has here become superseded by the body of

Blitzowski, the tramp. And as for the author called Mark Twain, Twain tells his readers that no one any longer recalls this man's full name, though it is thought it may have been Mike Burbank Twain, an agriculturalist from California. At any rate, years ago this inconsequential fellow murdered a man and was executed for it. Clearly, at this point in his life there was for Twain as there had been for Huck at the end of his book, "Nothing More to Write."

II

❧ THE OFFENSE WHICH PROVOKED the man known as Mike Burbank Twain to murder was another tired pun on his name. It had so enraged him that he shot the offender five times. The Twain who wrote this bleak fantasy in 1905 had for a long time been weary of his enormous celebrity. Yet he had sedulously courted it through the years—and on occasion courted it still. In our age of celebrity the phenomenon of such ambivalence is familiar: the rock star or athlete who has lusted after recognition but who, casually encountered at supermarket or restaurant, warns you off with snarling mien. But perhaps here as in other ways Twain was a pilot charting unknown waters. In his time others had had their brief hours of adulation: Grant, Jenny Lind; by the time of his death in 1885, Victor Hugo was a French national hero, his funeral the occasion for recitations of his poetry by the crowds massed about his bier at the Arc de Triomphe. But who, year after year, through more than three decades enjoyed and endured the celebrity of Mark Twain?

His novels of Tom and Huck and his book on river life had conferred celebrity on Hannibal, too, and the year after his death in 1910 the town began to memorialize its famous son. Led by a great Twain admirer, George Mahan, Hannibal took gradual but irreversible steps to advertise itself as Mark Twain's home town, "the Most Famous Small Town in the World," "Everybody's Home Town." Today after more than three-quarters of a century of this the man's stamp is inescapable anywhere within a fifteen-mile radius of his boyhood home on Hill Street, and signs of him are in evidence even farther away. When you get within fifty miles of Hannibal you begin to pick up Mark Twain diners and trailer camps. In the town itself you are assaulted by Injun Joe Campgrounds, Clemens Landing, Huckleberry Heights, the Mark Twain Motor Inn, Huck 'n Tom's Motel, Mark Twain Avenue, Mark Twain Savings & Loan, Mark Twain School,

Mark Twain Beagle Club. . . . In a small river town left behind by the shifting currents of commerce the Twain industry provides a much-needed source of revenue.

Naturally, this is resented by some of its citizens in the same way citizens of any tourist place come to resent the transient gawkers and litterbugs who are here and gone, leaving behind their money and their waste. For the locals this is a curious kind of thralldom, condemned by the inscrutable designs of history to stay put and service others of seemingly greater leisure. Talking with John Lyng, a two-term mayor of Hannibal and in 1985 the Executive Director of the town's Mark Twain Sesquicentennial Celebration, I learned there was a significant community split between those who wanted to vigorously promote the Twain association and those who, as Lyng put it, "say, 'Why are we worrying about Mark Twain when we ought to be out there chasing smokestacks?'"

We were talking at the Sesquicentennial headquarters on Main Street on a dank October afternoon, the litter of the spent celebration all about us—stacked posters, boxes of unsold t-shirts, piles of brochures—and a skeleton crew moving through the big room on the one-way errands of closure. The celebration, John Lyng was saying, had not been all it could have been, in part because of this community split. There had been, he told me, an unfortunate amount of resistance to the Sesqui committee's efforts. "You know," he went on in a tired voice, "in every community there is always a group that thinks that everything wonderful is happening somewhere else—Chicago, wherever. Here, you'll find some people who think Mark Twain is a curse on this town." (This surely was an irony since Twain himself had felt under a sort of curse from Hannibal.)

"Maybe so," I said, "but what would Hannibal be without Mark Twain?" Big laughs here from the crew moving around us. A quarter of a million people come through here every year, Lyng said, because of the Twain association.

"And without that we'd be just like Ste. Genevieve, Chillicothe, or Louisiana [all Missouri towns of smaller size]. I'd guess the annual percentage of the economy we get from this would only be about 10 percent, but without Twain we wouldn't be known beyond a hundred miles. This year, because of the celebration, people came from all over the world. Every one of our 500 motel rooms was booked, and trailer parks were overflowing all the way to Quincy." He went on to describe the celebration's highlights: three grand parades, concerts by the Beach Boys, Al Hirt, and the Oak Ridge Boys, literary seminars featuring luminaries like Justin

Kaplan and Stanley Elkin (these, Lyng said, had not been so well attended as other festivities). But his recitation had a hollow sound to it, and you couldn't help but sense some crucial failure here. Indeed, Lyng's own prefatory remarks rather prepared you to sense this. Maybe in some weird and oblique way the famous son's ambivalence about his home town had been a hidden part of his bequest of fame to it. If so, it surely would have seemed to Twain himself fully in keeping with the way things were in a world where the unsuspected consequence was in reality a commonplace. Musing on this, I drove out from Hannibal to Florida, going back to where it all began before Twain was Twain.

It was only forty-one miles now along a highway lined with fields of milo, corn, and soybeans, but at the end of the 1830s the distance was measured in the vast difference between life in the woods and life on the river. At Monroe City Route 24 branches southwest toward Florida and Paris. Monroe City wore a dusty look about its agricultural stores, antique shops, and video rental outlets. It did, however, have a tanning parlor, an ironic comment on changing styles here in this borderland of the South where it used to be evidence of your elevated class status if your face was milky white, untouched by the bronzing effects of outdoor labor. Wishing to make clear that Colonel Grangerford is a thoroughbred and a gentleman, Huck Finn tells his readers that the Colonel's complexion was pale with no trace of red in it. Now to wear a tan is a sign of membership in the leisure class, but no one I saw on the streets of Monroe City had this look. I stopped here anyway, hoping for some of the flavor of what used to be the backcountry. At the B & B Restaurant over a Fish Basket and coffee I re-read the relevant portions of Twain's autobiography.

"Recently," he had written there, "some one in Missouri has sent me a picture of the house I was born in. Heretofore I have always stated that it was a palace, but I shall be more guarded now." He characterized the village of Florida as "almost invisible," and if anything it is more nearly invisible now than it was in 1835. Running southwest out of Monroe City the stranger went to Florida on faith, for it was no longer on the map, and there were few road signs. When you reached it you found a dismal, stranded settlement of two restored nineteenth-century cabins, one or two homes currently inhabited, and a few trailers rusting down in the weeds. The John and Jane Clemens house that once stood on one of these streets had been removed to a more auspicious setting, a hilltop above the naked shores of the artificial Mark Twain Lake.

Here you once again encountered evidence of the failed Sesquicentennial since the development of the lake into a recreational site and marina was to have played a major role in the celebration's revenue strategy. Nothing like this happened, though, and the lake, lonely in its artificial newness, simply sits there in sullen mockery of human schemes. The Clemens cabin, however, was in itself worth the trip, preserving in its uneven structure and general ramshackleness a sense of that thin, hard margin of hope on which the family subsisted in its Florida years.

That sense was intensified in the Florida cemetery, for in this little clearing, surrounded by hardwoods just then in their late ochres and russets, were the slanting slate and marble markers of the casualties of frontier life:

STILL BORN
DAUGHTER OF
J & E THOMAS
MARCH 15
1853

CYNTHEA ANN
DAUGHTER OF
J & E THOMAS
DIED
JAN. 25, 1850
AGED 1 YR
9 MOS 20 DS.

SARAH C.
DAUGHTER OF
J & E THOMAS
DIED
NOV. 30, 1844
AGED 10 DS.

GEORGE W.
SON OF
J & E THOMAS
DIED
MAY 5, 1841
AGED 3 MOS
8 DS.

ELIZABETH
WIFE OF
JOHN THOMAS
DIED
MARCH 26, 1853
AGED
36 YRS. 1 MO.
23 DS.

After this, what? Did John Thomas give up trying to found a family here and move on to another town and another wife, leaving behind in this grove love and grief?

Also here:

SACRED

TO THE

MEMORY OF

MARGARET L. CLEMENS

WHO DIED

AUGT. 17TH A.D. 1839

IN THE

TENTH YEAR OF HER AGE

Standing in front of this tablet, I contemplated the vistas of the then and the now: the mother and father lowering their child into this soft soil in late summer's haze, surrounded by the woods' brittle green, beyond which you now could see the waters of that lake created and named for the little boy who was perhaps with them on that August day. Here was the place to begin an exploration of Twain's childhood landscape, though his subsequent fame has mitigated against this: almost inevitably you are drawn first to Hannibal beside the river. But with this utterly forgotten hamlet to ground your other experiences you might come to an appreciation of what "home" meant to Twain, an appreciation that the fun and folderol of the modern-day Hannibal prevented.

The center and axis of that fun was the boyhood home on Hill Street in Hannibal (actually but one of several the family lived in). Next to it was a well-done museum which included among its holdings some interesting material on Twain's valedictory visit to Hannibal in 1902. (At a banquet held his last night in town the great man interrupted one of his famous after-dinner discourses by breaking into sobs.) Across from the house and museum was the Haunted House, tenanted by effigies of Twain characters. The wax figures, the manager told me, were done by a Mr. and Mrs. Curzon. "She puts human hair in one strand at a time," Wilma Arthaud told me, "all except the black man [Jim]. He was too hard. They couldn't afford to do him that way. That wax is special, too: it won't melt until 200 degrees compared to 130 for your ordinary candle wax." She smiled at me knowingly. "Now the eyes, the eyes come from a clinic where they make 'em for humans." Beyond the effigies the haunting commenced. In dark passageways skulls pounced at you, coffins creaked opened, a funny tombstone glowed in lurid light.

Outside the Haunted House were benches on which you could sit if you needed to recover a bit. They faced the Twain home and its famous stretch of board fence, said to be the very one Tom Sawyer was to have whitewashed. If you sat very long there, you were treated to some fine left-handed compliments to Twain's work, most of them coming from women. Their men were mostly silent, had both hands plunged into their pockets, and wore the resigned look of those in a coffle. "I never did read that," a woman said in passing, "but I saw the movie, and I thought that little boy was just as cute as he could be."

"You know," said another, "Mark Twain doesn't do a *thing* for me. We had to read him in school, but it was just boring then, and it's boring now. I'd much rather, you know, see Graceland." So would Twain.

Kin to the Haunted House was the Mark Twain Cave a mile and a half south of town. Because of its sheer size and geological reality the cave was more than an underground version of the Haunted House, but its owners had done their best to achieve parity with that more centrally located attraction, and so there was a family resemblance. The narrow passages were dramatically lit, and some of the more interesting formations had colored lights to relieve the monochromatic yellow of the natural limestone. A vertical column was styled "The Devil's Backbone" and was lit from behind a hell-fire red. In a blue-tinted grotto a mechanized pump splashed water down into a little pool. Another grotto was "Jesse James's Hideout," an imposture recalling those practiced on the pilgrims in *The Innocents Abroad.*

Our touring group was small and mostly silent, and we went through the lighted ways with our hands held out as if to ward off injury or sudden assault. At one of the grottos a woman said to her husband, "Tom, take a picture of this, and hurry up." Tom hurried. "Stand here," she said, marking the spot with a small stamp of her boot. As I moved on I heard the plastic snap of the Instamatic. When we got to a large open space our guide, Tammy, seated herself opposite and slightly above us and delivered a lecturette on Tom and Becky in that sing-song cadence used on children, patients, and tourists. When she said that it was "right there that Tom sat Becky down and tried to comfort her," all our heads turned to the indicated spot in tribute to the power of Twain's work or Tammy's delivery, or the labyrinthine reality of the cave itself.

After being shown where Injun Joe expired, the spellbound group wound its way toward daylight. As we did so I sidled up to the golden-haired Tammy. "Say," I said, "I read in one of Twain's books that there was a dead girl down here buried in a copper casket. Where is she now? I wanted to see her face."

"Oh," said Tammy sweetly, "I expect she's wandered off somewhere," smiling at me as at a mischievous boy. "Actually," she went on, raising her voice to group level, "I was just about to tell that story." Which then she did, drawing on Twain's remarks in *Life* about the eccentric doctor who did indeed bury his daughter down here, preserved in a copper casket filled with formaldehyde. Tammy told us the dead girl had been dragged up for inspection so often that at last she was rewarded with a more conventional burial.

Outside in the fluorescent light of the lobby I waylaid Tammy again, this time wanting to know something of her training for this work, how she'd been able to master a spiel of such length. Was it, I asked, a kind of oral formulaic in which certain spots in the cave were used as mnemonic devices? "Basically," she said, "it's a two-day course. They take you through twice. And then you have to learn where the [light] switches are." She giggled shortly, then added, "They just sort of throw you in there."

She was sitting so far from me on the bench that I was foolishly provoked to move a little closer. "Seriously," I said, lowering my voice, "do you ever get asked about that dead girl? I was just sort of kidding you."

"Oh, yes, sometimes."

"Hold on just a minute there," a voice commanded loudly from the cash register behind us. "Who are you, and why're you asking all these questions?" My explanation began more haltingly than I could have wished, and I seemed to be hearing myself making a defense against the suspicion that I might be a necrophiliac when a beefy, red-faced woman hove into view and took up a spread-eagle stance close to me. Her eyes meant business.

"Can I help you," she said without the slightest interrogatory rise. Before I could launch once more into a defense of my character and intentions Tammy was gone. But the company had not been diminished: the cash-register lady had closed in from behind, as if to block retreat. I tried to summon what I imagined was a literary demeanor, keeping my notebook open, pen at the ready, the writer at work.

"How many people," I asked hoping to gain the initiative with what I supposed might be a neutral enough subject, "do you folks average through here in a normal year? Must be quite a few."

"That's privileged information," the beefy woman said. "We don't tell anyone that."

"Well, why not? I'm not the I.R.S."

"Because we don't, that's all. We're not a public concern, this is a private company. You're on private property right now." I said other organizations sometimes told how many people used their facilities, the neighboring St. Louis Cardinals, for instance. Now her face approximated the red of a Cardinals cap, and she tugged viciously at her company blazer, wrapping it like a sheath of virtue about her trunk.

"I don't care *what* the Cardinals or anybody else does. *We* don't give out such information." I asked then if she would at least tell me her name and position here. She would not.

"I don't have to tell you a . . . (a pause here while briefly contemplating strong language) thing. Who are you to be coming around here asking questions?" Writers, she said, had credentials, and where were mine? "If you really were a writer, you'd have a blue card from the Hannibal Information Center, and then you'd get the royal treatment. Instead of that we find you sneaking around the back way and start in with our girl. I don't mind saying I don't like a sneak."

"Here now, what's all this about?" asked a man, also in company blazer.

"Says he's a writer. I think he's a sneak."

The man poured an unguent voice over our troubled waters, and shortly I learned from him that he was Mr. Bogart, that the woman was Mrs. Bogart, and that the cave had been in his family since 1923. All this without a Blue Card. When, shortly after, I left, Mr. Bogart kindly saw me to the parking lot and called after me, "You have a good day now."

Late in the afternoon of my last day in Hannibal I stood at the foot of North Street, the river just behind me, the rumble of cars and trucks overhead on the Mark Twain Bridge. I had been down here talking with passengers on the steamer, the *Mississippi Queen,* which had stopped on its Fall Foliage Tour from St. Paul to St. Louis. These folks were a hardy lot, smiling through the mist and the rain that had been following them for days. One couple, retired to Winter Park, Florida, said they'd signed on because they'd missed the hardwoods so. "We're originally from Minnesota," Mrs. Louis Champeau told me, "and you have no idea how much you miss them until you're away for a few years."

Leaving the boat landing, you passed a plaque at the foot of North Street marking the site of the old town jail, memorialized, the plaque read, as the place where Muff Potter was imprisoned for the murder of Dr. Robinson in *Tom Sawyer*. To Twain, the jail was memorable for another reason, one he tried to remedy in his fiction by having Tom save Muff with his testimony against the real murderer, Injun Joe. Crossing the highway that goes over the bridge to Illinois, I climbed the steep steps of Holliday's Hill, passing the handsome bronze statues of Twain's immortal boys with their straw hats and fishing poles. I wanted to end my visit up on the hill where Twain had begun his eventful one in the spring of 1882 when he had climbed here on a quiet Sunday morning to muse over the sleeping town and the passage of the years. In that passage in *Life* he made light of the experience, but there is little doubt he had been much moved by it at the time. He could see the town spread out below him and mentally blot out its new and unfamiliar features, could see Main Street directly below him running south to the trees of Bear Creek and beyond the high bluffs. He could see the house on the corner of Hill and Main where Owsley shot Smarr and dozens of other sites associated with his childhood and youth.

He was forty-seven then, and it must have seemed to him that he had lived several lifetimes since those days. Sitting here it seemed to me I, too, had lived at least a couple of lives since I was eight and my father read *Tom Sawyer* to me and my brother. I could see now my father's youthful, spare figure, sitting on the couch of our Chicago living room, shirt sleeves furled after a day at the office, holding the red-bound book in his lap. And through his reading of Twain's words, I was perfectly able to picture the little white town and the great river that now I gazed on. The lights of shops and cars were becoming more distinct down there as a misty dusk settled. The *Mississippi Queen* had departed, and the landing was empty. For a moment past and present met in the Timeless Now.

A car pulled into the small parking oval beneath the hill's crest. I heard the slam of doors, the crunch of gravel, voices. Presently a man appeared at the top of the steps. Tall and gaunt, he might have been anywhere between forty and seventy, his cheeks caved in over missing molars, his skin a dirty, unhealthy yellow. "Haven't been up here in years," he panted, then stuffed a cigarette into his mouth and puffed. "We used to go to the roller rink Sat'day, pick up a couple of girls, then come up here and fool around." He took me further into confidence with a wink. "You know how you do when you're young." I knew and said so.

"But," he said, exhaling a gray sigh of smoke and gesturing outward toward the town, the wooded bluffs, the river, "time sure changes everything, don't it? Even us. You comin' up, Maw?"

From the lot below the unseen Maw hollered back. "I can see everything I want to right where I'm at."

He looked at me, and this time I winked first.

FOUR

COMMUNITY OF SIN: GEORGE WASHINGTON CABLE'S OLD CREOLE DAYS *and* THE GRANDISSIMES

I

~ AS TWAIN NEARED NEW ORLEANS on his investigative pilgrimage of April 1882, word came that Ralph Waldo Emerson had died in Concord. Twain had once subjected the old man—and Holmes and Longfellow at the same time—to an outburst of the savage humor of the Old Southwest in a speech before them at the Whittier birthday dinner and afterward had abjectly apologized. But possibly Emerson had missed the whole performance anyway, observing it, said William Dean Howells, "with a sort of Jovian oblivion of this nether world in that lapse of memory which saved him in those later years from so much bother." Despite Emerson's long, dim decline, Twain could hardly have failed to know that Emerson had cared more, and more deeply, about the health of the national letters than anyone of his time, and that he, Twain, and all the others were permanently in Emerson's debt.

But whatever misgivings the news of the death may have stirred in the manic man were quickly enough repressed by his interest in spending some time in New Orleans with a contemporary author Twain had come very much to admire. This was George Washington Cable, whose *Old Creole Days* (1879) and *The Grandissimes* (1880) appeared to Twain to herald a Southern literary renaissance after the cultural dislocation of the Civil War. His New Orleans visit, during which he was subjected to reports of the recent Mardi Gras, would eventually prompt Twain to write an almost hysterical condemnation of Southern culture in the steamboating book he was now struggling to piece together. The South, he would write, was dwarfed and retarded by its love of Walter Scott-style chivalry and

105

other "brainless" forms of "girly-girly romance" like Mardi Gras. But in the midst of all this outmoded rubbish there was Cable, and Twain regarded his work as a sign of hope.

They spent several pleasant days together. With Cable acting as Twain's "eyes" they toured the French Quarter and rode out to West End and Lake Pontchartrain for a magnificent pompano dinner. On his own Twain attended a cockfight and a mule race, then went down-river to view former governor Warmoth's model sugar plantation at Magnolia. Joel Chandler Harris came down from Atlanta and Cable entertained both writers at his home in the Garden District. Harris was too shy to do so, but both Cable and Twain gave informal public readings of their works and Harris's too. Perhaps even then Twain's restlessly fertile brain was meditating a scheme for a reading tour with Cable, one which in fact they undertook in late 1884.

What initially drew Twain to Cable's work was the latter's use of Creole dialect. Knowing so well his own passion for talk, for folk-say that carried the very flavor of a locality and its style of living, and knowing too the monstrous difficulty in faithfully reproducing dialect on the page, Twain extravagantly admired Cable's achievements in *Old Creole Days* and *The Grandissimes.* In several reminiscences Howells recalled Twain's delight in reading aloud from Cable's work, especially his favorite story from *Old Creole Days,* "Jean-ah Poquelin." But in New Orleans Twain came to new levels of appreciation of Cable. He heard for himself large amounts of Creole talk and so more fully understood Cable's skill in reproducing it. Then on their tour of the French Quarter Twain marveled at the faithfulness and brilliant execution of Cable's settings. Even more, he loved the way the little man had of imaginatively opening those closed, battened shutters to give you a glimpse of interior lives, his ability to push aside heavy iron gates into hidden courtyards, the way he could with a few well-turned words peel back the layers of the years, revealing the secrets of an old city. Twain found the newer, Americanized portions of the city uninteresting as literary territory, and had he not had Cable and his books as guides, the old part, the Vieux Carré, lying in shadow on the north side of Canal Street, might have seemed to him merely a shabby reminder of that quaint past the bustling commercial metropolis had outgrown, as unpromising for literature in its dirt and disrepair as the American portions were in their undistinguished, spanking modernity. But *Old Creole Days* and *The Grandissimes* had made the French Quarter blossom for him into a species of holy ground, peopled with Cable's characters, alive with those old buildings and streets that formed the settings of the fictions. "With Mr. Cable along to see for you," he was to write in *Life on the Mississippi,* "and describe and explain and illuminate, a jog through that old quarter is a vivid pleasure. And you have a vivid *sense* as of unseen or dimly seen things—vivid, and yet fitful and darkling; you glimpse salient features, but lose the fine shades or catch them imperfectly through the vision of the imagination: a case, as it were, of ignorant near-sighted stranger traversing the rim of wide vague horizons of Alps with an inspired and enlightened long-sighted native."

The more Twain was in the New Orleans area, the more he was struck by the singularity of his host. The man appeared to be sui generis, of his own kind and without either artistic ancestors or current colleagues, working his literary vein in a solemn isolation at once noble and awesome. Despite its embrace of modern commercialism, the city and the state, Twain came to feel, were still attached to the feudalistic thinking of the ante-bellum era, and only Cable was writing and thinking like a modern. The proverb, he was to write in *Life,* "Many men, many minds," did not yet apply here where there still reigned a sluggardly uniformity of thought and feeling. Another literary friend of Cable's from the north, H. H. Boyesen, said of the South at this time that it was a "literary Sahara." Twain would have agreed and would never for a moment have wished to exchange places with Cable. Yet somehow the latter seemed to be finding imaginative sustenance here.

When we look at the biographies of American writers we are struck by how often it appears as if they arrived at literary careers by accident. Twain himself is a good example, and so is Cable. At age fourteen Cable was forced to drop out of school when his father died. In 1863 at age nineteen, a refugee from Union-occupied New Orleans, he enlisted in the Fourth Mississippi Cavalry of the Confederates and two years thereafter returned home penniless, owning no other clothing than the tattered uniform he wore, bearing the unhealed scars of two combat wounds. Soon he was a clerk in a cotton brokerage, laboring doggedly to support his widowed mother and sisters. When he married a New Orleans woman in 1869 his financial obligations became even heavier. Perhaps nothing then could have seemed to him more wildly unlikely than a literary career. There was no literary tradition at all in his family, where the reading of fiction was frowned upon by Cable's mother whose New England background held within it a strong vein of old-time Calvinist piety. Apparently

Cable read little belles lettres as a child, except for a permitted indulgence in Walter Scott, Twain's bête noire. As a young man his taste in literature ran distinctly to the factual: he liked history and geography, read considerably in the literature of exploration, particularly in *Jesuit Relations.*

But there was the city itself, and from early childhood Cable was intensely interested in it. The city's sights, its quaint gallimaufry of architectural styles, its sounds and the peculiar way they carried so far—street songs of the neighborhood vendors, the singing of black laborers in the cotton press near his childhood home, the music of the marching bands—the pungent odors of orange trees, coffee, the mingled smells of the open air markets, all this entranced the diminutive boy. He loved to cruise the waterfront at the foot of Canal where the steamboats tied up and along Tchoupitoulas Street, where the flatboats lay that had come down the Ohio and the Mississippi with their boisterous crews. At five in the afternoon— with a grand show of blackest smoke, the pent-up shouting of the whistles, and churning paddles—the up-river steamboats would back out into the Mississippi as a group, leaving behind an exhilarated crowd gathered for the spectacle. Congo Square on the westernmost skirt of town was another captivating place, though considering Rebecca Cable's stiff religious scruples it is difficult to imagine the family attending the slave dances held there on Sundays. But perhaps at least their terrific, booming sounds carried through the New Orleans air to the Cable home, and the boy caught the excitement of the massed drummers and gourd rattlers, the shouts of the dancers. In any case, he had vivid memories of Congo Square and wrote of it in *The Grandissimes* and in a magazine piece in 1886. He loved, too, exploring along the city's drainage canals, following them out through reclaimed swamps that once had been indigo fields. With his family he took fishing expeditions to Lake Pontchartrain.

Perhaps best were the open-air markets. Down at what was probably the French Market on the waterfront the boy heard what he later recalled as a "bewildering chatter of all the world talking at once, mostly in German and French, a calling and hallooing, a pounding of cleavers, a smell of raw meat, of parsley and potatoes, of fish, onions, pineapples, garlics, oranges, shrimps, and crabs, of hot loaves, coffee, milk, sausages, and curds, a rattling of tins, a whetting of knives, a sawing of bones, a whistling of opera airs, a singing of the folksongs of Gascony and Italia, a flutter of fowls, prattling and guffawing of Negroes, mules braying, carts rumbling. . . ." It was all "great fun," he said, and indeed the whole city was great fun for a boy. Even after the harrowing of the war, after the occupation and the smoldering, mutinous years of Reconstruction, and now too Cable's impoverished circumstances and burdensome family obligations, the city continued to intrigue, even romance him. There was a richness, a density, to the texture of daily life here that the young man was in love with. But to him it appeared there were astonishingly few New Orleanians who cared about their city's uniqueness, its almost opulently rich past. Historic buildings were allowed to fall into irreparable condition or were blithely demolished in the name of Progress. Recollections of the past were allowed to pass unremarked and unrecorded into oblivion. In 1870 there was not a single literary society in the city. Years later when the literary historian Fred Lewis Pattee asked what had motivated him to begin his researches and Creole stories, he replied that there was such a wealth of perishable material lying about unused it seemed to him a pity it should go to waste. A quiet sense of urgency invaded and possessed the man who spent his hours in the counting room of a cotton brokerage. He wanted somehow to be able to say something about this city and its past.

In those days New Orleans newspapers made a practice of opening their pages to unsolicited material, and for Cable, casting about for some avenues of expression, here was the obvious one. In company with a few like-minded friends, he proposed they contribute a weekly column on city life to the *Picayune,* but after the first column the others lost heart, and Cable had to decide whether to let it go or try to produce all the copy himself. For the next eighteen months he wrote the weekly column under the by-line of "Drop Shot." In ninety columns on subjects as varied as baseball, women's rights, city streetcars, and yellow fever, Cable purposefully explored the surfaces and recesses of his native place, digging into its forgotten ruins, turning over its beguiling idiosyncrasies, remarking on the significance of current events. In the course of his investigations he learned how vast a treasure-house of oral lore its populace still retained, learned too that he had a talent for ferreting out folklore and legend as well as for probing into the alleys and abandoned byways of the heedlessly growing city. The city itself, he came to see, was a palimpsest, continuously, restlessly writing over itself, burying its past and those stories beneath successive layers. In short, during the months of his "Drop Shot" apprenticeship he learned that he was a writer and that he had on hand a great subject. In the last column he wrote, he made a remark that was both a public announcement of vocation as well as a challenge to himself. Echoing John Neal's earlier remark about the unrecognized riches of local soil, Cable wrote that Louisiana's history was a rich mine awaiting the writer who could see what was there. "Here," he wrote, "lie the gems, like those new diamonds in Africa, right on top of the ground. The mines are virgin." Only one man, he continued, had been at work here, and that was the Louisiana historian Charles Gayarré, but Cable ventured that the half of the story had not been told. He would begin trying to do so.

Still, at this point he could by no means think of cutting loose from the counting house into full-time authorship (nor was he ever to be wholly free from nagging financial worries until his old age when Andrew Carnegie's will settled an annual $5,000 on him for the balance of his years). He now had children of his own to support as well as his continuing obligation to his mother who lived with him in a series of crowded, quietly impoverished households around the Garden District. But in what moments he could spare, he wrote. In the fall of 1871 he sent *Scribner's* a slender manuscript culled from his "Drop Shot" columns and augmented with some new poetry and a narrative piece. It was rejected, but Cable remained determined, writing the publisher in response to their negative reaction that he was going to keep on writing until he had produced something that was publishable.

When *Scribner's Monthly Magazine* sent the writer Edward King south in the spring of 1873 to work on a series of articles to be called "The Great South," Cable was one of those local New Orleans writers King wanted to see. King and his illustrator spent several delightful days in Cable's company, and Cable read them several of the stories he'd been working on. When he left, King took with him a couple of Cable's manuscripts, and in July he notified Cable that one of them, "'Sieur George,'" had been accepted. Then in the year and a half following, Cable placed "Belles Demoiselles Plantation," "'Tite Poulette," "Jean-ah Poquelin," "Madame Délicieuse," and "Café des Exilés" with *Scribner's*. "Posson Jone'" ran in *Appleton's Journal*. So by the middle of 1876 the stories that would make up *Old Creole Days* had all appeared in print; in editions subsequent to the first one Cable would add "Madame Delphine," a novella published serially in 1881.

Quite as much as he'd intended it, *Old Creole Days* is a testament to its author's fascination with the city. The book is so filled with the facts and physical qualities of New Orleans that it could have been read at the time as a sort of guidebook—and indeed during the Cotton Exposition of 1884–85 it was so used by numbers of visitors, so concrete and specific was Cable about his settings. Partly this scrupulous fidelity to fact was the result of Cable's relatively shallow acquaintance with fiction; partly also of his inherited, religiously based uneasiness with fictive forms; and partly, too, the consequence of his apprenticeship as a journalist where he was bound to discover and respect facts. But more than any of these matters it was simply the fact of the city itself, which to Cable obviated the need to invent much. New Orleans itself had done all the inventing; the writer need do little more than construct a fairly plausible narrative framework and then let the city speak. So close to history was he working that at first he thought of assigning dates to the stories and then arranging them chronologically when they appeared in book form. Finally he did not, but every one of the stories is securely, precisely grounded in the facts of its time and place.

Each of the stories begins with a beautifully rendered description of place: the heavy-timbered cypress house of Jean-ah Poquelin sitting in the marshy midst of abandoned indigo fields above Canal Street; the three blank-looking buildings on Dumaine where the drama of "'Tite Poulette" is played out; the corner of the Rue Royale and Conti where the wastrel Jules St.-Ange hangs out, surrounded by balconies beneath which sit showy shops, rows of peeling shutters, and the great groaning key that is the locksmith's sign; the magnificent, doomed Belles Desmoiselles plantation with its immense veranda, its front steps "spreading broadly downward as we open arms to a child," presiding over its fecund fields along the roiling and unruly river. "Madame Delphine" opens in the newer section of New Orleans but in its second paragraph passes across Canal Street and into a past symbolized by the Rue Royale. Huddled in the shadows are the shabby buildings that have settled down into a "long sabbath of decay," their iron balcony railings rust-eaten, their streetward windows nailed shut, their alleys revealing a "squalor almost oriental." Yet, Cable writes,

> beauty lingers here. To say nothing of the picturesque, sometimes you get sight of comfort, sometimes of opulence, through the unlatched wicket in some porte-cochere—red-painted brick pavement, foliage of dark palm or pale banana, marble or granite masonry and blooming parterres; or through a chink between some pair of heavy batten window-shutters, opened with an almost reptile wariness, your eye gets a glimpse of lace and brocade upholstery, silver and bronze, and much similar rich antiquity.

Clearly, the writer was one who spent hours haunting the streets of the old quarter, taking more from the purposefully blank old facades than they ever wished to give—or than others cared to notice.

Such lush and leisurely descriptions as these of *Old Creole Days* have a tendency to make Cable's plots and characters look pallid and superficial by contrast, and reading the book through for a first time you feel the real subject is New Orleans as a physical and historical fact. If in retrospect this places Cable squarely within the tradition of post-bellum, local-color writing with its too often sedulous attention to place, at the same time there are other aspects of *Old Creole Days* that set it both apart from and above the work of the local colorists whose work filled the magazines of the 1870s and '80's. Cable was the only one among them to take on the challenge of the city. From the beginnings of local-color writing in the 1830s, the settings had been determinedly rural, and in the post-bellum years they remained so. With the explosive entrance of the city into the mental landscape of America after the war, local-color depictions of backwater retreats and unspoiled byways took on a nostalgic, sentimental quality that in fact constituted a chief attraction.

There is nostalgia in *Old Creole Days,* too. In fact, Cable's preferred title for the book was *Jadis*—once upon a time, formerly, of old—and none of the stories is set later than 1850. There is sentiment here, too, some sappy romantic stuff, and that chronic Southern adoration of the "fairer sex" that drove Twain wild with rage. But nostalgia in *Old Creole Days* is more than a way of avoiding the unlovely and unsung facts of the present. The use of the past, its insistent, brooding, prodigious presence in the stories is used to suggest that the present is the past in its present shape and that the past is thus inescapable. With an almost Hawthornian sense Cable tells us that in old places where continuous living has built up a kind of cultural humus, it is an illusion to think you can escape from the past into a bright, new present. This is what surely saves *Old Creole Days* from the characteristic defects of the local-color genre.

And there is authentic bite to these stories, too, that continues to recommend *Old Creole Days* to the attentions of succeeding generations of readers. For in accepting the challenge of his native place Cable was compelled to write of its darker aspects, that heavy counterpoise to its soft tropical light and airy ironwork that makes the place like none other in America. It would never be his style in fiction, or in life either, to revel in the darker aspects—he was no Dreiser, nor yet Frank Norris—but he recognized their existence and their force, and in these first published stories he suggested more than a few of them. In several of the stories we get sharp, darting glimpses of that historic underworld that made New Orleans the most notorious American city until well into the next century when Chicago would become the crime capital. In "'Tite Poulette" the dandy gloved-and-cane-carrying caller on the widowed Madame John and her gorgeous daughter is actually the manager of a nefarious dance hall where "arrangements" may be made between the white patrons and the quadroon dance-hall girls. When Madame John's phlegmatic neighbor across the street attempts to intercede on behalf of the widow and her daughter, he is immediately stabbed, caned, and left for dead by the manager and his bravos. Only the ignorant and romantic, Cable makes clear, would think to go up against organized crime. In the surprise ending to "Café des Exilés," we catch just a fleeting glimpse of New Orleans as the historic staging point it had long been for all manner of revolutionaries, smugglers and gun-runners, filibusterers, and assorted interventionists. This story also suggests the violence endemic to the city where so many habitually carried knives, pistols, or sword canes and were ready to use them, and where the duel was an ever-present possibility as a way of settling differences—as it is in "Madame Délicieuse." "Posson Jone'," in most respects a very funny story, also glances into the shadowed world of gambling dens, con men, and low doggeries serving rotgut whiskey to slaves in defiance of prohibitions against the practice. "Madame Delphine" features a wanted man, a pirate trying to go straight; the male characters in "Belles Demoiselles Plantation" are what we would now style "white-collar criminals" whose violent, dissolute ways have here the sanction of long custom. To us "'Sieur George," Cable's first published story, might be the most seemingly contemporary one in the book with its depiction of isolated, unremarked decadence and decay, the lone, unknown man, even his name a mystery, accepting the guardianship of a girl, then through the years squandering her inheritance on a lottery. Finally, all 'Sieur George has for the girl who has always called him "father" is the obscene offer of marriage. She flees to a convent and at the story's end sees him against a sunset, creeping westward out of the city to find a night's lodging in the tall grass. Or perhaps Twain's favorite, "Jean-ah Poquelin," would most recommend itself to our attention. The old man, living in his huge broody house on the outskirts of town and attended by none but a mute African slave, fiercely resists the efforts at municipal improvement that would drain his poisonous marsh and drive a thoroughfare close by his front gate. Rumors and whispers have followed him for years. What does he do out there in that ruin of a house, anyway, and where is that half-brother, once so beloved but vanished now these many years? What is the source of his resistance to the new street? The city's bureaucrats try in vain to reason with old Jean, and the city's street children tag along behind him, jeering at his bent back, flinging clods of mud at him. One night a drunken deputation decides to give Jean a charivari in an attempt to persuade him to allow the improvements. But when they arrive at his gate, clanging their pots, bellowing their chants, they are themselves met with a small, solemn deputation: it is the mute slave leading a bull and cart. On the chart lies the shrouded body of Jean-ah Poquelin, and behind it walks the long-vanished half-brother, a whited leper. The crowd falls into an awed silence, then parts to allow the cortege to make its way into the *Terre aux Lépreux* in the swamps behind the city. No one ever sees them again. "Jean-ah Poquelin" is with the reader long afterward like a bruise, and you feel the justice of Twain's admiration for it. In its tone and setting, and in its theme of the mystery of human bonds and bondage, it anticipates Faulkner and Tennessee Williams.

In writing of such things Cable had to overcome his own internal inhibitions. Like Rebecca Cable and his wife Louise he was an intensely religious person, involved all his life in various kinds of Christian uplift. The sordid and the unseemly were hardly pleasant for him, and when one considers the fact that the prevailing taste of the national magazines was strenuously opposed to fictional depictions of life's grimmer dimensions, Cable's courage is even more remarkable: "Posson Jone'," with its descriptions of drunkenness and gambling, was rejected by that prim guardian of America's Genteel Tradition, Richard Watson Gilder of *Scribner's;* it was rejected also by the *New York Times,* the *Galaxy,* and *Harper's,* whose editor in returning the story spoke for all when he said that it was unsuitable for his pages because it emphasized the "disagreeable aspects of human nature." Stories in *Harper's,* he went on, "must be of a pleasant character and, as a rule, must be love tales."

But the strength of Cable's commitment to a faithful depiction of the spirit of New Orleans is even more evident in his decision to deal with the fantastically tangled issue of race there. When he'd begun writing his "Drop Shot" columns for the *Picayune,* Cable by his own admission was no different in his views on racial matters than the great majority of his fellow whites in the city. Like many of them he had refused to take the oath of allegiance when the Yankees had occupied New Orleans, and with his mother (who had declared herself an enemy of the United States) had fled to Mississippi. He had fought gallantly in the war and had his own red badges of courage to prove it. And like many of the city's whites he was subsequently aghast at the venality and corruption of carpetbagger government in the state. New Orleans from 1865 into the 1880s was a powder keg of racial tensions, and these often exploded in isolated racial incidents, mass meetings of outraged white citizens, formations of segregationist organizations, and occasionally in riots: there was a severe one in 1866, a lesser one the year following when blacks stopped whites-only streetcars and beat up their drivers; in 1874 the White League fought a pitched battle with the city's police force and defeated it. Cable witnessed most of these events with the eyes of a white supremacist. Assigned by the *Picayune* in 1871 to cover a teacher's convention called by the Republican-appointed superintendent of schools, he reported with indignation that the white participants had been forced to sit next to their colored colleagues. When other newspapers took up the cry their ferocity alarmed Cable, and he dropped the issue. Subsequently he lost favor with his editor.

Possibly the incident set the conscientious man to thinking about the whole justification for slavery. If so, it must have been a staggeringly revolutionary process for one who had all his life simply accepted the natural supremacy of his race, the inferiority of the black race, and the divine and natural sanctions for slavery. His own family had owned slaves—as many as eight at a time, by one account—had bought and sold them like so many goods. These terrifically turbulent days, though, outsiders were telling Southerners that slavery had been wrong all along, that the whole superstructure of divine and natural sanctions supporting the practice of it and its kindred distinctions of caste were at best specious. Ever a searcher after recondite truths, he now went back to the republic's founding documents, the Declaration of Independence and the Constitution, and read them for what they had to say on the question of equality. He read, too, in the Bible for its alleged support of the master/slave hierarchy. He found his old prejudices shaken. And they were more seriously shaken in 1872 when he did research for a *Picayune* series on New Orleans churches and charities. Digging into the city's past, its crumbling archives and yellowing newspapers, he began to see that just as the physical city was founded on land-fill that raised it above the encroachments of the river, so too was the culture of the city founded on skin discriminations that were cruelly unjust. The old *Code Noir* was the most flagrant manifestation of this, a document that, superseded though it had long been, still so blazed with the viciousness of its Draconian measures that Cable felt forever branded by his tacit participation in the system it enforced. He never recovered from his primitive encounter with the document. It remained a reference point for the rest of his years; the very idea that a white master might send his chattel to the Calabosa for a specified number of lashes without at the same time specifying the offense committed signifying for him the whole mindless and violent tyranny of the system.

How then could he presume to write of this place without writing of those distinctions that had built the city, that still remained much in force, manifestly and covertly? He might as well try to portray the city without mentioning the river. As Cable looked about him in the early 1870s he saw clearly for the first time how indebted his city was and had always been to the workings of the *Code Noir,* to the systematically enforced peonage of its colored peoples, to the isolation and humiliation of its many mixed-blood citizens whose women had long been the preferred concubines of the white gentry. The very conformations of the buildings with their now-modified slave quarters behind them, the famous wrought-iron railings and fences forged by black hands, even whole neighborhoods such as the Marigny District with its neat cottages, homes to quadroon mistresses—all this was daily testimony to the operations of the color code. If the spirit of New Orleans was a wonderful gumbo of street songs, orange blossoms, steamboat whistles, and the half-forgotten tales of the past, beneath all, tingeing all was the history of the separation of the races, the subjugation of the colored peoples.

In this highly charged personal context, made the more intense by the prevailing racial atmosphere of the city, Cable wrote his first piece of fiction. It was a legend of slavery told him by one of the city's blacks, though the tale was then apparently widely known among New Orleans's colored population, and he may have had more than one source.* The man who told Cable the story of "Bibi," the African king brought in chains to New Orleans, was a porter in the cotton brokerage house in which Cable worked. In essence, the legend told how the monarch had fallen into the hands of the slavers, how he'd been brought to New Orleans and put to work on a nearby plantation. But the black giant rebelled at this violent change in his status and escaped to the swamps where he became a terror to those hunters and fishermen unlucky enough to encounter him. Out of the swampy mists rumor wafted, crediting the fugitive with superhuman powers, with organizing that nightmare of the whites, a slave rebellion.

---

*The great New Orleans jazz artist Sidney Bechet (born in 1897) knew the legend and in dictating his autobiography in the late 1940s retold it as a part of his own family history. Bechet, *Treat It Gentle* (New York: Hill and Wang, 1960).

Heavily armed posses pursued him and fired on him, then saw him laugh scornfully and stalk back into the swamps. Once, some said, he was wounded in the arm and amputated it himself, thus his name, "Bras Coupé," though others maintained he had never been hit and that "The Arm Cut Off" was a crude translation of his African name, which signified that in losing him his people would be losing their right arm. Whatever, Bras Coupé came to a fatally appointed end, betrayed by one he trusted, and clubbed to death as he slept in his swampland hut.

It is not known what version of the legend Cable used in writing his story, or whether he used several and added materials of his own. But in whatever form the story met a stony wall of rejection when it went north. Gilder at *Scribner's* rejected it as did other magazine editors in that supposedly liberal latitude, George Parsons Lathrop (Hawthorne's son-in-law) writing Cable that his "Bibi" produced an "unmitigatedly distressful effect". Well. So he could not write this glaringly about racial matters. Yet somehow he must find a way of talking about them, of showing their pervasive influence. And he did so. "'Tite Poulette" is a story about the plight of the city's quadroom women, and it is also a sketch for the longer "Madame Delphine." These two stories treat the subject frankly. But in other stories, "Posson Jone'" and "Jean-ah Poquelin," the issue of race is there less obviously, but it is there nonetheless, as racial matters would be there in the daily conduct of life in the city. To the visitor they might not be plainly visible, yet they would be there all the same, and the native would know it, whether showing his visitor the ruins of the old St. Louis hotel (as Cable had Twain) wherein were held the city's most significant slave auctions; or simply in passing in one of the streets of the old quarter some veiled, handsome woman whose skin barely betrayed the fatal tinge of the tar brush.

In "'Tite Poulette" the widowed Madame John is the quadroon mistress of a white man who in dying bequeaths her his handsome raised cottage in Dumaine. She sells the house and places the proceeds in a bank, "which made haste to fail." Thereafter she must work at the Salle de Condé, one of the places where the city's famous quadroon balls are held. It is a demeaning, humiliating existence for this still-handsome woman with her "nearly straight" hair and her skin so pale you "would hardly have thought of her being 'colored,'" and yet what else was she to do? In her youth she had been singled out by a white gentleman at one of the balls, had become his well-kept mistress (in New Orleans she would have been known by the politer term, *placée)*, and had the bank not failed would have had the money from her house as her life legacy. So worked the system of *plaçage* there. Thus there is good reason for her to accept the work at the Salle de Condé despite its conditions, for she is a half-caste outcast without other means of support and with none to look forward to. Besides, she has her daughter, 'Tite Poulette, a girl of striking appearance in whose fine features one could detect no least sign of the colored race, and Madame John would do anything to spare 'Tite Poulette what she herself has had to endure. She wishes aloud to 'Tite Poulette that they were both "real white" so that some white gentleman might legitimately ask for her little chick's hand and make an honorable marriage. Eventually, in a too-abrupt denouement, it is revealed that 'Tite Poulette is not really Madame John's daughter but instead the orphan of a Spanish couple, both of whom had died of yellow fever within hours of one another. This frees the way for true—and legitimate, white-to-white—love to triumph when the woman's neighbor across the street, the young Dutchman, Kristian Koppig, asks for 'Tite Poulette's hand.

Subsequent to publishing this story in *Scribner's* Cable received an anonymous letter from a quadroon woman who had read it and now begged him to rewrite it. "If you have a whole heart," she wrote, "for the cruel case of us poor quadroons, change the story even yet, and tell the inmost truth of it. Madame John lied! The girl was her own daughter; but like many and many a real quadroon mother, as you surely know, Madame John perjured her own soul to win for her child a legal and honorable alliance with the love-mate and life-mate of her choice." Cable did rewrite "'Tite Poulette" in accordance with his correspondent's plea and in accordance with that inmost truth he knew. "Madame Delphine" is the story of a quadroon mother who successfully passes her daughter as white and so wins for her an honorable marriage.* To accomplish this monstrous breach of New Orleans racial law she has to have the help of a sympathetic priest, Père Jerome, and it is he who speaks finally for Cable and Cable's sense of his community's common history. "We all participate in one another's sins," says Père Jerome. "There is a community of responsibility attaching to every misdeed. No human since Adam—nay, nor Adam himself—ever sinned entirely to himself. And so I never am called to contemplate a crime or a criminal but I feel my conscience pointing at me as one of the accessories."

The stories that were to make up *Old Creole Days* apparently aroused

---

Passe-a-blanc was the local term for the practice.

neither suspicion nor animosity in New Orleans or elsewhere in the South. In the North Cable was greeted as an innovator and genius of his region. Still he felt he had not half told the story of his place. The stories were all well enough: he enjoyed writing them, and they had allowed him to say some things about New Orleans he believed needed to be said, captured, before the stories—and even the buildings as well, maybe—passed into oblivion. But his experiences with the magazines had taught him the limitations of that medium, and he felt he needed both more physical space and more literary latitude to capture the languidly phosphorescent spirit of his city. Editors up north and friends, too—Lafcadio Hearn and H.H. Boyesen, particularly—were urging him toward deeper, more extended treatments of New Orleans. Sometime in late 1877, Cable began sketching out the plot of *The Grandissimes,* which he thought of as a *roman à clef de la ville.*

By this time he had come out on racial matters in the city, the occasion being the expulsion of colored students from the Girls' High School in 1875. The school was located in the old Lalaurie mansion at the corner of Royal and Governor Nicholls, and the building's lurid history was symbolic of that tragic continuity Cable saw between the slave-holding past and the troubled present, for in 1834 it was accidentally discovered that Madame Lalaurie had been using her home as a torture chamber for her slaves, several of whom had been found manacled to the floor in an upstairs chamber. Now in 1875 a mob had come to the school and forcibly removed those pupils who could not definitively prove themselves lily-white. Cable was there to see the spectacle and wrote an impassioned letter to the *New Orleans Bulletin,* which he signed "A Southern White Man." Much later he would write about the house and its history in *Strange True Stories of Louisiana* (1889). But for now he felt there was little to be lost as well as much to be gained by writing openly in a novel about what all New Orleans whites knew to be true but refused to acknowledge. His indignation and passion were such, though, that the early drafts of the novel he submitted to *Scribner's* were more tract than fiction, and his editors there, first Gilder, then Robert Underwood Johnson, had to work with him page by page toward a work of art and away from a thinly veiled sermon.

The Grandissimes, flawed though it turned out to be, is one of the great novels of nineteenth-century American fiction, a haunting, haunted portrait of a place as it was in 1803, as it in many ways remained in 1880. At first blush the novel seems almost impossibly cluttered (it would defy even the reductive powers of the staff at Cliff's Notes). There are several

subplots, an unwieldy cast of characters, and a webbing of interconnec-
tions that is baffling: Agricola Fusilier, the old Creole lion, is connected
by past dealings to Aurora Nancanou who is connected by her affections
to Honoré Grandissime, old Agricola's nephew; Honoré is himself half-
brother to a half-caste of the same name who loves Palmyre Philosophe,
a slave in Fusilier's household and who, in turn, loves the white Honoré,
et cetera. But whatever demands this intricate webbing makes on the reader
there is a strategy at work here, and ultimately it is effective. For what
the intricate webbing makes clear is that community of sin of which Cable
had gradually become aware and wrote of in "Madame Delphine." All of
the characters—white, black, mixed-blood—share in the system that the
city's racial codes have made, and there is no escape possible: not denial,
nor the perpetuation of the old wrong, nor violent rebellion against it;
not even a retreat to Europe, which the colored Honoré Grandissime and
Palmyre Philosophe attempt. Signifying this is the revenant story of "Bibi,"
here called "The Story of Bras-Coupé" and placed like a black marble
monolith squarely in the center of the book. All the characters, all the action finally must relate to the story of the king in chains, and whatever
version Cable may have tried on the magazine editors in the early 1870s
it can hardly have been more unsparing, more powerful, more condem-
natory than the one he now used. Here, for instance, is his spare description of Bras-Coupé's Middle Passage, as poetic and tragic in its language
and import as Melville at his best:

> Of the voyage little is recorded—here below; the less the better. Part
> of the living merchandise failed to keep; the weather was rough, the
> cargo large, the vessel small. However, the captain discovered there
> was room over the side, and there—all flesh is grass—from time to
> time during the voyage he jettisoned the unmerchantable.

Because of his unquenchably fiery spirit Bras-Coupé is punished according to the provisions of the *Code Noir*. ("We have a *Code Noir* now,"
Cable writes in an aside he couldn't bear to give up, "but the new one
is a mental reservation, not an enactment.") And Cable spells out those
provisions: for first-time runaways, their ears to be shorn and one shoulder
branded; for the second such offense, hamstringing and branding of the
other shoulder; for striking a master with sufficient force to produce a
bruise, death. Bras-Coupé's master is lenient by the standards of the day
and spares his huge chattel, though Bras-Coupé had nearly struck him dead

with a thunderous blow of his fist. But the other provisions are enforced after he is captured at the Congo Square dances. Still, mutilated and disfigured as he is, the king retains his power, for Bras-Coupé is a voudou sorcerer, and he has pronounced a curse on his master's land. In phrasing the author surely borrowed from his close acquaintance with the Old Testament, Cable has Bras-Coupé point his long black finger through the open window of the big house and pronounce a curse on its lands. "May its fields not know the plough nor nourish the cattle that overrun it." And "May weeds cover the ground until the air is full of their odor and the wild beasts of the forest come and lie down under their cover." And so it came to pass:

> The plough went not out; the herds wandered through broken hedges from field to field and came up with staring bones and shrunken sides; a frenzied mob of weeds and thorns wrestled and throttled each other in a struggle for standing room—rag-weed, smart-weed, sneeze-weed, bind-weed, iron-weed—until the burning skies of mid-summer checked their growth and crowned their unshorn tops with rank and dingy flowers.

On his deathbed Bras-Coupé is persuaded to lift this particular curse— one of the very few doves Cable lets fly in these somber pages—but he makes it clear that the curse of race yet hangs over the city like some moral miasma, infecting even the most charming of its scenes. He dares hope there are signs that New Orleans, the state, and perhaps even the South as a region are beginning to understand the necessity of a critical confrontation with the past, and here four of his younger characters find reconciliation and love, though he does not tell us of their ultimate fortunes. But as for the majority, they still lie under the "shadow of the Ethiopian," convinced that change is fatal to the order of things. Murder: of old Agricola by his family unacknowledged half-caste member, Honoré Grandissime, f.m.c. (free man of color). Murder: of old Clemence, the *marchande des calas,* who is caught planting a voudou fetish on the Grandissime estate and is summarily executed by the clan's younger members, the fat old woman frantically scuttling away through the tall grass shouting placatory phrases ("'We de happies' people in de God's worl'!'") while her judges stand there coolly until one of them shoots her in the back. Suicide: Honoré Grandissime, f.m.c., who had vainly sought Palmyre's love, knowing she preferred his white half-brother. Exile: Palmyre Philosophe, the voudou queen, who escapes to Bordeaux, there to live on the blood-money regularly sent her by the white Honoré who knows the dimensions of his debt and endeavors to pay some portion of it.

In the last chapter Cable gives the reader a parting glimpse of the beautiful city seen from its old center, the Place d'Armes. Here again are the manicured gravel walks and verdant shrubs, the skimming swallows and dancing butterflies, the sedate strollers beneath parasols, the gentle-manly riders. And beyond rolls the peaceful river. But we know too much now about what this beauty has cost in human blood and sacrifice, in blighted aspirations, in cruel vanity and the callous perpetuation of palpable wrongs for us to be beguiled by it as we might well have been when Cable described the same scene in the novel's opening pages. And we know also what it is that truly sustains the great Grandissime home on Esplanade, its white pillars holding the house high above the "reeking ground," its veranda so immense "twenty Creole girls might walk abreast" across it. Joseph Frowenfeld, the young American apothecary who witnesses so much of the novel's action and struggles to assess its significance, says at one point he feels as if he's reading a book called the "Community of New Orleans," one whose pages are dusty, torn, misplaced, obscure. *The Grandissimes* is hardly such a musty old tome, but it is the book of New Orleans in which Cable has pieced together the story of his community in all its dark vividness, its unacknowledged continuities.

II

❧ CABLE'S NAME IS pretty much a dead letter in the New Orleans of to-day. Kenneth Holditch, a professor of English at the University of New Orleans and a professional guide to the French Quarter, told me that no one asks about the sites of Cable's work anymore. We were at a sidewalk table of the Napoleon House on Chartres Street. Across from us loomed the rear of the huge Royal Orleans Hotel, and Holditch pointed out that a few of its cornices were remnants of the famous St. Louis Hotel under whose rotunda the great slave auctions were once held. Holditch said his students didn't know Cable's name, let alone his work, and that even among his colleagues at the university Cable was not much esteemed. "One of them actually told me the other day that Adrien Rouquette was a better writer," he said, rolling his eyes in amazement at this assessment. But in a left-handed, ironic way it seemed appropriate that in New Orleans some might really find Rouquette superior since it had been this poet/priest who commenced the community savaging of Cable's reputation shortly after publication of *The Grandissimes*. Rouquette was the designated hitter for white New Orleans's bitterness and outrage at what they conceived to be betrayal by one of their own. Eventually the attacks and public snubs that followed the publication of Rouquette's anonymous pamphlet made the city intolerable to Cable and his family, and when he took them north for the summer of 1884, it was for good. Up there, cut off from a daily, vitalizing contact with his native place, Cable failed to fulfill the literary promise of his first two books and became instead more a reformer than a literary artist. He did much good in this way—perhaps more than he ever could have in literature—but by the time of his death in 1925, he was remembered, if at all, as merely a local colorist of the nineteenth century whose best work had been done many years before.

In New Orleans itself, they shed few tears for the little man who had made himself so obnoxious at the beginning of the 1880s. Oh, they'd forgiven him—sort of—by the time he'd begun to revisit the city in the early years of this century. By then many of his fiercest antagonists were dead, and perhaps it had been to show him what Southern manners really were that the Louisiana Historical Society invited Cable to read at the historic Cabildo. But that honor did not signal a real and final forgiveness. In 1921, the illustrator Joseph Pennell wrote to Cable from New Orleans, saying that he had found Cable "remembered from the old days and not forgotten—or forgiven." Nor does he appear forgiven now when a silent, studious neglect seems to cover his name and early achievements.

Under Holditch's guidance we walked the Quarter, talking of Cable and literature, Holditch pausing now and again to point out some inconspicuous architectural feature that otherwise would have forever escaped my attention: the much modified remains of the slave quarters, for instance, that at many addresses have been integrated with the streetward buildings once inviolately separate from them. We also saw and paused before places lived in and frequented by Faulkner and Faulkner's early mentor, Sherwood Anderson; by several residences of Tennessee Williams, including the building in which he'd finished *A Streetcar Named Desire* (one of the old trolleys is now parked in the grass down by the U.S. Mint building, its faded green scroll bearing the superseded destination, "Desire"); and by homes once tenanted by Roark Bradford and Lyle Saxon.

In 1883 Lafcadio Hearn easily identified the scenes of *Old Creole Days* and *The Grandissimes* in an article for *Century Magazine*, and ever since then his leads have been followed, for those still interested. All but two of the scenes lie within the French Quarter, many of them along Royal or a half-block off it, and Holditch took me to them. Walking from the roar of Canal Street into the comparative quiet of Royal, you pass the "'Sieur George" house at the corner of Royal and St. Peter, a shabby, peeling, leaning thing, apparently much as it was when Cable wrote of it. Locally known as the "first skyscraper," it is still a tall building for the Quarter, where structures have remained on a human scale. Three blocks past it brings you to Dumaine. If you turn to the right, you're at the handsome raised cottage Cable identified as that legacy that slipped away from Madame John and left her dependent on the mercies of the manager of a quadroon ballroom. If you turn left out of Royal, you come to that high blank building to which Madame John and 'Tite Poulette were forced to move and across from which sits the low cottage of their redeemer, Kristian Koppig, its doors and windows painted a shiny, ballpark green. Farther along Royal at Governor Nicholls three women were snapping camera shots of the "Haunted House." When Cable wrote of the house in his *Strange True Stories of Louisiana* (1889), he ruefully says that though no one on Royal would be able to point out the sites of any of his fictions, everybody could point out the "Haunted House" and tell any number of conflicting, wild stories about it. If this was true in 1889, and it probably was, it is even truer now, for when I asked the three photographers if they knew the significance of the house, they chorused, "No! We only know it's supposed to be haunted!" There's a law at work here: the significance of legendary places gradually becoming blurred through the years until at last only the grossest outlines remain.

If you walk all the way to the end of Royal, you'll pass the sites associated with "Madame Délicieuse" and "Madame Delphine," and by the time you arrive at the site of the latter you're on the edge of both the old quadroon quarter and the former turf of the white Creole aristocracy. Their intimate proximity is suggestive. In *The Grandissimes* Cable writes that the angle between Esplanade Avenue and Champs Elysee once held some of the grandest homes of the old aristocracy but that they had been mostly torn down, including the Grandissime house. Along Esplanade there are still some grand old places, though now most of them are multi-family dwellings and are generally in some disrepair. At the corners of this once fashionable avenue are small groceries and convenience stores with winking neon signs. The grassy median that divides the avenue looks a bit threadbare beneath its canopy of live oaks, and it's clear that the city's iron has moved elsewhere.

We turned right off Esplanade into the Marigny District, so named after the opulent Creole family de Marigny, who once owned all the land between Marigny Street and the Inner Harbor Navigation Canal. After Bernard de Marigny had squandered much of his family's vast fortune shooting craps, he divided up the ancestral plantation into suburban lots and sold them off. He named one street the Rue de Craps; later it was changed to the present-day Burgundy (pronounced here "Bur-GUN-dy"). The Marigny District is currently undergoing gentrification, but however much this will change its complexion it will probably not change the rows of little shotgun houses that are in their own subdued way historic monuments: to the custom of selective breeding, caste discrimination, and enforced concubinage that were peculiar to this city. Here are the *petites maisons* of the *placées* where they were set up by their white gentlemen in what often amounted to the white man's second household, complete with children and servants. This system was another casualty of the war, but its effects lingered: in the sizeable numbers of people of mixed blood; in the postbellum hatred the city's whites now bore the mixed-bloods for their allegedly uppity ways (and for other reasons as well); and at last in the silent accusations these cottages continued to make. Cable, poking about in these moral shadows in the 1870s, asked whether a particular house was inhabited. A neighbor said simply, "Dey's quadroons," and that explained everything: the little house, its air of tight-shuttered anonymity, its place there at the wrong end of Royal.

"Marigny is a long way from being at the wrong end of anything nowadays," Kenneth Holditch was saying as we sat in the cool of his book-cluttered study in the Marigny District. His house looked out on tiny Washington Park with its avenues of out-sized live oaks. "I was lucky enough to buy this place years before the neighborhood became so popular with Yuppies and out-of-towners. The ways things have become I couldn't near afford to buy it now." He took down several editions of Cable's work, and his talk returned to the writer's bright promise there at the outset of the 1880s: "No one, not Twain, not Howells, not Henry James, was writing better prose than he was then." While I gently turned the browning pages, looking at the illustrations, it seemed somehow a justice to be doing so in this part of the city Cable had greatly loved.

Later, having said goodbye, I sat on a bench under one of the great blacksnake live oaks, its sinuous folds alive with the twitters and restive movements of late afternoon's birds. I was alone in the park, yet there wasn't the least sense of solitude. It wasn't the birds that kept me company, nor the rumble of the homeward-bound traffic along what's now called Elysian Fields. It was the past, Cable's feel for it. It sat beside me on the bench, stared out from the cottages on the park's far side, wrapped me as in some heavy velvet cloak. That, I thought then, was his special gift, to make you feel the presence of the past. There is its softness, as in the softness of sleep and dreams and reminiscence. And there is it terrific weight, and Cable makes you feel that also.

Sundays are ever unpromising in their beginnings, and this one in its sullen drizzle was no different. Yet as I set out to find the approximate spot of Jean-ah Poquelin's house, I could feel the wind freshening, see the skies shifting toward brightness. Sundays call for your powers of improvisation, so to rescue the day from blue laws and other forms of piety, and what better way to spend the Sabbath here than to go on a sort of spirit hunt along paths blazed by Cable's words?*

In his article on the scenes of Cable's stories, Hearn was finally able to identify the site of Jean-ah Poquelin's house as having been near the junction of Poydras and Freret streets, an area that in Cable's day was in transition. The indigo and cane fields had long been abandoned there and had gone back to swampland, and the city was draining these lands and driving roads through them toward outlying areas as yet uninhabited but surely coveted for future expansion. A mid-nineteenth-century map shows this particular area uninhabited and under the designation, "Cypress Swamp[,] Timber mostly felled." Out here a certain Dr. Gravier once lived alone in a house surrounded by swamp, and he served as Cable's model for the story's old slaver who takes his beloved half-brother with him on a slave-gathering voyage to the Guinea coast. It is from this voyage that young Jacques returns with leprosy—Cable's powerhouse symbol for the moral contagion that the slave trade carried—to live afterward a whited, rotting inmate of Jean's one-man asylum. And it is Jacques that the secretary of the Building and Improvement Company sees one night while spying on Jean, trying thus to discover the secret of the man's resistance to the proposed road: first, a strange, sickening odor on the night air, and

---

*On their joint reading tour of late 1884, Cable's Sunday piety drove Twain to exasperation. His scrupulous Sabbath observances, Twain wrote Howells, had taught him to "abhor and detest" that day and to "hunt up new and troublesome ways to dishonor it."

then the white form of a man slowly and painfully materializing out of the gloom and dragging himself up the rearward stairs of the house.

Poydras is now a fashionable and economically significant thoroughfare dominated by One Shell Square, the very hub of the new city that has grown up out of the huddled remains of the old one of Cable's early work. At Poydras and Freret where Dr. Gravier's house once stood there is on one hand the bloated hulk of the Superdome, once called locally the "Suckerdome" when it appeared it was to be a permanently losing proposition for the city. And on the other hand there is the high gleam of the NBC building. Oddly, on a Sunday morning with the wind whipping about the bases of these giant structures, the empty streets puddled, there was something that in a new way was as lonesome as that "grim, solid, and spiritless" cypress pile of which Cable had written, that house that looked "like a gigantic ammunition-wagon stuck in the mud and abandoned by some retreating army."

Turning northward on Claiborne, you're kept company by the freeway, and under it, out of the rain or the sun as may be, are the Sunday sidewalk mechanics, sheets draped over auto fenders as they tinker with planned obsolescence. About St. Louis Street you begin to encounter the first of the black nightclubs of the old Downtown section. So early on a Sunday they are rank, the refuse of a Saturday night strewn about the banquettes of their entrances. Tucked in between them are beauty parlors, auto-parts stores, and take-out restaurants. At St. Ann Street an ancient black woman, her head close-wrapped in a turban, waited for a bus to take her to worship. Then Esplanade again for another more leisurely inspection of the quondam high road of Creole fashion; on this day a century and more ago there would have been a parade of Creole carriages, of horsemen bound out for the lake, of promenaders under parasols. With the memory of *The Grandissimes* in your head and perhaps a copy of it with you for ready reference, you can see them still, forging a phantom way through the present's morning shoppers in shorts and t-shirts, out for the Sunday paper and a pack of cigarettes.

I had my copies of the novel and *Old Creole Days* in a book bag, dipping into them at one spot and another, then turning down Royal where business was beginning again with tourists browsing the antique shops and curio stores, a few swollen-faced drinkers already glimpsed in the dimness of the bars. At the "'Sieur George" house, I stopped to chat with an artist setting up his little canvases of French Quarter scenes in the archway of the house. I read him Cable's description of the place as it had been in 1850, and it made him howl with laughter. "It's still the same place," he gasped between laughs. "They haven't fixed up a thing!" No, there weren's any singular old gents living there now, but there were characters aplenty, he said. Like the woman whose lover had left her for a man and who in desperation drank a can of Drano. She survived, but each night, the artist claimed, she appeared on her inner balcony to hurl imprecations after his departed form in what shreds remained of her vocal cords.

A friend had given me a key to rooms he owned in the building, and simulating 'Sieur George's lonely, weary climb, I went up the courtyard stairs to the second floor. There was nothing in the scabby rooms but an old cypress gambling table and a couple of captain's chairs. Dragging one of these to the window that looked out on the corner of Royal and St. Peter, I mused on the passing show beneath, a seduction of any elevated vantage, and then from that drifted into thoughts of the Drano woman somewhere else in the building, and finally to her fictional kin, 'Sieur George, both of them chasing into the future chances that belonged to the past.

I wanted to meet a Creole, a white one, and through the help of local friends, it was arranged for me to pay an afternoon call on Mrs. LeBeaux at her home in the Garden District. She apologized for having taken so long to answer the door. "These old houses," she explained, "are truly wonderful, but they're so long, front to back, that it just takes a long time to get anywhere in them. And they do soak up sound so." A tiny, frail woman, her back as bent as a comma, she led me to a sitting room, seated herself slowly, smoothed her dress, and looked across at the visitor through glasses that hugely magnified her eyes. She was ninety-five, but the eyes, dark and piercingly intelligent, told you instantly that she was very much in control here, that this would be no generous social call on a lonesome old woman. She knew just who she was, and if she'd agreed to talk to me about George Washington Cable, it would be on her terms.

"We're not really Creoles," she said at once. "A man was doing a genealogical project years ago, and he came to us [she used the collective to refer to her family] and said, 'Now I've found my perfect Creole subjects.'" She laughed shortly. "We had to tell him we had Belgian, English, and Scotch in us as well. I think we disappointed him a good deal." She laughed again, her eyes disappearing into the deep creases of her face.

"Anyway, you were asking what a Creole was? Well, a Creole is someone born in the colonies of either French or Spanish blood. Nowadays, of course, there aren't any pure Creoles left. There's no such thing any longer as a Creole subculture in this city. I couldn't introduce you to a Creole if I wanted to. They're all gone. Many years ago there was a tourist here—I think he may have been a historian—and he spoke to a gentleman on a street corner. 'I'd really like to meet a Creole,' he said to the gentleman. 'Do you know how I might do so?' The gentleman drew himself up straight as an arrow. 'My dear sir,' he said, '*I* am a Creole.' Oh, they were *proud*. But that was a long time ago.

"How long ago? Well, I'm ninety-five. Back before I was born they used to say, 'Nous sommes français, mais français comme le drapeau blanc, pas comme le drapeau tricouleur!'" She laughed again at the pride, and maybe even the lofty pretensions of the old Creoles, then went deftly into a description of her family and the Creoles from just before the war until the present. The war, she explained, always meant the Civil War. "Why, do you know," she said, leaning toward me for emphasis, "as a sixteen-year old boy my father fought at Shiloh? Think of that!" The war had changed everything. After it the Creoles had moved, almost en masse, from the Esplanade area into the Garden District. But, she wanted me to know, it wasn't the war that made the Creoles so bitter: it was Reconstruction. "It was Silver Spoon Butler. My grandmother was told that if she refused to obey the orders of the trooper, she'd be treated like a woman of the streets." Indignation and a hurt surprise still tinged her voice here. "The Creoles weren't bigoted until Reconstruction. There was, of course," she said, glancing down into her gnarled hands that lay so lightly in her lap, "always the matter of race. . . . " Her voice trailed off. "And when the colored people who'd worked for the Creoles began calling themselves Creoles—ohhh, they resented that terribly.

"Yes, they were proud, as Cable said they were. But you must understand that part of their resentment of him was because of what had happened to them in the war and especially afterward. They simply lost everything. And then, they found themselves deleted from society! No wonder they were bitter. Roger Baudier, a wonderful historian here, told a story about those years that may help you to understand this. Every evening, he said, the wife of a Creole family he knew would make a great rattling of dishes around the dinner hour: she wanted the neighbors to think the family was about to have its grand dinner, just as in the old days. But inside they were dining on red beans and rice." For Cable to hold them up to censure and even ridicule, she said, angered them profoundly. She said Cable really didn't know much about the inner lives of the Creoles, and that, too, had angered them. He wasn't, she wanted me to understand, an educated gentleman, whereas many of his subjects had been educated abroad and could speak a number of foreign languages. "His dialect was all wrong. These were highly educated people I'm speaking of, and his dialect made them sound like *Negroes*!

"Then Cable makes them seem so terribly hard-headed and narrow. They weren't really, as I've said. They had a good sense of humor, and they were good businessmen, contrary to what Mr. Cable wrote. And they were so fond of children." She paused and smiled inwardly. "We [here lapsing for a moment into a closer identification with the Creoles than she had admitted to at the outset] never took ourselves so seriously as Mr. Cable claimed. Not at all like New Englanders who always take themselves so seriously. *They* have to have a cause, it seems. And Cable's identification with them was resented. Oh, I didn't actually read Cable when I was a young girl. You were told, 'Don't read him.' And you weren't to read *Uncle Tom's Cabin,* either. But when I was older, I read him. Oh, I read everything then."

She talked about other aspects of Creole culture and history, about her own career as a teacher at a local college, about the accuracy of her memory at this stage in her life. Interviewers, she said, "come here so often. But my sister says, 'What you don't know, you invent.' And you know, it may be so. I don't have anyone left to check my reminiscences with. They're all dead." She threw up her hands in a gesture of cosmic helplessness, then brightened and asked if I'd like a glass of sherry with her while I waited for my cab. We went down the long, dark hall to the dining room, its table piled high with books, magazines, and correspondence. Mrs. LeBeaux rummaged in the mahogany hutch for appropriate glasses, then broke off, laughing, "Oh, well, it's just cheap cocktail sherry, you know, and I suppose it'll taste just as good—or bad—out of these."

I poured for us, and she toasted me—"A votre santé"—and as we sipped and chatted I suddenly became aware of a ghost-white apparition materializing slowly, inch by inch, from the darkness of the hall: first the propped feet, then the blanket-shrouded legs, and last the torso and head: a powder-pale woman with hair screwed into a bun at the side of her head, wheeled slowly in by a black nurse in starched whites. It was Mrs. LeBeaux's last surviving sister, it was explained. Mrs. LeBeaux spoke brightly to her, "I'm afraid we can't offer you some sherry, my dear. The doctor says you're not to have any until you go off your medication." There was no visible response from the figure in the chair. Behind her the black face of the nurse was implacable.

"This is Mr. Turner," Mrs. LeBeaux explained. "He's come to talk with me about George Washington Cable. You recall his books, don't you?" This time the sister looked at me, with what regard I couldn't fathom, but I could not help thinking of Jean-ah Poquelin and his invalid brother so tenderly cared for through the years. And then the present intruded on the peremptory honks of a cab's horn. I had to say my hasty goodbyes, and then retreat down the hall, out into a driving rain and away, leaving the sisters there with their dense history, their memories of a culture now present only in memory and preferences, their clear convictions about Mr. Cable.

I told John Scott a story of Mrs. LeBeaux's. We were sitting in a studio at Xavier University where Scott teaches sculpture. A tall black man with prematurely white hair and strong, tensile hands that shape ideas while he talks, he listened carefully with a slight smile while I told him about the black maid who'd worked for her family when they lived on Barracks Street, between Burgundy and Dauphine. The maid's name had been Alice, and Mrs. LeBeaux remembered she'd had a little boy to whom the girls had given tips when he carried packages for them. "When he'd saved up enough money, what do you suppose he asked his mother to get for him?" Mrs. LeBeaux had asked me. "A clarinet. And she did. Years later, when my sister and I went to London there were five hundred people waiting at Victoria Station to see that boy. He turned out to be George Lewis. But there was no one waiting for us!" She had laughed then at this irony of history. John Scott merely smiled a bit more, for the irony hit him differently. Lewis was one of those whose full humanity had been studiously denied by the city's whites, and even after the evidence of that humanity had burst from the bell of his clarinet it had still been denied by New Orleans whites who thought of jazz as jungle music and were actually ashamed that it had its origins in their city's backyards: there would not be five hundred admiring fans awaiting George Lewis on his return to this town.

No one knows these facts of New Orleans life better than John Scott. He had nothing to say about Cable and his clashes with the Creoles, but he had plenty to say about the city, its past, its special spirit, and about the changes that have come upon it since Dutch Morial was elected as the city's first mixed-blood mayor. Born on a farm owned by the Kolb family (Kolb's restaurant just off Canal is one of the city's fine old establishments), then raised in the Uptown section in some poverty, Scott is a nationally recognized sculptor who stays in his hometown because he wants his life and work there to make a statement. And he stays because he finds that New Orleans and its past—all of it—continues to nurture him. "When I'm on campus here," he said carefully, "I want to be saying every day to these [black] students, 'You can, too.' I'm not an artist. Not yet. But I'm striving to become one. My mother told me once, 'You can't control your coming or your leaving in this world. But only a fool won't try to control what's in the middle.' That's what I'm trying to do, in my life, in my work. If I can do this, I'll feel justified.

"I could leave, of course. I could go someplace where maybe the environment might be a little easier for a black artist. I could go to New York or Los Angeles and work on my reputation." He paused here, searching for the right way to convey his sense of New Orleans's spirit, what it gave him. "This," he said at last, "is a town that talks to you. The sidewalks talk to you, so do the shapes of the buildings. Everything here's telling you a story if you know how to listen, if you want that story. Some don't. The iron work in the Quarter: you know the old work there was all done by black artisans, and you can just hear the hammers ringing on it, something all the way from Egypt to here. You walk around in the Treme District, you've got to think of Jelly [Jelly Roll Morton], you hear all of his sound, all of the life that went into that sound, in the shapes of the buildings, the street games. This whole city is built of survivals and adaptations, if you know what to look for. You'll see bead work in the Zulu parades [of Mardi Gras] that's in every essential absolutely identical with that of the Yoruba culture. And these are people who for the most part have never even heard the word 'Yoruba.'

"Then there are, of course, all the memories of slavery and the caste system. For me, they're utterly inescapable in the older portions of the city. The chants of the street vendors of my childhood—the strawberry woman, the rag man, the three vegetable men—these go back to slavery times, to field hollers and work songs. For me, this city is a mirror of the past as well as the present, a mirror in which I might one day see myself. I've *got* to stay."

And then, too, there are the political developments of recent years, ones that have changed, perhaps irrevocably, the way the city runs and the way it thinks of itself. Scott thinks Dutch Morial was "one of the best mayors this city has ever had." His administration, Scott said, had done more for the arts and for the city's cultural life than any before. Nor was this all. His election and the success of his administration had broken the old racial alignments that had kept New Orleans locked in patterns and preferences that went back to slavery times. The mayoralty contest of 1984 was between two "colored" candidates. "Morial's administration," Scott said as we stood in the sun outside his studio, the students streaming by at the change of classes, "and [Sidney] Barthelemy's that's just beginning are telling us something we need to hear: they're telling us that this city's not one thing or another: it's not French, or Spanish, or black. It's a gumbo. And your spice isn't going to make the stew if mine isn't in the pot with it."

In 1886, well after he'd been forced out of town, Cable returned imaginatively to the scenes of his youth and the promise of his early career in an article for *Century Magazine* on the dances once held at Congo Square. In it he shows readers the cultural significance of the view from Congo Square down Orleans Street to St. Louis Cathedral at the other end. Congo Square, he tells us, was one world, the world of the blacks, out there at the very fringe of the city, while the Cathedral was the world of the white man, surrounded by the seats of government and law enforcement.

Also down at the white end of Orleans Street was the Orleans Ballroom, one of the sites of the quadroon balls. These days it's called the Bourbon Orleans Hotel, but the building is the same one, and the ballroom still exists on the second floor. If for Cable the plight of the quadroons symbolized that community of sin he found beneath the city's rich and gamey history, so the Orleans Ballroom symbolized the whole system of racism and caste that once made New Orleans so unique and that still, in muffled but real ways, is a part of this city. You heard it said, for instance that in the 1984 mayoralty, it was really the blacks against the colored Creoles, and that Sidney Barthelemy's election was due to the whites' preference for the colored Creoles, their fears of what might happen if State Senator William Jefferson (who is black) were to win. It might not be true: John Scott said it wasn't. But the fact that such things can still be soberly—or even snidely—said tells you again and in a new way what Cable said more than a century ago: the past lives here, intensely so. I wanted to see it living on in the grand ballroom and to make this my last stop in the city.

The day was bright and sunny, the subtropical sun on this late spring morning still withholding its full powers, the air soft under a blanket of blue. The Bourbon Orleans sits at the corner of Royal and Orleans, and along the iron railings of the garden behind the cathedral you could see the sidewalk artists setting up their pitiful paintings. A shop a door off the corner had its doors flung wide, its tape player up, and Bessie Smith shouted blues into the heedless sun of the banquette. Inside the hotel I was told by the manager I'd have to wait until a meeting of AT&T broke up in the ballroom above. I occupied myself by copying the text of the historical plaque outside the main entrance. "Former site," it read, "of Holy Family Sisters' Convent, the old Orleans Ballroom, built in 1817, served a number of purposes over the decades." There wasn't enough space, apparently, to list them all, no mention being made of the quadroon balls held here. It did mention that the property was purchased for the convent for colored nuns through the philanthropy of Thomy Lafon. It didn't say Lafon was a Creole of color, child of a French father and Haitian mother, like Cable's dark Honoré Grandissime an f.m.c.

When the manager summoned me with the news that I could go up, he told a bellhop, Richard Thomas, to show me the way. A muscular young man in his stiff-collared uniform jacket with a bushy Afro, Thomas bounded up the curving stairs two at a time ahead of me, then waited patiently at the door of the ballroom, a high-ceilinged affair about half as long as a football field. We walked in and stood in a moment's silence amidst the neat clutter of the just-departed telephone executives who were now happily at some nearby restaurant, perhaps savoring the heady reward of that first luncheon drink. Thomas told me he was New Orleans born, and so I asked him what he knew of the ballroom's history. He told me it had been built by the nineteenth-century financier John Davis at a then astronomical price of $60,000 and that it was "the place where men bought their mistresses." After that, he said, it had been a convent.

"You must yourself be of mixed blood," I said then, "to judge from your looks."

"I am," he said with a smile. "I'm what we call here a Creole." I asked him then what it felt like to work in this place, knowing what he did of its history, what it felt like to show this room to a stranger and have him ask about it.

"Well," he said without much pause, "if you want to know whether I feel bad about it, the answer is no. I feel like I fit in here, you know? I feel like I understand this. It gives me a good feeling to be working here, some way.

"Say," he said suddenly, "how'd you like to go out on the balcony there to get a feel of the place?" I hesitated. "Come on," he laughed, "you're not that old. You can climb this sill. Come on." And he preceded me, clambering nimbly up onto the broad, waist-high sill, then dropping down through the open window to the balcony. I followed.

Now we were once again into the bright soft air with the cathedral garden on our left, the green of Congo Square down the street on our right, and beneath us the busyness of a spring day's noontime. "They used to go over there to settle up when they had arguments about their mistresses," Richard Thomas was saying, nodding in the direction of the small garden. The sounds of Bessie's blues filled in the silence as we regarded the scene.

"Well," I said shortly, "up at the other end there is Congo Square." And we looked squarely at each other then, two strangers united in being Americans and so sharing, however obliquely, a history. "And here we are," I resumed, "on a balcony of a place where they used to sell quadroons. And the last two mayors have been of mixed blood like you."

He laughed a full, throaty laugh. "Yeah. It just goes to show." What, I wondered aloud, did it show? "It shows," Thomas said forcefully but without heat, "that what you look like doesn't count anymore. It's what's inside that counts. It's an individual thing." I said I thought that was the right point to make and we went in then, crossed the empty ballroom, and, descending the graceful stairs, reentered the rush of the present in which the past so often seems but fiction.

THE BISHOP'S FACE: WILLA CATHER'S <u>DEATH COMES</u> <u>FOR THE</u> <u>ARCHBISHOP</u>

I

  YOU COULD GET AN ARGUMENT ABOUT THIS from residents of other parts of the state, but it would still be reasonable to claim that the hub of New Mexico's thriving tourist industry is the lobby of Santa Fe's La Fonda hotel. There in a cool, high-ceilinged dimness you may see the smart-looking strangers—silver concha belts, scarves, broad-brimmed hats—whisking in and out on errands of pleasure. Occasionally, but not often, they rest in the deep-cushioned lobby chairs above which are the framed prints of Gerald Cassidy that depict the very lineaments of the state's vaunted enchantment.

If they are looking bookishly inclined, there is the surprisingly literary gift shop, and on its racks, snugged in between Louis L'Amour and John Nichols, is Willa Cather's *Death Comes for the Archbishop*. Its presence amidst the gaudier currency of contemporary Southwestern fiction might be taken for another bit of local color, the literary equivalent, say, of a chile pepper magnet, which the gift shop also offers. It is not. Just as for many Gerald Cassidy has caught the image of the Southwest's magic, so for thousands of readers Cather's New Mexico novel has captured the human emotions evoked by the region's landscape. The shop clerk at La Fonda said it was "probably our number-one seller," a status confirmed at other bookstores down San Francisco Street. At Los Llanos Book Store they told me that only John Nichols outsells Cather's novel about Father Latour, and farther west on the street, Lynne Moor, owner of Collected Works, said that *Death Comes for the Archbishop* was "probably our most consistent seller." Hardly a day goes by, she continued, "that we don't sell at least one copy."

Many tourists do not wait to buy Cather's novel until they get to Santa Fe but read it in preparation for a New Mexican vacation. One middle-aged New England couple told their Santa Fe hosts they'd read passages of it to one another as they drove across the country. For such travelers *Death Comes for the Archbishop* is far more than a fine work of fiction. It is at once a guidebook, an authoritative work of history, and, deeper, a legend: a narrative in which a place and all its history are imaginatively expressed.

The immediate inspiration for the book came to Cather in the summer of 1925 when she and her companion, Edith Lewis, decided to spend that season in New Mexico. Though she had long since made the acquaintance of the New Mexican countryside, Cather felt far more comfortable in the city of Santa Fe, then in its halcyon days as a writers' and artists' colony. She and Lewis set themselves up at La Fonda, a brand-new establishment that had replaced the historic Exchange Hotel on the same southeastern corner of the Plaza. Less than 200 yards east on San Francisco Street rose the golden facade of St. Francis Cathedral, and through its mismatched bell towers you could see the red hills dotted with dark green clumps of piñon and juniper and beyond, the fathomless blue of a high-plains sky. In front of the cathedral, beneath a locust tree, stood the life-size bronze statue of Bishop Jean Lamy who had dreamed the cathdral into solid reality. Adjacent to the cathedral were the Bishop's old residence and the gardens he had created and lovingly tended. The branches of the imported fruit and nut trees nodded over the adobe walls at walkers in the dusty streets.

Among those who passed beneath the stern gaze of the bronze Bishop none was more alive to the potential implications of the figure in its setting than the little woman in long full skirts who more and more frequently passed that way on her town walks. The Bishop's face interested Cather. It seemed, she later declared, "something fearless and fine and very, very well-bred—something that spoke of race." She wondered more and more purposively about him, then came across a biography of Joseph P. Machebeuf, Lamy's vicar general, which gave in rich and suggestive detail the lives of these two Auvergnats who in the three decades from 1851 changed the diocese of New Mexico from a paper fiction to a functioning reality. One hot night the writer stayed up until dawn reading Father Howlett's *The Life of the Right Reverend Joseph P. Machebeuf,* and it would not be too much to suppose that the next day she looked upon

Santa Fe and its environs with new eyes. For more than ten years she'd wanted to write a book about this region, and now she had a subject around which her feeling for New Mexico and the Southwest might cohere, as a French Impressionist landscape coheres around a girl's diaphanous parasol or a red rooftop.

In that summer of 1925 Santa Fe was hardly the tiny, low-slung adobe village Lamy and Machebeuf had first seen in August, 1851. Architecturally the town was in transition, vestiges of Western frontier frame buildings vying with bulky Victorian and Queen Anne structures and with the vogue of Spanish–Pueblo Revival. Across the shade-dappled plaza from La Fonda, for instance, La Casa de los Conquistadores, a large movie theatre and garage complex in the Spanish–Pueblo style, stood surrounded by undistinguished four-square brick buildings. Still, for all the substantial changes of three-quarters of a century, there remained around the city's margins plenty of hints of the place as Lamy had known it.

In search of these myself, since I wanted to understand what Cather had to work with as she began to think about her book, I called on Margaret Pond Church, a noted New Mexican writer. In 1925, she was just a twenty-year-old girl from the country who came to Santa Fe for supplies and socializing. She had warned me on the phone that she hadn't ever met Willa Cather and didn't think much of her work. But she agreed to see me—as much, I suspected, to set me straight about Cather as anything else.

When I asked her what Santa Fe was like in 1925, she arose from the table at which we were seated, went into another room of her apartment, and returned with a loose-leaf notebook. "I can't answer your question better," she said at last, "than to read you a poem I wrote then. It's called 'Warning,' and it'll give you a sense of what this place looked like and felt like then."

Indeed it did. Of the advent of spring in Santa Fe, Church wrote:

> Orchards had awakened
> On every valley-slanted hill; through every field
> Wild plum ran riverward like a blown fire;
> And over the walls of our ancient gardens
> Pear trees lifted candle-white spires of bloom
> Toward the sky. Our narrow crooked streets
> Were flinging before us around every curve
> Some unexpected beauty. Petals rippled
> Along the mad rush of the acequia;

And there were children entering the cathedral
With crisp white frocks, and blossoms in their hair.

"It was rural," she said when she'd finished reading, glossing her own lines. "There were orchards everywhere. There were little adobe houses with real people living in them, tending their gardens and fields and goat herds. Upper Canyon Road was alfalfa fields on both sides. Acequia Madre ran then because it was necessary as an irrigation ditch. Palace and Don Gaspar were the only paved streets, and the altitude and the population were just about the same: 7,000. Now when I want to wow people, I tell them I used to ride my horse down Palace Avenue and sell vegetables on the Plaza. Can you imagine that?" She smiled indulgently at me as at a child. "Now," she added with brave sigh, "I live in the heart of a tourist ghetto. But I look out at the wall of that house there and pretend it's a canyon wall."

So, despite all the changes from the Bishop's time to her own, it appeared as if Cather had had that distinctly rural feel to start with. She also had, of course, the cathedral and its mountainous backdrop, the Bishop's residence and gardens, and his beautiful little retirement residence in the Tesuque hills. Best of all, she had the largely unaltered landscape through which Lamy had ridden, carrying with him the burden of his lonesome responsibilities in this unknown outpost.

At some point during the summer of 1925 Cather received an invitation from Mabel Dodge Luhan to visit her up in Taos, and though Cather could be bluntly unsocial, this time she accepted. Luhan gave Cather and Lewis the Pink House on her property, decorated with a drawing of a phoenix by its former inhabitant, D. H. Lawrence. Cather had said she'd only spend a day or so but ended up spending two weeks. The reason was Tony Luhan, Mabel's Pueblo Indian husband. In him Cather found the guide she wanted to the northern New Mexico country. Tony Luhan drove Cather and Lewis west into the country associated with Lamy's rebellious priest, Father Martinez. He drove them east into the high, remote Spanish villages of the Sangre de Cristos. He showed them the rich farming lands of the Española Valley. And everywhere he took them he had some small, laconic, but revealing thing to say about place and the life lived in it; everywhere he provided introductions to native dwellers who told their own stories to the writer avid for the fullest sense of the land through which her hero had moved.

Cather and Lewis came back the next summer, again establishing themselves at La Fonda and again going north to Taos and Tony Luhan. Cather subsequently told her friend, Elizabeth Shepley Sergeant, that in the summer of 1926 she'd left the manuscript of her new novel behind in a New York vault. But the writer Mary Austin claimed Cather finished it this summer at her house in Santa Fe.* Certainly Cather visited Austin often in these months, walking from La Fonda to Austin's rambling adobe home at the foot of what was then called "Cinco Pintores Hill" (Five Painters' Hill). And there is an autographed copy of the first edition of *Death Comes for the Archbishop* (1927) inscribed by the author to Mary Austin, "in whose lovely study I wrote the last chapters of this book." Wherever she finished it, the book was for Cather a positive joy to write, and both Elizabeth Sergeant and Edith Lewis testified to the seeming ease of its composition, the serene mood in which Cather worked on it. Willa Cather herself called the writing of it "a happy vacation from life, a return to childhood, to early memories."

On its surface that remark seems an odd one: Cather's childhood had nothing physically to do with New Mexico or the Southwest. Actually, the remark is paradoxical and leads us, as perhaps Cather intended, to an understanding of the significance of this region to her life and work.

As a writer Cather's supreme strength is her sensitivity to place. Indeed, it was said of her by a long-time friend that she cared as much for places as she did for people, a trait that might help to explain that rudeness she could occasionally display. The sensitivity was the result of a childhood dislocation: in 1883 in her tenth year Cather was taken from the ancestral home near Winchester, Virginia, to south-central Nebraska where her father joined his parents and brother who had emigrated there. The contrast between the old home and the new one was shocking to the girl, and suddenly all the smallest details of the home place came indelibly to mind while at the same time she was made acutely aware of all those new, alien features of the prairie. It was as if she were now overwhelmed by a feeling for two landscapes.

The new country, she recalled in afteryears, "was mostly wild pasture and as naked as the back of your hand. I was little and homesick and lonely

---

*Father Angelico Chavez, historian of St. Francis Cathedral and author of *My Penitente Land*, told me he'd always heard that Cather was staying with Austin at one point and that when the latter went on vacation, Cather stayed on at her house and appropriated for her novel some notes Austin had been gathering on Bishop Lamy.

and my mother was homesick and nobody was paying any attention to us." "I would not know," she said on another occasion, "how much a child's life is bound up in the woods and hills and meadows around it, if I had not been jerked away from all these and thrown out into a country as bare as a piece of sheet iron." In what many regard as her masterwork, *My Ántonia* (1918), Cather writes of the advent of a Virginia-born child to the Nebraska prairies. It isn't a country at all, young Jimmy Burden thinks, "but the material out of which countries are made." He feels as if he has left the world behind him and is now in some other, foreign sphere, without markers, without the comforting old feature of mountain ridges against the sky, without resident spirits. Between the empty earth and the equally empty sky, the boy feels "erased, blotted out."

Since the latter part of the eighteenth century Cather's people, the Cathers and Boaks, had farmed the Back Creek country of Virginia. Willa was born in her grandmother's house, and all about her in her youngest years were the signs and symbols of tradition, of things in place, of the old ways. The earth itself had been prepared by her ancestors, and their bones, gone back to that earth, enriched the life of succeeding generations. A sense of history—familial, communal, regional—was part of the air she breathed. History had happened here and helped to tell the natives who they were: Stonewall Jackson, for instance, had ridden these very roads. Then suddenly, as it must have seemed to the ten-year-old girl, all this was gone. Suddenly she was in the middle of a vast expanse of rough red grass across which the west wind blew with a wearing constancy. In place of the ancestral home there was now a lone-standing farmhouse with no near neighbors. There was no graveyard with its familiar slabs telling the old story of generations and continuity. There was only the tiny post office christened Catherton to remind the newcomers of civilizations as they had known it. In that first Nebraska year the Cather family did not even live in the little village of Red Cloud—where the pangs of transition might have been somewhat eased—but in the unopened country to the northwest.

Here was a land that, as Cather's character puts it, wasn't even a land yet, by which the author meant that to the newcomers it had no history, no associations, no legends they knew. It was a land seemingly without a language, without its indigenous songs, and it was so terrifically different from Virginia that neither the Cathers nor the small colony of transplanted Virginians around Catherton could bring the old songs to bear on the problems of life in a new place: they simply didn't fit. By Babylonian rivers, sang the Psalmist, we wept when we remembered Zion, and though our captors required a song of us, we asked, "how shall we sing the Lord's song in a strange land?" For the Charles Cather family with its strong Christian background, this passage might often have come to mind in those first Nebraska months.

But, of course, in a large way the predicament was hardly peculiar to the Cather family. It was a microcosmic recapitulation of the American pioneering experience: how to sing the old songs in a strange land. Some never learned, while others never acknowledged that they cared, and a later Nebraska writer, Mari Sandoz, would brilliantly assess the effects of life without song, calculate with her own life the cost of the effort to sing. But for Willa Cather life without song was unendurable, a mean and brutish existence that could hardly be called human. Besides, this land, whatever else it was, was simply too big and powerful to be met without the mediating presence of a language, and she set out early to devise her own.

At the very least, she could set herself against the new land and sing a song of stubborn resistance, refusing to be blotted out, erased between sky and unmarked horizon of grass. So, early on she identified herself with the village rebels of Red Cloud, those dreamers and poetic types who in their own ways were trying to sing. She became one of them, a mannishly dressed intellectual, critical of local xenophobic ways and apt to champion unpopular causes such as scientific rationalism and vivisection. When she began publishing fiction, a theme of a number of the early stories is the tension between the new, unsung land and its would-be singers.

In *The Troll Garden* (1905), Cather's first collection of stories, there appeared "A Wagner Matinée," a tale of two artistic types on the Nebraska frontier. Clark, the narrator, has escaped Nebraska for Boston, but a letter from his aunt, who has not, brings back to him

> the tall, naked house on the prairie, black and grim as a wooden fortress; the black pond where I had learned to swim, its margin pitted with sun-dried cattle tracks; the rain-gullied clay banks about the naked house, the four dwarf ash seedlings where the dish-cloths were always hung to dry before the kitchen door. The world there was the flat world of the ancients; to the east, a cornfield that stretched to daybreak; to the west, a corral that reached to sunset; between, the conquests of peace, dearer-bought than those of war.

Clark remembers himself in that unrelieved setting, a gangling farmboy, tutored and patiently encouraged by his aunt, sitting at her parlor organ and fumbling through his scales with stiff, reddened fingers. Subsequently, when Aunt Georgiana pays him a visit in Boston, they attend a Wagner matinée. At its conclusion, as the crowd flies chattering and laughing from the bright hall, Aunt Georgiana bursts into tears, sobbing to Clark that she doesn't want to leave, for to do so is to go back—back to the black pond, the cattle-tracked bluffs, the tall, warping house that to her lies just beyond the concert hall.

Even grimmer is "The Sculptor's Funeral." At some point in their friendship, Cather told Elizabeth Sergeant that when she went back to visit her family in Nebraska she could never stay very long: she was afraid of dying in a cornfield. Sergeant thought that a peculiar phobia and said so. You wouldn't understand, Cather wrote her, because you haven't seen those miles of fields. There is no place to hide in Nebraska, she said. In "The Sculptor's Funeral" Cather allows herself to imagine a kind of corn-field death for a sculptor who has apparently escaped his prairie origins and has, indeed, achieved fame in the East. Yet in death he is brought back to stranded little Sand City where the townspeople who hated him in his escape and success file past his defenseless face in ultimate triumph and judgment. Yet Cather can't quite let them have that triumph, pass that judgment. She creates a dipsomaniac town lawyer who will pass the final judgment—and not on the dead artist but on the town that hated him. "There was only one boy," the lawyer says over the coffin,

> ever raised in this borderland between ruffianism and civilization who didn't come to grief, and you hated Harvey Merrick more for winning out than you hated all the other boys who got under the wheels.

Then he essays a sort of Sand City Anthology, describing the early, failed ends of those boys of bright promise who became drunks or arsonists or who were gunned down in gambling houses.

These early fictions suggest a truth about their author: that she had yet to come to truly creative terms with her home territory. True, she loved the great, shaggy plains, stained as if with wine. But there was a good deal of resentment, even hatred, mixed with that love, a resentment of the meanness of small-town life, of the endemic xenophobia, the hardness of living that the land exacted of its pioneers. Her life and attentions were still very much elsewhere—the East, Europe—and she could not yet write about home with any empathy.

In 1908 she met someone who had. She was introduced to Sarah Orne Jewett whose stories of small-town life in southern Maine were models of what could be done with such apparently unpromising material. The older writer must have sensed in Cather and her work a familiar artistic predicament, that of a talent in danger through neglect of its roots. In December of that year she wrote Cather an extraordinary letter, telling her that her talent would wither and die if she didn't dig deeper into her own life and memories and write out of that secure, unassailable place.

It is tempting—and dangerous—to say the letter was decisive for Cather. Such a conclusion makes neat literary history but specious biography. Still, it is a suggestive coincidence that the very next year Cather published a story that drew in a fundamentally different way on her Red Cloud years. In "The Enchanted Bluff" Cather recalled the long-ago days when with her brothers she would lie out on a small sandy spit in the Republican River and talk of legends, of faraway places, and of future travels out into the great, waiting world. A consistent subject of talk in those lazy but intense days was Coronado, who some thought had traveled this far into Nebraska seeking his gilded cities. Cather chose to think he had done so, had in fact passed along this very watercourse, and in the story this speculation leads to the legend of New Mexico's Enchanted Bluff and of a tribal tragedy that occurred there in the prehistoric past. None of the story's characters has ever seen the Enchanted Bluff, but the legend is so compelling that all swear they'll visit it someday. At the story's end the narrator describes his return to the town of his youth and a talk he has with one of those youthful companions who swore to go to New Mexico. None of us, the narrator concludes, "has ever climbed the Enchanted Bluff," and though there is here a whisper of the familiar, failed ambitions of the small town characters that Cather had written of before, this is not the dominant mood of the piece. Instead, the writer has gone beneath the old resentments, reaching toward a more generously inclusive memory of a time and a place, that quiet place of which Jewett had written. The mood is gentle and somberly reflective, and in the midst of the story, like a brilliant spot of sunshine, is that legend of the Southwest.

She first saw the region in 1912. Like the characters in her story, Cather had been longing to see that sun-filtered land out of which the misguided Coronado had come to die of a broken heart on the great plains. Her favorite brother, Douglass, who had dreamed with her of Coronado, was now working for the Santa Fe railroad and living in Winslow, Arizona. In April Willa went out for a two-month visit.

Winslow itself was a disappointment, a gritty little railroad town in which the only buildings of any distinction were the line's division head-quarters. The high winds of a Southwestern spring hurled sand, the black gravel of the yards, and all manner of rubbish through the unpaved streets. Douglass lived with two other men, and his sister took an immediate dislike to one of them, so living conditions in the smallish house were difficult. Cather began taking target practice with a pistol and wrote a friend she was beginning to fear she might turn the gun on Mr. Tooker, the man she disliked. The only compensation the town had to offer her was its Mexican section on the wrong side of the tracks, and Cather began going over there for relief and instruction: the looks of the residents there intrigued her as did their ways and music. To be sure, there was a romanticism and exoticism mixed with her interest in the Mexicans, but there was something more, for she saw that these people had achieved a kind of triumph over the trashy impoverishment of their circumstances. As always, she admired this.

But it was the countryside beyond Winslow that drew her far more than anything the town could offer. The land was in fact far more stunning than her grandest expectations of it. She rode out into it as often as she could, either on horseback or in horse and wagon. Sometimes she stayed out camping for days at a stretch. She followed the Little Colorado northwestward toward the San Francisco Mountains, looking for cliff-dweller ruins said to be somewhere along it. At Walnut Canyon in the San Franciscos she did see the Anasazi ruins, and the impact was as powerful as the land itself. Here was something that took you out of yourself and your petty concerns, much as did the cathedrals of the Old World. Here was history, more ancient even than the monuments of Europe. Together, the rocks and the rock city opened a depthless American dimension to the alert imagination.

At the end of her stay she crossed into New Mexico, found Albu-querque and its setting a more luminous version of the country between Marseilles and Nice, wandered up to Santo Domingo Pueblo. There the radical otherness of the country and its indigenous cultures hit her: all this had nothing to do with her; it was utterly foreign; she didn't belong here. Beside the Rio Grande a suddenly bitter wind blew through her, and she turned back toward the East. For the moment, anyway, the encounter with the Southwest had been too overwhelming.

But its effect on her proved decisive for her work. Seeing that magnificent landscape and the ancient cultures that fit into it like bones in a body, she saw her own Nebraska landscape in a new way, a perspective neither the East nor Europe had provided. She had fallen in love with the Southwest, and, as always, love inspired imagination. More, it had provoked memory. True, the Anasazi ruins and the living pueblos had nothing to do with her in a direct blood sense. But in a larger, generic way they had everything to do with her: they were a part of the history of the human race, and they spoke profoundly of the often difficult accommodations human cultures must make with their environments. These cultures were ancient songs in their land, and her admiration both for the land and its cultures thrust her back on her own past. "Life began for me," she said, "when I ceased to admire and began to remember." The Southwest taught Willa Cather that the "mission of man on earth is to remember. To remember to remember" (Henry Miller's words again). And so she turned back to her memories of Nebraska, to herself as she had been then, to recollections of childhood friends, to the stories of pioneer lives, the songs obscurely embedded in those lives.

In *O Pioneers!,* published the year after her first Southwestern encounter, the new feeling for Nebraska is expressed in the strong and enduring figure of Alexandra Bergson who is neither defeated nor embittered by the land but nurtured by it. "For the first time, perhaps," Cather writes of her, "since that land emerged from the waters of geologic ages, a human face was set toward it with love and yearning.

> It seemed beautiful to her, rich and strong and glorious. Her eyes drank in the breadth of it, until her tears blinded her. Then the Genius of the Divide, the great, free spirit which breathes across it, must have bent lower than it ever bent to a human will before.

That face is, of course, Alexandra Bergson's, but it was now also the author's. And just as Alexandra is successful in creating a bounty out of her portion of the prairie, so Willa Cather would now strive to create an enduring literary harvest out of her portion of the Nebraska earth: *Primus ego in patriam mecum . . . deducam Musas,* " 'for I shall be the first, if I live, to bring the Muse into my country.' " Thus *My Ántonia*'s Jim Burden translates the line from Virgil's *The Georgics,* and as if to make the point unmistakable, Cather has him go on to explain that "patria" here means, "not a nation or even a province, but the little rural neighborhood . . . where the poet was born." Virgil wants, Jim Burden says, to bring the Muse "to his own little 'country'; to his father's fields, 'sloping down to the river and to the old beech trees with broken tops.' " Nor was this a mere intellectual platitude, for in *My Ántonia* Cather indeed brings the Muse to her country, sings the songs of that time and place, illumines the sharp truths of prairie life. The justly famous passage in the novel describing a plow magnified by sunset into a heroic dimension is the fitting symbol for her large and ambitious intention. No longer resentfully estranged from the prairies, Cather now felt absorbed by them and had discovered that this was a unique species of happiness.

So, she owed the Southwest much, and it may be that almost from the first she had intended to honor the region in a work of art. Perhaps conscious of the need to steep herself in the land, its history and moods, she went back to it again and again—1914, 1915, 1916, and then again in the significant summers of '25 and '26. And during these years the region kept appearing more and more insistently in her work, like the spread of sun across the landscape: a reference to old Mexico in *O Pioneers!;* a long section inspired by the Walnut Canyon ruins in *The Song of the Lark* in which the heroine's temporary salvation is to be absorbed by the land and its ancient history, a telling reversal of what Cather had once feared from the Nebraska landscape; speculations about Coronado in *My Ántonia;* and the wonderful set-piece, "Tom Outland's Story," in *The Professor's House.* By this time she was wholly ready for *Death Comes for the Archbishop,* which would help to explain the ease of its composition.

To be sure, not all are as pleased with Cather's Southwestern novel as she apparently was and as are those thousands of tourists who rely on it so trustingly. Some critics find it considerably inferior to *My Ántonia, O Pioneers!,* or *The Professor's House,* precisely because it is so loving a tribute to the region. Cather here, the arguments runs, was so in love with the Southwest that she surrendered her imagination to history and geography, and the result is an inferior work of art. At the History Library of the Museum of New Mexico in Santa Fe, I talked about this with Orlando Romero, the research librarian. "I love Willa Cather's other work," he said, shaking his head, "but not that. It's as if she became enamored of the 'Blue Sky' school of writing—all that romanticism about the country and the quaint customs of the natives." Romero also vigorously objected to the geographical and historical inaccuracies of the novel and led me to a stinging attack on these in a back issue of the *New Mexico Quarterly*.

There are, in truth, several instances where Cather took liberties with history and geography. In the passage, for instance, where the Bishop and Father Vaillant visit the site from which the Bishop will quarry the stone for his cathedral, Cather provides directions that would almost certainly place the site at La Bajada, the high ridge south of Santa Fe on the Albuquerque road. In fact, as Cather well knew, the stone was taken from a hill near the little village named after Bishop Lamy. There is the lone yellow hill from which the stone was taken, and atop its steep slopes near the westward prow are some of the great golden blocks, stacked and still waiting for the trip down the cut to the railroad siding below. Cather had to pass the hill each time she arrived in or left New Mexico since Lamy was the nearest train station to Santa Fe. The point is obvious: Cather transported the hill from its relatively prosaic setting to a far more dramatic one at La Bajada because as a novelist it suited her to do so. The gain for her was artistic, the loss in geographic literalism unimportant.

There are a few other instances of this sort in the novel, but they are as insignificant, and those who place their trust in *Death Comes for the Archbishop* as a guide to the region are right to do so, for far more important than the few liberties she takes with geography is Cather's ability to evoke the land, its spirit, and the effects of these on travelers and natives alike. Her description of the landscape on the Bishop's return to Santa Fe from a springtime missionary trip into Navajo country, for instance, remains one of the most superb and sharply detailed in the large canon of Southwestern literature. The sky, she writes here, was as "full of motion and change as the desert beneath it was monotonous and still"—precisely that contrast that causes the traveler in the Southwest to stand at emotional attention. And, she goes on,

> there was so much sky, more than at sea, more than anywhere else in the world. The plain was there, under one's feet, but what one saw when one looked about was that brilliant blue world of stinging air and moving cloud. Even the mountains were mere ant-hills under it. Elsewhere the sky is the roof of the world; but here the earth was the floor of the sky. The landscape one longed for when one was far away, the thing all about one, the world one actually lived in, was the sky, the sky!

Montana may currently advertise itself as "Big Sky Country," and the sky is big there. But in the Southwest the sky sometimes seems all there is and everything below somehow trivial. Even the mountains and mesas appear adrift in it. *Death Comes for the Archbishop* is full of this sense of sky: sunsets, cloud shapes, the constant brush of wind. Here is a felt, home truth about this place, and beneath this, Cather's few geographical liberties are also trivial.

But what of the novel's alleged historical inaccuracies? If we assume here that even the unfriendly critics would grant the novelist license to invent and embroider character and incident, what could these inaccuracies amount to? That Cather altered both the content and tone of some of the incidents of the Bishop's life is undeniable, as is the fact that she invented others. In the main, however, she stayed faithful to such sources as she had, chiefly Father Howlett. But if *Death Comes for the Archbishop* is truly that place legend many believe, it ought to be so true that it is beyond both history and historiography; it ought to include these and raise them to another power.

The essence of the claims against the novel's historical accuracy is that the Bishop was never the hero Cather made him. Mary Austin was the first to say it. That doyenne of Santa Fe's artistic and literary culture of the 1920s was offended that Cather should have given her talent and allegiance to a man who arrogantly trampled on the Indian and Hispanic cultures he found when he assumed his New Mexican duties. The symbol of this, Austin felt, was St. Francis Cathedral, a French Romanesque structure utterly out of character with its setting. The cathedral, according to Austin, was a constant affront, and had the effect of intimidating native New Mexicans. She was not the last to say this, and to those holding this view, it is apt that the Bishop's statue—the same that so inspired the novelist—should literally be standing atop the rubble of the old parish church that served as the Bishop's seat when he first entered Santa Fe in 1851.

There evidently was something imperious in Lamy's manner, something haughty and disdainful in his attitude toward local ways. Paul Horgan, his biographer, gives Lamy's fairly brutal census of his diocese as he found it in 1851: 68,000 Catholics, 2,000 heretics, 30,000–40,000 infidels. Lamy judged Hispanics to be congenitally inferior to Anglos in intellect and incredibly lax in their sexual conduct, and his treatment of the rebellious Taos priest, Father Antonio Jose Martinez, smacks of this prejudice. The native styles in architecture he found primitive and ugly, and almost from the first he dreamed of a French-style cathedral to replace the Santa Fe parish church he thought little more than a "stable of Bethlehem."*

Just as Cather could not have missed knowing where the cathedral's stone came from, so she had to have known of Lamy's imperiousness. Yet when she suddenly saw her Southwestern book, she couldn't help but see him as its hero, for in his life she saw once again the drama of the artistic sensibility isolated on the frontier from nurturing sources and forced to improvise ways by which it might sustain itself. This, of course, was her own story as she saw it and as she had been compulsively rewriting it in all the years since she'd left Red Cloud for the wider world. No wonder then that she gave her allegiance to the French priest, sympathized with his dreams, his monument, and ignored much of the local controversy that clung to his robes.

But had this been all, *Death Comes for the Archbishop* would not be the book it is. The difference between the book we have and the book that might have been is simply—and wonderfully—that just as the author surrendered to the spirit of the Southwest, so does her hero. True, he is occasionally imperious, and he candidly announces that his mission in the great diocese is to reform the lax and scandalous practices of both clergy and laity. There are also some practices of Indians and Hispanics that the Bishop finds impenetrably mysterious, even vaguely repellant. But Cather makes the Bishop assent finally to the indigenous peoples and, more, to their land of living.

He finds a special sort of joy in the wooden *santos* of the Hispanics and develops a genuine respect for their faith that remained steadfast through all the years when the region was almost wholly cut off from any sustaining contact with the Church. He respects, too, the ancient traditions of the Indians, senses that behind his friends, Jacinto and Eusabio, there are long, rich histories he will never know but which inform their every action. He admires their ability to accommodate themselves to any situation, an ability less sympathetic observers mistake for indolence but which the Bishop sees as "inherited caution and respect." It was as if, the Bishop thinks, "the great country were asleep, and they wished to carry

---

*It may be that Lamy has been posthumously paid back for such attitudes. Though his statue guards his cathedral, the diocese does not make a cult figure of him, and his tomb within the cathedral is no more prominent than those of his successors. Recently the church gave its approval to the destruction of the remains of the large formal gardens the Bishop brought into being behind the cathedral, and his episcopal residence has been torn down.

on their lives without awakening it; or as if the spirits of earth and air and water were things not to antagonize and arouse."

We are told this characteristically formal Frenchman finds the odor of burning piñon delicious, the hand-shaped contours of adobe walls and sills pleasant and reassuring, the cloud formations and ceaseless sky-change awe-inspiring. Climbing up to Acoma, the Sky City, the Bishop is caught in a thundershower. From beneath a sheltering rock, he looks out over the "great plain spotted with mesas and glittering with rain sheets" and thinks that the "first morning of Creation might have looked like this. . . ."

These are not, it hardly need be urged, the sentiments of one who holds himself stiffly aloof from the life of his place of work, and though they are perhaps not truly the sentiments of Bishop Lamy, they are a significant source of the enduring appeal of the novel Cather wrote. And, as if to seal her hero's surrender to the land, Cather devotes considerable space to his years of retirement when he lived and worshiped in his little residence in the Tesuque hills, surrounded by his fruit trees and with an amazing view of the blue Jemez Mountains to the west.

He might have retired to his beloved Auvergne and died there, Cather tells us, gathered to his fathers. Instead, he chose to stay on in what was still very much a rough frontier. The reason, Cather says, is that the land wouldn't let him go. Indeed, he had tried to go back, but "in the Old World he found himself homesick for the New. It was a feeling he could not explain; a feeling that old age did not weigh so heavily upon a man in New Mexico as in Puy-de-Dôme." In New Mexico, Cather continues,

> he always awoke a young man; not until he rose and began to shave did he realize that he was growing older. His first consciousness was a sense of the light dry wind blowing through the windows, with the fragrance of hot sun and sage-brush and sweet clover; a wind that made one's body feel light and one's heart cry "To-day, to-day," like a child's.

There may somewhere be a finer single description of the special feel and quality of this place, but to those fortunate enough to have experienced the thrilling promise of morning on the high plains of the Southwest, this will do.

II

❧ NEW MEXICO AND EASTERN ARIZONA are plentifully dotted with sites associated with *Death Comes for the Archbishop.* Most are quite specific, though a few are only generally identifiable, like that verdant village, Aqua Secreta, the Bishop stumbles upon in the red hills below Albuquerque in the novel's opening scene. William Lumpkins, a New Mexico native, recalled driving cows along the Rio Grande south of Albuquerque more than half a century ago when there were a number of tiny settlements down that way. "That may be one," he said, "that's just disappeared." But if you wanted to experience the hot sensory deprivation that can come from an encounter with an unfamiliar stretch of semi-desert, you could on a summer's day get off on a side road around Socorro or Belen and wander a bit through the red dirt, junipers, and piñons.

There are a few narrow, twisting roads through the Sangre de Cristos, and these would give you a sense of the rainy ride the Bishop and Father Vaillant take toward the village of Mora. The region is still as lonely and sparsely populated as when Cather described it, and the fierce, hard-bitten mountain villages up there appear to the traveler the appropriate setting for the Penitentes whose practices the Bishop wisely lets alone.

At Pecos there are the ruins of the ancient pueblo where the Bishop and Jacinto spend an anxious night on a mission to rescue Father Vaillant, ill in a mountain village; and a bit farther north, up the canyon cut by the Pecos River, there is a cave at the tiny post office of Terrero that may well have given Cather the idea for the secret cavern that shelters Latour and Jacinto when they are caught in a blinding snowstorm.

And, of course, there is the Santa Fe area, seat and center of the Bishop's career. The cathedral gardens the Bishop so assiduously planned and cultivated are almost gone now, and in place of the pond he created by diverting a portion of the Santa Fe River there is nothing but a dry, wheel-rutted parking lot. Pointing out to me the original layout of the place, Brother Rick of St. Francis Cathedral said the bulldozing of the gardens had been a sort of temporal godsend. "They'd become a weed-patch," he told me as we stood there in the bright morning, "and over against that wall was the hangout for the winos." A small square of enclosed lawn is all that remains of the gardens. Within it Brother Rick showed me the remnant trees of old husbandry: almond, apricot, walnut, pear, apple, and a twisted old quince leaning wearily against a wall. In a corner of the lawn, shrouded by grasses and plants, are the stones of the old *parroquia* (parish church) that made way for the Bishop's Romanesque cathedral.

The back road north from Santa Fe is called the Bishop's Lodge Road and leads to his retirement residence in the Tesuque hills. Once, Cather writes, the Bishop was riding in that direction and followed a stream through the hills until he came out beneath a rocky ledge into a perfect little bowl, filled with warm sun and crowned with an apricot tree of "such size as he had never seen before. It had two trunks, each of them thicker than a man's body, and though evidently very old, it was full of fruit." The site is now part of the Bishop's Lodge resort where the Thorpe family, which owns the resort, has carefully preserved the old tree though frost killed it several years ago. One afternoon there Mrs. Thorpe stood on the stone patio of the Bishop's little residence and chapel and said she sometimes liked to imagine what the view from this place must have been like before the resort's buildings blocked all but a beguiling blue glimpse of the Jemez Mountains forty miles to the west. You can't ignore the buildings, it's true. But if you were to follow the stream a bit eastward along its course through the boulders, scrub oak, junipers, and grand cottonwoods and then retrace your steps, you could easily imagine Cather's hero picking his meditative way along here, then being suddenly surprised by the sunny bowl and its great tree. Out in that same sun, surrounded by the blue air and the birds' songs, and with the Bishop's trees in leaf, you yourself might be pardoned for exclaiming aloud, "Today, today," like a child.

One spring, I took the Bishop's long trip into Navajo country, described in the section called "The Great Diocese." In terms of the novel's design and verisimilitude, Cather evidently wrote this chapter to give readers a vivid sense of the vastness of the Bishop's diocese and of the corresponding vastness of the administrative and spiritual challenges he faced as Vicar Apostolic. In writing it she drew on her formative encounter with the Southwest in the spring of 1912—more than reason enough for me to follow the trail of author and character out from the Bishop's cathedral to that stand of cottonwoods on the Little Colorado in Arizona where Father Latour spends several days as the guest of his Navajo friend, Eusabio.

E.K. Brown, one of Cather's more perceptive critics, points out that many of the novel's significant scenes occur near sunset, as if Cather too, like so many other travelers in the Southwest, found that hour's colors the distillation of all the region's visual treasures. So on a Sunday at sunset I stood under the Bishop's stern face as it caught the dust-yellowed light winnowing up San Francisco Street, and then in my car I swung out of town heading for La Bajada hill where Cather makes the Bishop find the stone for this cathedral. Spring had come late this year, preceded by day after day of high winds, snow squalls, raw weather, but once on the road beyond the city I could see the sure, small signs of the new season on the land: green shoots beneath the gray smother of winter, the shocking chartreuse of willows along the watercourses, runoff in the arroyos.

Albuquerque had delighted Cather, but the enthusiastic descriptions she wrote of it to friends seemed now to refer to some other place than the ugly sprawl with its toy metropolitan center mocked by the long thrust of the Sandia Mountains, and I was happy to clear the city to the west and point toward Navajo country. Here the road lifted into a long, gradual rise, and over its crest, the city dropped from view and memory both, the country suddenly got bigger: the land was spread into ever more generous folds, the vistas lengthened, and there was less and less of the human-made to suggest scale. Twenty miles west of Albuquerque you would have no suspicion that so large a city existed on the other side of a hill, and in this new amplitude I felt truly on my way, beginning to take a traveler's simple, direct pleasure in the swiftly lengthening shadows, the sharp geometry of the mesas that stood against an amethyst sky like the designs of Indian blankets. By full dark I was in Navajo country at Gallup, that place of terminal sorrow, and looking for a motel on the western edge of town I caught in my headlamps the inevitable, twisted forms of Indian drunks outside the Crazy Horse Bar.

The next morning was one of those classic New Mexican mornings: a bladelike cleanness to everything and a hundred miles of windless blue in any direction you looked. Rolling toward Winslow my thoughts were all of Cather, of her seeing this for the first time. Crossing the Little Colorado just east of Winslow, my eye briefly followed its braid of tamarack, Russian willow, and cottonwoods, wondering how far up that stream Cather had gone to where she'd encountered that memorable cottonwood grove within which she places the Bishop in the hogan.

Winslow itself was still what it was in Cather's time, a railroad town, and the Santa Fe line's division headquarters were still the handsomest buildings around. In their airy offices I cajoled the superintendent's assistant into looking through what are called, without the slightest irony, the company's "skeleton records" for anything on Douglass Cather. The assistant was skeptical, a whey-faced, slump-shouldered young man with scant use for history or literature, especially such arcane reaches of them as my errand evidently involved. He obliged me, but I felt as I watched his quick rifflings through the brittle brown cards that he wasn't going to be extravagantly thorough. Maybe he surprised us both by turning up the skeletal remains of Douglass Cather, discharged, so the file said, for having overstayed a leave of absence. But there was no address on the card nor any other information, and subsequent searches at the library and town hall failed to reveal records that might have told me where Douglass had been living when his sister came out for that visit in 1912. In any case, for me as for Cather the real lure lay north of town on the reservation.

In the main office of the Bureau of Indian Affairs' Leupp Boarding School, a broad-backed Navajo woman mumbled aloud the alien words of a memo she was typing, and waiting to speak with the principal, I had the leisure to read over her shoulder the faculty names of the Chart of Academic Locations: Bedoni, Yazzie, Cody, Begay, Tsinijinnie, Nrbitsi, Redhair, Becenti, Wilson. . . . Next to it on the wall was a carefully lettered sign on apple-green construction paper that read: "Whether we are educators or students the success of our educational efforts depend [sic] on our belief in our own excellence and our willingness to transform our beliefs into action."

The principal, an Anglo man in his thirties, had, like Cather before him, been captivated by this land and its native people. Influenced by the spirit of social uplift of the American Sixties, he had come out here to "share the dream of a bilingual, bicultural education," as he put it to me while we ate lunch with the students in the school cafeteria. But he had not, he said, been prepared for the visual and spiritual lure of this place. The land had seduced him, he said, taken him in, but even more than this, he had come to feel the spirit of the place, expressed in the various aspects of native religious practices, including those of the Native American Church, which is a force among reservation Navajos. Even the shape and traditional ceiling pattern of the hogan had finally spoken to him, had taken him into a sort of large and impersonal embrace in which he felt his personal problems subsumed, as indeed Cather's Bishop comes to feel when he brings his burdens to Navajo country and feels them drop away in the presence of the people, their herds, their great, silent place. "I was troubled when I came out here," the young man was telling me, "I see no reason now to deny that. But when I finally began living in a hogan seventeen miles from school, things began to come around. I'd get out there [to the hogan] and lie on my back and look at that peaceful spiral of ceiling logs, and I'd think, 'Well, everything's really all right. Everything's the way it's supposed to be.' Now I'm building a hogan for my family.''

He had kindly arranged a tour for me of that portion of the reservation likely to have been visited by Cather in 1912, and I arose early that morning feeling like a hunter bright for his prey. The TV was predicting sharply soaring temperatures—upper 90s in the valleys—and it seemed as if summer might come to these high plains without benefit of spring. But it wasn't so. Opening the door I found the morning mild and bright with that light, dry quality to its wind that made the old Bishop feel a young man again.

My guide for the day was Mal, a weather-winkled man in his late fifties with alert eyes and hair bright as bird feathers. A native of the Pacific Northwest, he'd eventually tired of the eternal weep of those skies and had moved south to work on the reservation. Mal had graduated from high school, but in the Korean War he had time on his hands and discovered that he really couldn't read. He taught himself, laboring, Conrad-like, through a few pages of literature each day, gradually becoming not only a proficient reader but a voracious one, devouring all of Steinbeck ("my favorite") and Theodore White ("completely changed our views of politics") as well as a great miscellaneous lot of history and fiction. No Cather here, but he was sympathetic to my desire to trace her probable movements and in particular to find that grove of cottonwoods along the river. He thought he knew the spot. But first, *inter alia*, he wanted to give me a rolling tour of this corner of the reservation; he knew much of its recent history and thought I'd find it absorbing.

Indeed I did, though perhaps not quite in the way my guide had intended, beginning with the moment when we jounced in Mal's pickup into a wash on the west bank of the Little Colorado and arrived at the desolate site of a World War II Japanese internment camp. Mal was eager to show me this exotic bit of the local past, point out the tatterdemalion cemetery from which over the years almost all the bodies had been removed by relatives. Those left behind without kin to care had no markers to their unmerited misery. He was eager, too, to show me the outlines of their clever ditching and diking system whereby they'd diverted the river into their rice paddies. And still, he told me, in the local monsoon season (a Southwestern joke for the infrequent, tempestuous cloudbursts of July and August) the green shoots would come up here and there, as though they would be the slender markers of what many would just as soon forget.

When I was moved by the desolation, the melancholy aspect of this place—the two of us standing in the wind that thrashed the cemetery's rubbish—to observe that such a place was a blot on our history and a betrayal of our most cherished national ideals, Mal shrugged. He'd seen "gooks" in Korea, and his shrug and his eyes said he'd seen a lot more besides. Well, he came back, "War makes people do funny things. I guess they [the Americans] felt the threat." He, too, seemed suddenly saddened, but not by the significance of the place. It had been five years since he'd been down here, "and just look at it! They [the Navajo] don't care. They come down here to get stone from the old buildings to use at home. Seems to me they ought to have the decency to leave it like it was."

Whether this was a preservationist's sentiment or a comment on the Navajo—or both—was unclear, but an oblique amplification came a bit later. We were headed now for the stretch of the Little Colorado Mal felt confident must be the place Cather had in mind for her character's stay among the Navajo. "I was a real liberal years ago," Mal was saying, "but boy, am I ever a radical conservative now! They say it's age does that to you, and maybe it is. But [down-shifting here with a small grunt to meet a rocky hump] I'll tell you what: when I got down here and saw what that give-away program was doing to these people—that's when I changed my thinking.

"We've ruined these people we've made them so dependent. Why, do you know, they're so dependent on us they'll actually come into the school office and ask us to make their phone calls for 'em! They're really just like children. You have kids: how long did it take them to learn to make their own calls? My solution? Turn 'em loose. Make 'em Americans."

On our way down to the river we had to skirt several Navajo ranches, each with its hogan, a few animals, the inevitable pickup, and, occasionally, a couple of teenagers, the boys in ball caps. Here and there as if hurled from a great height were red outcrops of sandstone, some as fantastic and twisted as constructions poured by a seaside child. Under the red rocks at noon the shadows were as blue as air. In previous days as a school administrator, Mal used to come this way looking for runaways. On this corner of the reservation it was the logical place to look since just below on the river there were two good swimming holes. "We'd just go down there and wait," he said with twinkling eye, "and if they weren't there then, they would be sometime. Now they don't even bother to go swimming."

We arrived at the river and the cottonwood grove. "There aren't too many stands like this along the river hereabouts," Mal said, "so I'd have to guess this is the one she was thinking of. Read me that passage again." And I did so while we sat in the truck, reading how Father Latour and his guide, Jacinto, come down here to visit Eusabio who lived with his "relatives and dependents, in a group of hogans on the Colorado Chiquito; to the west and south and north his kinsmen herded his great flocks." Eusabio gives his guest a solitary hogan sitting apart from the group, and there the Bishop spends several meditative days. Above the hogan, Cather writes, towered the cottonwoods, "trees of great antiquity and size—so large that they seemed to belong to a by-gone age." The trees "rose out of the ground at a slant, and forty or fifty feet above the earth all these white, dry trunks changed their direction, grew back over their base line. Some split into great forks which arched almost to the ground; some did not fork at all, but the main trunk dipped downward in a strong curve, as if drawn by a bow-string. . . . The grove looked like a winter wood of giant trees, with clusters of mistletoe growing among the bare boughs."

As a girl living on the treeless prairie outside Red Cloud, Cather would anxiously scan the distance when the family made periodic trips into town for supplies, searching for the first dark spires that were the Lombardy poplars planted as windbreaks and the cottonwoods that grew along the Republican. Here again in another treeless expanse she met the cottonwoods, and surely they must have spoken to her of the other place where she had dreamed of the Southwest and bright-armored Coronado.

The passage finished, we walked out into the warm sun and toward the grove, Mal warning me to watch for rattlers. Here had been the site of the first school on the reservation, and its remnant walls still poked up stubbornly here and there amid a slowly dissolving litter of cookstoves, wheel hubs, shattered glass. In summer in such a grove the leaves would make a fine, constant clatter in the wind that sweeps unimpeded over the plains to the west, but now there was silence under boughs that showed only the sprouts of spring. Mal told me he and his girlfriend, Jan, often came down here on summer afternoons to sit in the shade of the big, old trees and listen to them. "There's a peace here," he said at last with what I found a surprising softness. And I thought again of Cather, escaping the gritty little town down-river, and of her Bishop, too, both finding here a somber sort of peace, "favorable for reflection, for recalling the past and planning the future." Here the Bishop writes long letters to old friends in France, sitting in his hogan that was "isolated like a ship's cabin on the ocean, with the murmuring of the great winds about it" and the air without "the turbid yellow light of sandstorms." This house, Cather writes, "was so frail a shelter that one seemed to be sitting in the heart of a world made of dusty earth and moving air."

Cather based her narrative of the Bishop's long trip on a similar one Bishop Lamy had made in fall 1863 and has her hero come down to the Navajo country from the north, through the Hopi village of Oraibi. She doesn't say how he goes back to Santa Fe, but I assumed he retraces his route, and so in an early afternoon I climbed the newly paved Oraibi road that follows the ancient trail northward. Looking back from the crest, I could see what the travel-stained Bishop might have: the river, the faraway mountains, the table-flat plain. I had had my own brief encounter below with the land, the reservation, with Cather, too, and her Bishop who at moments had seemed as palpable as the people I'd met. Looking at the greening twist of the river through the dry land I saw again Cather's small, stout figure, her close-cropped hair, bonnet snugged hard against the blowing sand, driving on out of sound of Winslow's freight whistles and iron wheels to some rendezvous with the land and her own long imaginings of it.

Halfway to Oraibi I saw to the west a deserted hogan sitting off by itself and on impulse turned toward it. Unlike the one loaned the Bishop, this one wasn't in use. It had been abandoned by death or dislocation or some other cause I would never know, but its door was open, and I entered it thinking of both the Bishop and the young school principal finding a day's-end peace in regarding the slow swirl of ceiling poles about the skylight. The floor here was not earth but concrete, and there was half a child's bike along a wall, its frame a rusted canary yellow. Wasps hung about the skylight, their long, delicate legs dangling like parachute lines. Plainly, this hogan's service was over, and it was going on into another stage.

The concrete was cool beneath me as I lay gazing into that octagon of blue. Barred light fell in slats across my legs through the spaces in the roof poles. The wasps stayed up high, drifting against the peeling gray of the wood, then silhouetted against the open air.

What had gone on in here, what confused or purposeful efforts to stand fast against threatening changes, what tides of blood, memory had been contained by the slumping walls about me now, this little fortress in the midst of a world grown strange? And above it then as now was the incongruous sameness of a sky that looked on the old and new with a tolerant indifference. . . .

When I awakened, the blue octagon had grown paler, the barred light no longer fell across my legs but lay curled for evening at the eastward door, and arising, I saw the shadows arrowing away, the afternoon fled.

At Oraibi I gave a ride to a Laguna Pueblo man married to a Hopi woman and living out here now. This, he said, as we passed the miles in conversation about the weather, had been a dry spring. "I don't know what we're going to do this summer. I've got my garden in, but it if doesn't rain some more this spring, well. . . ." He said he'd dug a six-foot hole "for my trash, you know. And it was just dry all the way down." He shook his head as I let him out at his small cinderblock house hard by the road.

Eastward toward Window Rock there were junipers with trunks as thick as a man's waist; they appeared as files against a sea of sage that in the lengthening light was literally stunning as I whipped through mile after mile of it. There were a number of new hogans under construction, suggesting that hereabouts, at least, there might be a revival of traditional concerns. In the sunset I saw a ballgame on a diamond without grass, bases, or backstop, the lithe, high-shouldered figures of the players etched against the rough red dirt. At roadside, a young woman twirled a bright silk sash waiting to cross. Up ahead, a clubfooted boy waved down a car and caught a ride. Two small boys herded sheep in the waist-high sage, the sun coming now in a last blinding slant over the tops of the bushes and through the dust slowly rising from beneath the oily hooves of the flock. The boys trailed after the sheep, whacking occasionally at the sage with switches. Just outside Window Rock I was reminded of the prints in the offices of the Santa Fe at Winslow when I saw a Navajo woman in the traditional wide, full skirt, astride her horse and surrounded by her sheep.

Fast-food hamburgers and a six-pack kept me company from Gallup on the night ride to Santa Fe, and topping the long hill west of Albuquerque, I was given the illusion of a city simply jewelled in night and a thousand incandescences. Seeing it so, I was ready to believe it might have been Marseilles or Nice.

Back in Santa Fe and down again at the cathedral, I parked on a side street, and then, undetected, climbed the solid statue of the Bishop to pass my hand, wonderingly, over those features that had said so much to a writer more than half a century ago. Past midnight on a weekday there were but a few rumbling riders still abroad in the streets, and as I climbed down from the pedestal the lights of one of them played across me. And maybe I, too, for that instant excited a sort of wonder.

SIX

PLAINSONG: MARI SANDOZ'S OLD JULES

I

  ON THE EVE OF THE CIVIL WAR, Walt Whitman said he could hear America singing, the songs welling up from the common stock of the populace—mechanics, woodcutters, boatmen—and the whole beginning to form a national anthem. Twenty years later when he wrote *Democratic Vistas* he was obliged to admit the singers hadn't really appeared and that there was as yet no anthem, scarcely even what he could call a truly American literature. And a nation without its own radically native songs, he wrote there, is a nation without a soul. Looking at his America he found at its center a vast and desertlike void. "Never was there," he says, "more hollowness at heart than at present, and here in the United States. Genuine belief seems to have left us." In a mood as close to despair as he could permit himself, he claims that after a hundred years democracy had failed to prove equal to its great opportunity and that perhaps the only hope left was the rise of native songs out of the unsung spaces of the west or else out of those few unvisited pockets of rude vigor that remained east of the Mississippi. "To-day, doubtless," he wrote, "the infant genius of American poetic expression . . . lies sleeping far away . . . lies sleeping aside, unrecking itself, in some western idiom, or native Michigan or Tennessee repartee, or stump-speech—or in Kentucky or Georgia, or the Carolinas—or in some slang or local song or allusion of the Manhattan, Boston, Philadelphia or Baltimore mechanic—or up in the Maine woods—or off in the hut of the California miner, or crossing the Rocky Mountains, or along the Pacific railroad—or on the breasts of the young farmers of the northwest, or Canada, or the boatmen of the lakes. Rude and coarse nursing-beds, these; but only from such beginnings and stocks, indigenous here, may haply arrive, be grafted, and sprout, in time, flowers of genuine American aroma, and fruits truly and fully our own.''

A decade later he was still at this subject in a cloudy essay he called ''Poetry To-Day In America—Shakespere—the Future,'' writing that what the United States still required was ''native-born-and-bred'' writers who would express the silent lives of the broad, average mass of their countrymen. Here, claimed Whitman, is a ''magnificent mass of material, never before equaled on earth. It is this material, quite unexpress'd by literature or art, that in every respect insures the future of the republic.''

If few outside the domain of letters shared Whitman's view of the vital connection between a nation's literature and its spiritual health, the notion was after all an ancient one Whitman was reviving for his time and place. Most so-called primitive peoples believed they had a sacred responsibility to sing the songs of their lands, for the songs belonged to the lands, and in singing them the people came to know themselves. Joseph Campbell, for instance, in his survey of world mythology, quotes a description of a group of Australian aborigines retracing the footsteps of their mythological ancestor, then chanting the story of his wanderings over the landscape. The aborigines, Campbell tells us, knew the routes of the ancestral wanderings as Songlines, for wherever the ancestors went they sang the country into existence. The remembrance of this mythological period, called the Dream Time, in the ritual recitations of the descendants is what keeps those descendants in vitalizing touch with their origins in their lands. In recent years Bruce Chatwin discovered that for the Australian aborigines the entire landscape is a kind of spaghetti bowl of criss-crossing Songlines, a constellation of sacred spots once sung by the ancestors. So in his call for native singers Whitman was in a deeply grooved cut—as so often he was.

Still, he did not correctly anticipate the content of the songs and stories that would well out of the life of American lands, and the singers who began to emerge in the late 1870s and early 1880s gave voice to themes that only in a very special sense could be considered patriotic. Out of the defeated, flattened South in these years began to come the first rough whispers of what would be the black blues. Rooted in the daily round of rural work—sounds of boat horns, train whistles, mule laughs, the feel of a long and lonesome row—the blues were yet curiously, tellingly rootless, expressing little allegiance to place and still less to country, America being for these singers hardly a land of freedom. And at the same time there began to emerge from the great West, for which so much had been hoped by Whitman (and Emerson before him), Songlines made by the successors to the pioneers. The Western explorers had hardly known how to sing the land, so overwhelmed, disappointed, even cowed by it had they been. But following in time on their dusty, blood-tracked footsteps had come unlooked-for laments, a sort of white blues, into which entered a tally of the disappointing struggles of those who found how hard were harvests of life on the tamed lands left behind by the moving frontier. Neither Whitman nor anyone else had quite anticipated these emerging songs, and so when they began to appear out of the Middle West they were not recognized for what they really were: Edward Eggleston's *The Hoosier Schoolmaster* (1871), for example, with its unflattering portrait of backwoods Indiana culture. Here was hardly a sing of the broad-ax, the open road, or that hardy yeomanry Whitman was certain existed somewhere out there, but instead a song of something almost dwarfish, deformed by supersession, cultural isolation, and the scruffy circumstances of daily living, a mean village culture tyrannized by public opinion, by bullies, and schoolhouse louts. Or E.W. Howe's bleak if melodramatic *The Story of a County Town* (1883), a song of soured puritanism, of betrayals, divorces, murders, and suicides enacted in a drab land no longer frontier nor even the West anymore but simply a spot left behind. Then as the frontier plowed and threshed its way west, the trees thinning now and giving way to the tall grass prairies, came Joseph Kirkland and then Hamlin Garland whose song was of bowed backs, crabbed fingers, grasshoppers, usurious middlemen, blighted ambitions. In the early years of the twentieth century came a new crop of writers, the literary equivalents of Russian thistles as they seemed to those of the regnant Genteel Tradition, determined to write about those things the Genteel Tradition thought were better left unexpressed. Edgar Lee Masters, Sherwood Anderson, and Sinclair Lewis all had grown up in the heartland's towns and villages, and all felt compelled to perform literary vivisections on that heartland, relentlessly revealing the shabby inner truths that lay beneath the surface. Masters's *Spoon River Anthology* (1915) perhaps cut deepest. Drawing on boyhood memories of Petersburg and Lewistown, Illinois, he created a chorus of ghostly voices, calling up through the weeds and grasses of the village graveyard, telling the stories of their failed, twisted lives. "Franklin Jones" speaks:

> If I could have lived another year
> I could have finished my flying machine,
> And become rich and famous.

Hence it is fitting the workman
Who tried to chisel a dove for me
Made it look more like a chicken.
For what is it all but being hatched,
And running about the yard,
To the day of the block?
Save that a man has an angel's brain,
And sees the ax from the first!

Writing of Sinclair Lewis, Mark Schorer makes the point that comprehends Lewis, Masters, and the other voices of Middlewestern discontent. "Have we," Schorer asks in his biography of Lewis, "enjoyed a cultural irony more absurd than this, that it should have been Sinclair Lewis who emerged to fulfill Walt Whitman's cry for a 'literatus'? Whitman had called for a popular prophet, a maker of democratic images that would elevate and amplify the democratic life that was falling short of its promise. Sinclair Lewis emerged, no mystagogical orator of the sort that Whitman had in mind and he himself was, but an image maker certainly, stridently drawing for Americans the picture of their crass, petty, hypocritical, and barren lives."

Farther west where the frontier was still a vivid memory there was Cather, and as she found her voice in these early twentieth-century years it sounded a good deal like those of Howe, Kirkland, Garland, and Masters in its scabrous descriptions of small-town life—though she was from the very first a far better craftsman than any of these writers. But, as we have seen, by the time she had written *O Pioneers!* she had learned her relationship to her material could not remain as hostile as it had been. She makes her heroine, Alexandra Bergson, sing the land "like the larks over the plowed fields . . . singing and singing,/Out of the lips of silence,/Out of the earthy dusk."

The song of the lark stayed with her as a metaphor of her own task, appearing as the title of her next novel, one about a girl who must learn to sing out of the shabby circumstances of her life in Moonstone, Colorado (Red Cloud again). Thea Kronborg, monosyllabic as a child, is apparently stranded there like the rest of the townspeople in that little place, set down, so Cather writes, like Noah's Ark in the desert. And yet there are the rudiments of song even here: old Wunch, the battered, drifting piano tuner, who first teaches the girl to play; and Spanish Johnny, like Wunch a drunkard, but also a superb mandolin player. From such mentors and out of some finally mysterious personal reservoir of need and talent Thea improvises song—and an escape from Moonstone's deadening sandy wastes.

Writing from Canada years after the book's publication, Cather said that what she had cared about in writing it, and what she still cared about, was the girl's escape from "commonness." Having, as she had to have seen it, escaped herself from "commonness" in Red Cloud, from death in a cornfield, Cather after 1925 became progressively disinclined to consider the lives of those not so fortunate, somewhat blithely writing of Thea Kronborg and her escape that to persons of such vitality and honesty "fortunate accidents will always happen." But behind her, in her past, were those still struggling for some sort of grace on the Great Plains. There under the endless skies and in those equally endless pastures that once had so threatened the transplanted Virginia girl were stories waiting to be sung of endurance, hardship, protracted isolation, tragedy—a mass of silent material that Whitman had believed was magnificent stuff for some "divine literatus."

It hardly seemed so to those living out there. Survival was perhaps the best one might hope for; song was out of the question. For years Mari Sandoz seemed fated to be just another one of those whose lives would remain silent material, whose songs would remain unsung. But through a heroic persistence she eventually gave voice, not only to her own song, but to the songs of her home region, the upper Nebraska panhandle. In the course of what became belatedly a literary career, she wrote eighteen books, some of them, like *Crazy Horse* and *Cheyenne Autumn,* truly distinguished. But her great book, her unique contribution to the national letters, was her first, her heart's cry, *Old Jules* (1935), the story of an unforgiving man and an unforgiving place. Near the end of her career she wondered, somewhat like the aged Whitman had, why the Great Plains had not produced a truly great writer. Where, she asked, "is the writer with the gusto, the wild unmeasurable power, the wide horizons, the sense of the universal in man, and his infinite variety, that the Plains might have produced, should have produced?" If she herself was not quite that writer, she had at least written a book that was equal to the high, harsh tone and scale of her native place.

"Old Jules" was the not wholly affectionate soubriquette bestowed by his fellow settlers on Jules Ami Sandoz, Mari's father. It connoted a number of things: Jules's pioneering activities in the upper panhandle country in the early 1880s; the fact that out there settlers quickly took on the look of age far beyond their years; and Jules's reputation as a character. Watching him stalk the main street of Rushville, said a contemporary, you would have thought him either a freak or else one who had let the image of the Wild West go to his head, for he dressed like a filthy drifter and always carried a long-barreled rifle at the ready. Born in Neuchâtel, Switzerland, of thoroughly middle-class parents, young Jules was pointed by them toward a career in medicine. His father had definite ideas of how a young man ought to prepare himself for this, and in the summers he sent his son to the Jura Mountains where he learned planting, husbandry, and hunting from German farmers. Later, after he had entered medical school at the University of Zurich, Jules worked summers on a mail train running between Zurich and Florence. His father thought a doctor ought to know more than school and books. Yet when Jules announced to his parents that he had fallen in love with a fellow worker on the train and wished to marry her, his father was adamantly opposed. He had not foreseen such a complication: marriage at this stage would surely impede his son's progress toward his degree, and in any case this little Rosalie of Jules's was clearly beneath them in class and station. Hot-headed and willful from youth, Jules said he would marry Rosalie anyway and take her to America with him. But Rosalie herself was not in a temper, and the prospect of tearing up her roots held nothing but dread for her, so Jules went alone in 1881, vowing to send for Rosalie when he was settled.

Impetuous as his behavior was, it did not proceed *ex nihilo*. In these years of the late 1870s and early 1880s America was in the Swiss air, and a goodly number of those who could think of leaving the homeland were doing so. Swiss agriculture had been pummeled by a series of natural and economic blights since the 1840s, culminating in the international economic depression of 1873 and the collapse of commodity prices. At the same time the American West was swinging open like the door to a treasure-house now that the Indian menace had been effectively removed. For the young, the venturesome, the have-nots, prospects at home looked far less pleasing than those that might be encountered abroad. For Jules Sandoz, his neck hot against his parents and the stodgy ways of the tiny homeland, the American West looked like a good place to bet your life. He knew agriculture; he knew guns. He could make his way, and in a short while he could send for Rosalie.

It had long been established among immigrants that all the good farming country east of the Mississippi had been taken up, and so by the 1880s settlement had ventured cross the Missouri into eastern Nebraska—though the movement out onto the great plains was as much the result of a desperate optimism on the part of the settlers as of an accurate knowledge of the new country. Nebraska had been given a bad name even before it had been Nebraska, and there were still those who believed it fully deserved it. After the region had been acquired in the Louisiana Purchase, expeditions by Lewis and Clark and by Pike, Long, and Frémont had established the notion that Nebraska sat squarely in the midst of the Great American Desert. The cartographer of the Long expedition wrote "GREAT DESERT" across a large portion of what was to become the state, and this designation was still to be found on maps of the U.S. as late as the 1870s. Yet in the 1850s the hardy few who went across the Missouri found that Nebraska soil and climate were pretty much the same as that of western Iowa, and slowly settlement began near the river, then spread out along the valleys of the Platte, the Elkhorn, Logan, and Niobrara. Apparently the Great American Desert commenced somewhere farther west. Settlement was slowed by the Indian troubles of the late 1860s and by the great grasshopper plague of 1874, but by 1880 the tide was running again, and now it was widely said that the Great American Desert was a great hoax: there was no desert, at least not in Nebraska. Rainfall had been increasing steadily through the late 1870s, and the trend seemed likely to continue. Geologists explained why. The cultivation of prairie soils, they said, had a generally cooling effect on them, and this in turn called moisture out of the air as the winds blew across the state. The more soil in cultivation, the greater the amount of cooling and the more plenteous the rains. "Rain follows the plow" became Nebraska's marching song even as "Over the Hills" had once cheered the Garland family and other settlers in Wisconsin and Iowa. Settlement passed the hundredth meridian, beyond which, so Major John Wesley Powell had warned, the annual rainfall would not be sufficient for consistent crop production. In 1880 this too seemed erroneous.

Jules Sandoz chose Knox County for his claim, locating on a spot near the confluence of the Niobrara and the Missouri. As he began to farm the rainy cycle continued and he wrote hopefully back to friends in Switzerland, urging them to come on while good land still remained available. A few did so, but not his Rosalie, so Jules took a wife. He couldn't wait forever, not out here where a man needed a woman with a strong back, a silent tongue, and a passable hand in the kitchen. The woman he chose, however, failed to satisfy him on at least several counts, and on a day in 1884 he flew into one of his violent rages, flung the flour and sugar at the pigs, took the available cash, and left his wife behind in their tiny cabin. With his wagon and team of horses he pushed westward once more, following the Niobrara across the northern portion of the state, through the sandhills (which really did look like the Great American Desert), and out onto the hard land of northwestern Sheridan County up near the South Dakota border.

He came into the country at a propitious time. For some years the area had been a favored grazing place for cattle ranchers, but a combination of blizzards, rash speculation, and overcrowding had reduced the ranches around what had become known as Mirage Flats, and Jules Sandoz found he had things pretty much to himself. Sinking his spade into what settlers came to call the "nigger-wool" sod, he inspected the soil beneath and was satisfied crops would prosper here. Once more he wrote glowing come-on letters back to Switzerland. Rosalie still refused to come, but Rosalie or not, he was here to stay. There was nothing for him in the Old World. This land would soon become populous and valuable, and he would be its prophet. He made friends with the Sioux who still wandered from their reservation through what had lately been a homeland, and they taught him much about the country. He built a dugout, and as fall came on he began to dig himself a well with the help of two settler friends. Sixty-five feet down they struck water, and for Jules Sandoz the strike was also disaster: knowing too well his volatile temperament, his two friends could not resist teasing him, jerking him up and down as they hauled him out of the well, until they accidently dropped him all the way to the bottom, smashing his ankle and rendering him a cripple for life.

The accident did nothing to improve Jules Sandoz's temperament, but if anything it strengthened his determination to persist where he was, no matter how unpromising his patch of American earth might seem to others. He could not admit failure and limp home in his ragged clothes. The neighboring ranchers might jeer at him and his dreams of a fat and fertile land filled with hard-working Swiss and German farmers. He would show them all. He took another wife and moved from the Flats onto her homestead on the Niobrara. There he continued his tireless propagandizing for the panhandle's prospects in letters to friends and to newspapers in America and abroad. He located new settlers, began to experiment with horticulture, lobbied for a post office and got it. By this time, however, his second wife had wearied of his temper, his chronic feuds with all who lived in his vicinity, and of the beatings he was too happy to hand her. One day in 1891 she threw his belongings into the barnyard, then filed for divorce. Jules simply moved across the river, built a rough two-room house under the old Sioux landmark known as Indian Hill, and took a third wife. She lasted but two weeks before she fled in horror to friends in Rushville. The man was a vicious tyrant.

Yet he was oddly gifted, and out in the panhandle so far from any real center of civilization his abilities made him respected, if not liked. He knew guns, of course, could mold bullets, carve stocks, sight in a new gun. He was himself a crack shot and made sure the local cowhands knew it. He knew doctoring, too, could act as midwife to woman and cow, could administer either medicinals for illness or poison for wolves. He knew about the stars; on clear nights he could take his visitors out under the spangled heavens and tell them of constellations and the unimaginable spaces between. He knew natural history, too, and when so minded could talk of the ponderous creatures that had dragged themselves over the still-cooling earth in the ages before the advent of humankind. Above all, he knew agriculture, and he remained unshaken in his view that the panhandle would one day blossom as the rose.

That view was harder and harder to maintain as the 1890s came on, for Nebraska's rainy cycle had ended, and an equally pronounced (and more typical) period of drought had begun. The panhandle, where the average annual rainfall is only between fifteen and eighteen inches, was especially hard hit. Crops shriveled unripened on the stalk before merciless, fiery winds. The sandy soil, never the best conservator of water, cracked and yawned, rivers shrank into ironic deserts of baked sand and mud. 1890 was a severe drought year and 1893 an even worse one. In 1894 the entire state received but 13.54 inches of rain. July 26 of that year was the nadir for the panhandle county of Dawes when the temperature reached 105 degrees and the wind velocity forty miles an hour. The corn crop there was dead in three terrible days. Farmers began to leave by the hundreds, crawling back across the Missouri with the legend of defeat scrawled in axle grease on their wagons:

> In God we trusted.
> In Nebraska we busted.

Sky pilots and rainmakers appeared in the land, both promising deliverance from above. At Alkali Lake just south of Hay Springs on the Flats they held a grand revival in 1890. Jules Sandoz scoffed and said now was not the time for such foolishness, and still less was it a time for leaving: land was cheap now, and it was a good time to buy more. But others put their trust in the above rather than the land and gladly went to be doused in the stinking lake waters—perhaps a disguised effort at sympathetic magic. A week later and no relief, Mrs. Schmidt, one of the most enthusiastic communicants, was taken raving to the asylum, and a Bohemian hanged himself. A neighbor of Jules's hanged himself from his well curbing and swung in the wind for two days before he was noticed. Up on Box Butte a woman horsewhipped her neighbor for setting a prairie afire. A woman told her neighbor she'd had enough and was going back east. That night six wagons arrived at her house and shadowy figures began dismantling it for its lumber before the men realized the woman hadn't left yet. "Nebraska land," the sun-blackened survivors sang,

> Nebraska land,
> As on thy desert soil I stand
> And look away across the plains
> I wonder why it never rains.

Carved on the door of a deserted shack near Chadron, another legend of the time:

> 30 miles to water
> 20 miles to wood,
> 10 miles to hell
> And I gone there for good.

As for the theorists and quack practitioners, business was brisk. The theory that plowing caused rain was now flung aside as a tragic lie, and in its place came a plethora of new notions. Large prairie fires would cause rain. So would smoke from chimneys. Electricity carried in telegraph poles might cause showers. A prominent theory held that great concussions would break open the heavens, and in Old Jules country the citizens got up a thousand dollars to purchase gunpowder. Then on a torrid July day in 1894 charges were set off from Long Pine to Harrison at a pre-arranged signal. No rain followed. Nor did any but chance success attend the secret machinations of the numerous rain wizards who came through the towns and villages with long black boxes filled with rain-inducing chemicals.

The population of the panhandle continued to dwindle, but Old Jules stayed on, stubbornly tending his fruit trees, predicting the drought would soon break. In 1895 he married for the fourth and final time, this wife a Swiss-German immigrant. A year later Mary Sandoz had the first of the couple's six children, a daughter, also called Mary—later "Marie" to distinguish between mother and daughter.

From the first the girl was thin and sickly, and her father was in a constant temper at her whimperings. One summer's night her cries so enraged the man that he beat her blue with bruises. To save her daughter from maiming or even death, Mary ran with her into the night and stayed out till dawn. This was the first of the many beatings the girl was to endure, and it would not be the last time her father would come to the very verge of murdering her. She learned early to stay out of his way, to efface herself around the farm, to establish hiding places, to keep silent behind the kitchen stove, warily watching the broad, bearlike man as he cleaned his game, carelessly discarding the bloody entrails onto Mary's hard-scrubbed floor and spitting great gobs of tobacco juice everywhere. The girl was an utter stranger to affection: she was never hugged, held, or kissed, nor did she ever see such demonstrations pass between her parents—apparently not an altogether unusual situation out there where settler life was often reduced to a raw, crude level on which there was little to spare for grace notes. Courtship was the time for the exchange of affections, but after marriage it was understood that the business of life commenced in earnest, and when children came it seemed appropriate to raise them in a fashion that would realistically prepare them to take their places in fields and kitchens. Whippings were an accepted technique of child discipline. Wife-beating was not uncommon, and the only instances when it became a matter for community gossip were when it became chronic or when a man went too far: a man in the sandhills stopped his wife from running away by beating her to death with a metal-tipped wagon tongue, and this offended local opinion.

When the younger Sandoz children came—Jules, James, Fritz, Flora, Caroline—Marie became their surrogate mother, caring for them as best she could by day, sleeping with them in the attic room at night. Before the age of ten she had settled into the numbing routines of a farm woman: barnyard chores, housecleaning, cooking, babysitting. When Old Jules sent Rosalie a photo of the young Marie and the other children, Rosalie was struck by the prematurely aged and saddened look on the girl's face. "They are not like children here," she wrote Old Jules, "particularly the little Marie. . . ."

She could speak only a few words of English and had never seen the inside of a school. Because Old Jules was paranoid, constantly feuding with neighbors and on occasion even jailed for criminal actions against them, he saw to it that his women had little contact with the outside world. He discouraged visiting, never took Marie on his trips to Hay Springs or Rushville, and only rarely took Mary to these hamlets. Once when he did so he left her alone all day in a store while he talked politics with the men. When she remonstrated with him for his neglect, he laughed in her face and said he'd had so many wives he couldn't be expected to remember where this one was.

Thus Marie's world was a severely circumscribed one, consisting of the house, the Niobrara down the slope, Indian Hill with its summit blackened by the ashes of old Sioux signal fires. She had almost no contact with children her own age, and except for the random visits of the Sioux and Cheyenne who still came to visit their white friend at their old camping grounds, she had little human contact outside the members of her own family. And yet she found things in this small world that absorbed her. There was to begin and end with the natural world of river, hill, prairie. In her penal servitude she was able to find solace in the look of the stretching fields, the sound the constant wind made through the grasses, the burble of the river, the clouds scudding on toward unknowable destinations. Whenever weather permitted and she was freed for a few minutes from her heavy responsibilities she escaped to Indian Hill or the river. She was intrigued too by the Sioux and Cheyenne visitors. She could not speak their languages, but she could communicate a bit by a combination of sign language and the few words of English she knew. In this way and also from her father she learned something of Sioux and Cheyenne history. Some of the men still wore bits of their old-time, pre-reservation finery, and these relics gave her a whiff of the great days the men remembered before the coming of the all-conquering whites.

Despite all, she was interested in Old Jules himself. Vicious as he could be, selfish, slovenly, wholly unpredictable, still he was to her a figure of commanding significance. She was, of course, terribly afraid of him, but at the same time she sensed a power there, a force that drew her despite her fears. His breadth of knowledge was so obvious that even the inexperienced, sequestered girl could not but notice how often his neighbors were forced to come to him for advice about agriculture, politics, hunting, medicine, and land matters. As a woman she knew well that she was beneath his notice (and perhaps even his contempt), but there were occasions when in her presence he would talk of things that really mattered to him—usually when she accompanied him on hunts in the general vicinity of the farm. On these hunts she served the old man in the capacity of a retriever, and a man who once watched her and her brother Jule(s) scurry after quail Old Jules had downed said they were better than any dogs he'd ever seen. When it was just the two of them Old Jules would often stop to rest atop Indian Hill, the silent girl behind him, and then he might talk, almost to himself, of his student days in Zurich, of the Indians and their buffalo hunts and wars, of the way the country was when first he had looked on it. Once in a while he would lift up his voice in song. The girl liked the stories and songs, though she was afraid to say so, and such occasions seemed to suggest obscurely that somewhere in the wild fastness of his being Old Jules cared for Marie, found her worthy of overhearing his thoughts and reminiscences.

There was also the fact that he could make things grow in this inhospitable soil where so many others had failed and fled. His grandfather's seal had shown a man planting trees, a prophecy the grandson was now fulfilling. To be sure, he had suffered like the others through the awful dry years of the early 1890s, but he had survived where he was, and now that the full force of the drought had been broken at last his orchards began to prosper just as he had said they would. It was as if the man had willed them to bloom and bear, caused that life to spring out of the sandy soil. Perhaps here was the inmost source and seat of that power she sensed in him, the Life Force triumphing over his savage personality as over the panhandle climate. Like the solitary peasant of Jean Giono's classic fable, "The Man Who Planted Trees," Old Jules had brought shade and the juice of fruits into his portion of an arid and treeless land: plums, cherries, apples. Every spring he would give away wagonloads of shrubs to all who would promise to plant them, so to spread the green throughout the panhandle. In part he had been this successful with his trees because he had intuited early on that agricultural methods applicable in the more humid lands to the east were simply not going to work out here. Perhaps before dry farming became popular in the state in the mid-1890s he had worked out for himself some of its basic principles, such as the search for drought-resistant crops and the fact that compacted soil will retain its moisture longer than loose, worked soil. At any rate after the Nebraskan Hardy W. Campbell published his studies of dry farming Old Jules assiduously applied the new technique to his orchards, and his successes brought him the attention of American agriculturalists in the highest circles. How much of this the young girl understood is not known, but she surely had to marvel at the rows of sturdy trees that waved their boughs in the flats beside the river, and when Old Jules began to invite neighbors to come into the orchards and pick the ripe fruits for their homes it was a vindication of what he was.

He might have kept her at home indefinitely had not his snarled network of enmities caused someone to turn his name in to the local truant officer. Old Jules had long scorned the primitive efforts of the prairie schoolmarms he had known, saying his children could learn more from him than from such ignorant pedagogues—and he was probably right. Now, though, he had to obey the law and send Marie and little Jule. Mary was distressed because neither of the children had shoes, and it would be embarrassing for her to have the family's poverty exposed by sending them off to school barefoot. Somewhere shoe money was found, and the children went. Marie was almost nine as she entered first grade, and so in addition to the fact that her peers were two grades ahead of her and fluent in English she had to endure the taunts of those who told her she was the daughter of the local lunatic. Though she had to learn to read along with the rest of the first-graders, she picked it up quickly, and it was not many days before she discovered that printed words were made up of letters of an alphabet and that a certain word always looked the same on the page. Later, after she had become a well-established author, she often remarked that this was the greatest discovery of her life.

Soon enough it became known that Old Jules's Marie liked to read, and a neighbor, Charley Sears, told her he had a small library he would put at her disposal: Hawthorne, Dickens, Bill Nye, and a ten-year run of the woman's magazine *The Comfort*. Others contributed what they could, and Sandoz was later to recall that she had read Poe, Melville, and Hardy before she was out of grade school. Of these, Hardy made the most immediate and lasting impression, for his dramatization of the brooding, somber, and irresistible force of environment made a home sense to a girl already aware of how all those she saw about her had been shaped by this place. "So it was like that everywhere," she remembered thinking after reading Hardy. "Well, it was best to know."

Old Jules knew nothing of this. The creative arts were not for his family. Though he allowed them to enjoy such stray snatches of music as came to them—at occasional barn dances now in the late 1890s—to aspire to play an instrument or to become singers was out of the question. As for reading, some of it was well enough, but fiction was lies, the silly stuff that hired girls might indulge in, and Marie had to smuggle her treasures up the ladder to the attic room she shared with whoever happened to be the current baby. But however much reading she managed on the sly she could not get enough, and she began to compose her own stories, even daring to send one to the junior page of the *Omaha Daily News*. When the paper printed the story and Old Jules found it out, he gave Marie another thrashing and threw her down into the dark cellar. Later that day he cooled, let her out, and took her quail hunting.

In 1910 when Marie Sandoz was fourteen her father's itchy feet compelled him to relocate, this time on a section in the sandhills southeast of the Niobrara homestead. The region had been opened to settlement through the special provision of the Kinkaid Act (1904), and Old Jules was attracted by the challenge of it. He had long since created all the enemies he could on the Niobrara, and now in the sandhills he would be surrounded by the ranchers he so loved to fight. Besides, the soil there looked even more stubborn than that on the Niobrara place, and he wanted to make his seedlings thrust up through the short, spare grass, cactus, and poison ivy. To protect the new claim he sent Marie on ahead with James, now ten, and for months the two children lived alone in a tar-paper shanty without windows.

Like others who had formerly been exclusively farmers Old Jules decided that he should diversify and run some beef cattle on the sandhills claim. In May 1911, a blizzard struck the sandhills after an unusually mild and open winter. The next morning Mary discovered the cattle gone, strayed off into the great drifts where they would soon die. Marie and Jule were sent out on horseback to track them and bring in the ones they could. All that day of dazzling sun and snow the children worked, tracking through the wind-shifting drifts, finding cattle, freeing them, then leading the way back to the barn. That night they trailed into the barnyard with the small string they had been able to rescue. Utterly exhausted, they tumbled from their saddles and were carried into the house. Marie's head throbbed, her eyes burned. Why, she wondered aloud, didn't her parents at least have a lamp lit against the darkness. "A-ach," her mother said, making a strange noise in her throat, "the lamp is lit. You are blind."

Through the anguish of the pain-racked days and nights that followed the girl writhed and moaned in her bed, her eyeballs scalding, her skin falling away from her blistered face in gray rags. Old Jules administered morphine and predicted permanent blindness. Yet in a few weeks Marie was up again and about the house and barns on her chores. Not until she aimed a rifle and found she did not have to close her left eye did she discover it was blind.

The incident was a symbolic coda to her panhandle childhood. As if she had realized that little more could happen to her, she began to dare to stand up to Old Jules, even occasionally to criticize his ways. When at age seventeen she graduated from grammar school she sneaked into Rushville and in a daring act of emancipation took and passed the rural teacher's examination. She was splinter-thin, hatchet-faced, blind in one eye, and carried on the back of one hand a sharp white knob, token of one of Old Jules's beatings. She was shy and self-deprecating, but beneath these appearances lay a vein of determination as iron as any that lay in the breast of the old Cronus, her father. She went home after the exam to face his wrath, weathered it, then turned to face the long, hard-packed row that would be her own life. "Sticking feathers on a buzzard's neck won't make an eagle out of him," Charley Sears said laughingly of her new "career," "but maybe you can fool 'em." After he had calmed down Old Jules grumbled that he wouldn't tolerate any lazy schoolmarms around his place, but inwardly he was both amazed and pleased by this show of independence from Marie, and he helped her start a school in the Sandoz barn.

A year later Marie made another bold and surprising move toward independence when she married Wray Macumber, a twenty-seven-year-old farmer out from Iowa. Old Jules and Mary had always sneered at Marie's looks, Mary in particular often remarking on how skinny and gawky her daughter was, yet now Wray Macumber found her to his liking. Five years later when the marriage ended in divorce Marie would tell friends that her father had pushed her into it, anxious that another would take this daughter off his hands. Still later she said that Old Jules had at first approved of the match then later had seen it was a mistake and had helped her file for divorce. Perhaps here, too, he had secretly felt a larger affinity for this daughter than he could openly admit and had found in her marital troubles an echo of his own. Whatever the case, the years of her marriage remain a blank: in later years she never mentioned them, and her family learned never to bring the subject up. Most who knew her during her literary career never knew she had been married.

The divorce ended Marie Sandoz Macumber's panhandle years. Though Old Jules, long hardened against local opinion, appeared unperturbed by the divorce, Mary Sandoz felt humiliated by what she considered a disgrace. As for Marie herself, she would not slink home to face her mother's tears any more than her father years before would have gone back to Neuchâtel in limping defeat. Instead, she faced the other way and in the fall of 1919 moved to Lincoln (the opposite end of the state) where she enrolled in the Lincoln Business College for courses in spelling, composition, shorthand, and typing. For the next eighteen years she would live in Lincoln, a place she never felt was more than an exile but one that was at the same time a refuge. She would emerge out of those lonely, impoverished, and often profoundly discouraging years having succeeded at last in telling the story of Old Jules and of the place made mythic for her by the rude force of his personality. It took much of what she had as a person and a great deal of what she had as a writer, and in the spent aftermath of *Old Jules* you glimpse an author persistently looking back over her shoulder at the specter of that great achievement that had cost her so dearly.

In March 1935, the unknown writer who signed herself "Mari Sandoz" once again submitted her manuscript about her father to an east-coast publisher, this time entering it in the *Atlantic* nonfiction contest. It was the fourteenth time she had sent it out, and after each rejection she had revised it extensively, stubbornly convinced that this chronicle of a man and his place was worthy of far more than purely local notice.

She had begun in the mid-1920s without fully realizing what it was she had in mind as she went through old newspaper files in the dingy basement of the Nebraska State Historical Society where from time to time she was employed. She believed she was steeping herself in Nebraska history as potentially valuable background material for her fiction. Some of this material she began to use in the short stories she kept turning out—and which kept getting sent back by eastern magazines who found Nebraska the literary equivalent of the Great American Desert and the writer's tone dreary as Dreiser. One story utilizing a sandhills background won honorble mention in a short-story contest, but all she ever realized from this was a note from Old Jules who had seen notice of her award: "You know I consider writers and artists the maggots of society. [Signed] Jules Ami Sandoz." But then in 1928 Old Jules had died, and almost his last whispered words had been to his oldest daughter: "Why don't you write my life sometime?" Perhaps already she had been working at the idea, but now in the darkness of his death a light dawned for her, and she began in direct earnest on what she said was not only the biography of a man but also of a place.

From the outset it had been like hauling a wagon up a sandhill. Her mother resolutely refused to help despite her daughter's almost inquisitorial efforts and had gone so far as to throw out most of Old Jules's records. Sandoz was able to rescue some of them from the floor of the farm's meathouse where they had moldered in the sodden sawdust, but a good many were ruined. She tried to fill the gaps with her indefatigable researches at the State Historical Library, and she blanketed the upper Niobrara country with questionnaires. Only one was ever returned. By the fall of 1933, dejected, she decided to quit the project, which to her meant quitting as a writer. The year previous the manuscript had been returned by the judges of the *Atlantic* nonfiction contest with a penciled note saying it was a dull book about a dirty old man. Maybe it was. Maybe no one outside the state was interested in the Nebraska panhandle and its life, Whitmanian prophecies notwithstanding. The country as a whole was in the deeps of the Depression with fifteen million Americans out of work and farm prices sixty-three percent beneath their 1929 levels. Maybe escape, fantasy, top hats and tails dancing across silver screens were what was needed instead of the somber and ponderous serving of American soil she had been offering. Her mother had maintained her refusal to help, had scoffed at her daughter's hopes of making a living "without working," and maybe she was right, too. Mari Sandoz was now thirty-seven and had earned less than five hundred dollars for ten years of persistent literary effort. She called three friends to the backyard of her Lincoln apartment building and had them witness a bonfire of her manuscripts. That, she announced to them, was the end of Mari Sandoz as a writer. There was nothing left for her here in Lincoln; she was going back to the sandhills.

The sandhills were hardly welcoming, though; were, if anything, more depressing to her than her J Street apartment in Lincoln. If once she had looked on them as the locale of her book, now they mocked her as the recalcitrant scene of her humiliating failure. And how terrible they looked now! The whole country from Oklahoma to the Canadian border had become a giant drift of dust, and in November of that year with Sandoz back on the home place a dust storm swirled through and carried tons of Nebraska soil all the way to the east coast in an ironic harvest of Hardy W. Campbell's dry farming techniques.

She claimed she had left her literary aspirations behind her in Lincoln, that she had come back to the sandhills to find something else to do with her life. Yet her youngest sister, Caroline, recalled that Sandoz was not acting as though she had given up. "Oh," Caroline Sandoz Pifer said, "she talked a lot about giving up. She hadn't had much success, you know. But she must've brought about a dozen reams of paper out here with her, so it's pretty clear she wasn't really going to give up. You don't haul that much paper into the sandhills unless you think you're going to use it." Mary Sandoz watched in some wonderment as the grim prodigal went immediately to work, remodelling an old garage on the place. With hammer, nails, meat saw, and tar paper, Sandoz turned the building into a rough studio, its single window turned pointedly away from her mother's house, and there from dusk until dawn she worked at her writing. In the mornings, drawn with fatigue, she would go over for coffee with Mary. These seemingly casual visits bore unexpected fruit according to Caroline Sandoz Pifer because in that relaxed setting Mary Sandoz was willing to talk informally (and informatively) about Old Jules and the life she had lived with him. "In the past," the youngest sister said, "she [Mari] had always approached Mama the wrong way. You know, you don't approach these old country people and ask them to tell you this, tell you that. Now, if you want to take the time to *visit* with them, then you might find out what they have to tell you. But if you start in pumping, why, you'll never get anywhere. Anyway, it was during those morning visits that Marie got a whole bunch of valuable details for *Old Jules* as well as most of the information she used for *Slogum House*" (a novel published in 1937).

Through the rest of that fall and into early winter Sandoz worked simultaneously on *Old Jules,* using her precious new material to amplify and fill in gaps, and on her novel. Then when she was offered more work under New Deal relief legislation back at the State Historical Library in Lincoln she returned there. She had gotten what she had come back to the sandhills to get; now she went back to Lincoln in the belief she had everything she needed—or was ever going to get—to complete the book. In March of the following spring she thought *Old Jules* was in shape to try it again in the *Atlantic* contest. This time it won.

"Any landscape," writes Jacob Wasserman, "which somehow becomes part of our destiny generates a definite rhythm within us, an emotional rhythm and a rhythm of thought of which we usually remain unconscious and which is hence all the more decisive. It should be possible to recognize from the cadences of a writer's prose the landscape it covers as a fruit covers its kernel. . . . " This radical harmony between work and place was precisely what Whitman found lacking in the American literature of his time, quoting in sad concurrence in his "Poetry To-Day In America" essay the disparaging remarks of a British reviewer who said it was often impossible to tell from internal evidence whether an American poet was writing from the banks of the Hudson or the Thames. Reading *Old Jules* it is clear this book comes straight out of its place and is as native to it as the "nigger-wool" sod the first settlers turned up with their mold-board plows. This is a great, shaggy, rough-hewn book in which the reader hears the wearing whine of the wind, hears too the great lonesome distances, the silences of hot noons, the buried hush in the wake of blizzards. It is a long book, even in places a tedious one, and yet its length and occasional tediousness are felt to be wholly appropriate, as if in a shorter, more concise treatment something essential about the feel of the place and its life would have to have been left out. Here is a song, as Sandoz herself later wrote, of

> dun-colored sandhills crowding upon each other far into the horizon, wind singing in the red bunch grass or howling over the snow-whipped knobs of December, and the heat devils of July dancing over the hard land west of the hills. No Indian wars, few gun fights with bad men or wild animals—mostly it was just standing off the cold and scratching for grub. And lonesome! Dog owls, a few nesters in dugouts or soddies, dusty cow waddies loping over the hills, and time dragging at the heels—every day Monday.

Here, too, is a song of those who were trying to sing against a long historical background of failure and disappointment: the lost, dazed Coronado, searching forlornly for his shimmering cities; the American explorations of the early nineteenth century on which the subsequent desert legend of Nebraska depended; and then the settlers of the later years of that century, trekking ever westward in search of their green and promised paradise and with each mile of the state finding less and less chance of it. The panhandle? an old-timer asked incredulously when the young Jules said where he was headed. "Starve to death farming. Never rains, cold as blazes in winter. They brought a feller up from the lake country south of here last week. Both hands froze, fingers rotting off, crazy as a shitepoke."

In telling the story of this place and of the big, shambling man who moved ruthlessly through it, Sandoz withholds little. In spare, sharp-edged prose accented here and there with the angularity of second-language diction, she tells the story of Old Jules, beginning with his arrival at Mirage Flats, his first three marriages, his feuds with neighbors and ranchers, and his fourth marriage. She describes in unflinching detail her father's brutality to Mary and the children. She tells of the time she helped to save Old Jules from death by rattlesnake bite, of the blistering droughts of the 1890s, of the blizzard in which she was blinded. In the background, but fully there, are the comparable struggles of the other settlers, those few who hung on like Old Jules, and those many more who gave up and went back, or who got free rides to the asylum at Norfolk, or ended up crushed beneath the wheels of their wagons in drunken accidents. These people, she shows, broke the land and were themselves broken by it, so that no matter how mild their late days may have become the settlers carried always about them the stamp of great hardships undergone and endured: a certain ineffaceable gauntness of mien, work-corded hands and arms, sun-faded eyes that seemed forever scanning far horizons. As a prologue to the book she quotes the settler she calls "Big Andrew":

> One can go into a wild country and make it tame, but, like a coat and cap and mittens that he can never take off, he must always carry the look of the land as it was. He can drive the plough through the nigger-wool, make fields and roads go every way, build him a fine house and wear the stiff collar, and yet he will always look like the grass where the buffalo have eaten and smell of the new ground his feet have walked on.

Despite the incidents of violence she knew too well, the often mean and crude circumstances of settler panhandle life, there was something finally heroic about the whole phenomenon, Sandoz thought, something like a prairie sky that grandly transcended the incidents of the individual lives. Here, for good or ill, was the truth about America, how it was built and what that building had cost—vanished wildlife, the destruction of the tribes, the grief and madness of the unprepared pioneer women. Here were the real songs. And it was this deep and unconquerable impulse toward song, both as fact and as metaphor, that at last Sandoz most admired, the effort she had witnessed out here to find something to sing about. There were the earliest settlers on the Flats, sitting in young Jules's filthy dugout at night and consoling themselves with old airs recalled from home. There was that remarkably fortunate family that owned an organ and whose Sunday visitors gathered around it and sang all day until they had to board their wagons and creak home in darkness. There was the Christmas Old Jules ordered a phonograph and three hundred records, and the family stayed up all night listening. When the new telephone line came in there was a man who rang all the neighbors on it and yodeled for them. And there was at last her own impulse toward song, the silent, scared girl who became the thin, grimly-determined woman who shut herself away for days at a time in her J Street apartment, trying to make song out of silence and wind and grief. No one knew better than she the odds against song, how hard it was to lift your voice and sing, as Cather had written, "Out of the lips of silence/ Out of the earthy dusk."

II

NO ONE ENTERS A NEW LANDSCAPE without a full complement of mental baggage in tow, no matter how light he wishes to travel, and encountering a landscape made special to him by its literary associations he has his fond expectations and is glad when they are ratified by the look and feel of the land itself. Encouraged by Sandoz, I thought the Nebraska panhandle wore a somber, uncompromising beauty as I encountered it from northeastern Colorado. It was early July. The sky was white with heat and haze and clouds with a few ragged swatches of pale blue. Where the sun struck through to the wheat fields the effect was dazzling in contrast to the shadowed darkness surrounding. And that was all I saw for mile on mile—wheat fields nodding darkly under the clouds and wind. Then, way in the distance, the little towns—Gurley, Dalton, Angora—announced their existence with the white bulks of their grain elevators. They looked huddled and dusty as I drove through them: maybe a block of Main Street, two at most—Rexall, drygoods, hardware, supermarket with notices of meat specials plastered to the windows—and then on the nether side the route signs again. The towns grew smaller by the second in the rearview mirror, and then I was out once more into the fields.

Near Antioch I encountered the first of the alkali lakes Sandoz mentions in *Old Jules,* their borders reedy and abruptly marshy in the otherwise dry-looking fields. Antioch had a boom during World War I when it was discovered these lakes could supply the potash the Germans had cut off. When the war ended and the potash pipeline flowed again, Antioch shrank into an invisibility it had retained: only a few boarded-up buildings and, patiently waiting alongside the tracks, the broken and tilting remains of the potash factories. Like other phenomena of Great Plains life—gray, deserted farmhouses, the report of a murder or a suicide, a farm foreclosure auction, a solitary windmill—it is the setting that makes these things seem so stark, placed in such wide spaces that your attention is sharply compelled. This and the fact that there yet lingers in the national psyche a trace of the old pre-settlement image of a prairie idyll, some strange, unvisited amalgam of Virgil, Emerson, and Dwight Eisenhower. A rusted-through factory between Youngstown and Cleveland cannot occasion the regard tinged with awe that a gaunt, deserted farmhouse out here does, its spectral windmill whirring emptily.

The closer I got to Sandoz country proper the more the wheat fields gave way to choppy hills scantily covered with short red grass through which I could see sand patches. Old Jules had persistently maintained that the upper Niobrara country was prime farming land, and for him and his trees, so it proved. But unless there were farms placed well away from the highway, this looked like ranching country now. The old man had lived long enough to see it reverting to grazing land in his last years, and his daughter, listening at his deathbed in Alliance, quotes him as railing on in thin whispers about the desertions and betrayals by the farmers up there, how he would have to begin building all over again.

Caroline Sandoz Pifer was one of the old man's two surviving daughters, and she still lived in the heart of that country for which he had such grand hopes. I arrived at her snug home in the sandhills at evening, she having warned me earlier by telephone to eat dinner in Alliance. "I'm a rotten cook," she said cheerfully, "and I don't like to do it. You eat down there, then come on." Now she met me at the front door, a pleasant-faced woman in her late seventies. She offered me a beer, then said she had evening chores to do. "This is the nicest time of day out here," she told me over her shoulder as I followed her down the path to her garden. "The breeze comes up, and it isn't so hot." There was indeed a strong breeze out of the east, and the sun was beginning to die in the hills. I watched her moving among the slender slips of young fruit trees, her strong brown arms bare to the breeze and the mosquitoes, tending branches that were bitten and stripped by the hail that had raked the area two days before. It seemed obvious to ask whether her interest in horticulture could be traced back to her father. "Oh, I suppose so," she said casually, again over her shoulder. "And I guess it's the Swiss, too: always planting, planting. Once I went back to Switzerland, where Papa was from—Neuchâtel— and you could see it over there: everything, every inch of land, it seemed, was in cultivation. And—oh!—it was beautiful, I'll say that. More beautiful than here? Well, I don't know why anyone would want to leave such a place." Meadowlarks sang in the grass and the welcome evening breeze brought us the unmistakable odor of skunk, provoking Caroline Pifer to recall that she used to be asked all the time if Old Jules had really eaten skunk, as his enemies had alleged, and had used skunk grease for oil. (When Marie and Jule first went to school, this was one of the sharpest taunts the girls flung at her, telling her she stank like a skunk.)

"That used to make me laugh," Caroline said, "the fact that strangers would be so offended by the very idea. Well, if you know how to take that musk sac out of a skunk, there's no reason you shouldn't use the oil. But the idea that Papa and the family had to live on skunk. . . . " She left the thought unfinished since it seemed so clear to her that a hunter as good as her father would hardly have had to go after skunk.

When she had finished her garden chores we went up to the house and sat on its front porch, talking in the fading light, slapping occasionally at wind-borne mosquitoes. Caroline had put on a wide straw hat and a sweater, and when she turned toward me to answer some question the sun's last rays were prisms in the lenses of her glasses. But though she was certainly more than polite and candid in response to my queries, I had the feeling that what really engaged her attention in these last minutes of day was the old spectacle of sunset over the sandhills. She had watched it from here thousands of times, of course, and yet it seemed still to draw her, her face turned toward it in an attitude that suggested a sort of respectful, familiar attention.

At full dark we went in, and she showed me down to the cellar where she'd said I could spend the night. The cellar was, I quickly learned, a tribute to Mari (always spoken of as "Marie," though in the 1920s the writer had begun to call herself "Mari" on the advice of a numerologist who said it would bring luck). One room was a warehouse where Caroline stocked her sister's books, the high shelves ranked with copies of *Slogum House, Cheyenne Autumn, Crazy Horse, The Buffalo Hunters, Old Jules.* She said she was in effect the "regional distributor" of Marie's books, regularly stocking the region's drugstores, souvenir shops, and supermarkets. "They don't like to order from companies," she explained, "and so if I don't go out and bring them the books, they don't get on the racks. It's that simple." Another room was an informal memorial to the writer. There was her desk, her typing table, a cylinder holding pencils and pens she'd used. On the desk was an opened checkbook of hers. In a corner cabinet were some of her books, including a battered copy of *Betty Crocker's Picture Cook Book,* its flyleaf inscribed, "Mari Sandoz, 24 Barrow St., New York." Next to the cabinet was a big Blaupunkt radio Caroline said Sandoz had kept on constantly in her New York years. Its bands said it could pull in police and fire reports as well as Paris, Tokyo, Brazzaville. Caroline had thought I might like to sleep surrounded by Sandoz's effects, and I said I would. Yet when I was alone in the clammy darkness I found it vaguely unsettling. To me, the books on the other side of the wall were intensely alive, as good literature always is. But here I was surrounded by the effects of a dead writer—there was a difference.

Most of these mute things had been rescued by Caroline out of her sister's Greenwich Village apartment to which she had moved in 1943. Perhaps like her near neighbor there, Willa Cather, Sandoz had finally had enough of her home country and had felt she'd earned the right to the ease of distance. Perhaps, too, she felt that Nebraska had not appreciated her achievement as much as it should have, though by then honors had certainly come to her and would continue to do so through the remainder of her years. And then, with her honors, her steadily lengthening list of books, her comparative financial security, she had gotten cancer. Her youngest sister had talked at some length about this as we sat on the front porch, as if she were mulling over yet again the irony of it all. "At the end," Caroline had said, as much to herself as to her listener, "she was just in terrible pain. She was strapped down, and the nurse and I both had hold of her arms, and still she struggled so. It was all we could do to hold her in the bed. I kept asking the nurse, 'Isn't there something you can give her for the pain?!' But the nurse kept telling me, 'That's just the body's reaction. Marie herself doesn't feel it.' I wasn't so sure about that.

"In her last days Marie had wanted me to come to New York. She kept telling me it was urgent, said she wanted to talk to me about 'the book.' I didn't know what she meant, but of course I went. It turned out she wanted me to finish her autobiography for her. Well, I couldn't do that. How can someone else write your autobiography? Then it wouldn't be autobiography, would it? Anyway, I told her I didn't have her gift. I couldn't turn those phrases like she could. 'Don't worry about that,' she told me. 'After you work with my style awhile they won't be able to tell where I left off and you began.' She was still talking about this as long as she was conscious." Here Caroline had paused a long moment. "After they said she was clinically dead, whatever that means, her heart beat on another five minutes or so. It was awful."

I couldn't seem to get these mortal matters out of my head there amidst the daily materials of a life left behind, and some time near dawn I snapped on the bedside lamp and plugged in the old Blaupunkt, thinking that after all those years of neglect and isolation, those threadbare years in Lincoln where Sandoz had felt so cut off from both her panhandle roots and any really vitalizing contacts, this big radio might have seemed a species of magic, bringing her sound, song, the news of a wider world. There was no noise from it now, though, as if it had with Sandoz's life fulfilled its function.

At 5:00 a.m. there were sounds from above. Caroline was up and about, but I waited until I heard the latch at the top of the cellar stairs click open: she'd told she always locked the cellar door and that I'd have to wait for her to open it in the morning. I could hear her rattling about in the kitchen and behind that the hum of the TV. When I emerged there was sun on the hills, and the grass waved in a gentle early wind. It was close to the ghastly Fourth of July–Statue of Liberty extravaganza in New York, and on the screen there was a talk show personality and his toothy twaddle about "Lady Liberty."

If Caroline Pifer wasn't a good cook, she was a hearty one, and presently she served up a breakfast of sausage, fruit, cheese slices, and cereal, then wondered aloud at my meager appetite. Over the TV we talked of another Fourth of July, a hundred years ago, when 1500 of Red Cloud's Sioux had come into Chadron for the festivities and Old Jules had gone up into their camp to visit. "Goodness me!" Caroline exclaimed. "If 1500 Sioux were to show up in Chadron now, they'd scare every white clear out of the country!"

Then we were out into the morning on a tour of Old Jules country, I at the wheel of Caroline's pickup and the two of us jouncing down the three miles of sandy road to the highway that goes north to Gordon. We were on a portion of her ten-thousand-acre spread, and on either side of the road Herefords grazed the short grass. "I lease this part," she told me. "They know I like Herefords, so they graze them right alongside the road so I don't have to see those black ones [Angus]. They keep those up there [pointing to the hills northward]. I don't know whether that's too high a price to keep me happy. I guess they figure it isn't." She laughed shortly. On the highway to Gordon we passed hail-flattened fields on both sides of the road. As in the days of Old Jules, this section of the panhandle gets more hail storms—and more destructive ones—than anywhere else in the country, and now Caroline was shaking her head and clucking as she looked at the ruins. She muttered something about a rancher who "might not be able to make it another year with his hay crop destroyed like this." Farther on she exclaimed, "Oh. God! You wouldn't believe the way this field looked just the other day! Why, it was fat and ripe. Now just look at it!'' I wondered aloud whether Old Jules had been wrong after all about the upper Niobrara, that it was simply too harsh to be consistently productive farming country.

"Well, I don't think he was wrong," Caroline Pifer said slowly, "but it might take another hundred years to prove him right. The hail storms are just terrible, as you see. They've talked about seeding the clouds to get them to pour rain, you know, instead of this destructive hail. But then you only have a very few minutes to get out and do that. These storms hardly give you any warning. They say over in Chadron those folk'll have to redo all the insides of their houses: the wind and hail blew in all the west-facing windows and just covered the eastward walls with leaves and such.''

At Gordon long ago there used to be a sign that told the traveler how far it was to Old Jules's. Now there was a weathered green and white one that said this was the home of Mari Sandoz. That represented a triumph, I thought, for the silent, scared little girl over the bruising, egocentric father, though still like a stalking psychic shadow his name would be linked with hers. What I said was that it had been a wonderfully brave thing Marie had done in riding off to take and pass the teacher's exam in 1913. Looking straight ahead, Caroline smiled, then turned and looked at me across the cab. "Yes, it was. But do you know, she was still so much under Papa's thumb that she wouldn't dare to buy herself a new dress for the exam? No sir.'' Whether or not this was an invitation to ask more directly about family matters, I took it as such and asked Caroline about her own feelings for her father. Had she loved him, I wondered?

"Oh, you couldn't help but admire him," she said, looking at the road again. "He was so smart. He knew everything. But I don't think I loved him. My mother once told me I didn't love anything or anybody except myself. Maybe it was true.'' She gave another short laugh. "Did he hit me? Well, he was a cripple, you know, and when he told me to 'Come here,' I wasn't fool enough to do it. He had a bad temper, always grumpy about something. I learned pretty early to stay out of his way. We all did. Except Jule. He was just so stubborn. Finally Papa ordered him off the place, and Jule was gone two years. And when he came back, Papa screamed that Jule had cheated him out of two years of work!''

I asked then if her mother had loved Old Jules. "You have to understand," she began patiently, "that Mama came from a refined family. And then to get hooked up with someone like Papa!'' She paused, and a mile or two rolled by. "It . . . was . . . a . . . *jolt*. He never washed, and you know, out here it doesn't take too long to get pretty grimy. Oh, she had a lot of reasons to resent him. For one thing, he concealed his divorce from her. Then he pretended to be better off than he really was. The day after he married her she found out what he expected of her: he told her to massage his foot, and, of course, since he never washed, his feet smelled. And she could never get him to stop spitting on floors."

Still talking, we turned south onto a dirt road leading to the site of Old Jules's original dugout on Mirage Flats. On either side were fields of wheat, rye, and oats. Startled redwing blackbirds, killdeer, and meadowlarks scattered up before us, while behind a long funnel of dust drifted upward into the hot, bright air. "Stop here," Caroline said as we came to the crest of a long, gradual rise. "The Swiss always seemed to build against a slope if they could, and we figure this is the spot where he had his first place, the dugout." There was not a trace of any habitation on the grassy slope at the summit of which were three squat aluminum storage bins glinting in the sun. At the foot of the slope a marker was set in the grass, and the property owner had obligingly mowed around it so that you could read what it said. It was surmounted by a likeness of Old Jules in his fur cap, and beneath, the words,

SITE OF WELL ACCIDENT

'OLD JULES'

PAGE 100

ERECTED BY

BLANCHE

MRS. FRITZ SANDOZ

Upset by our intrusion, the redwing blackbirds sailed and jeered, just as they must have a little more than a century ago when the lone white man had sunk his spade into the earth here, turned it up for his inspection, and then begun to dream beyond the present that was then nothing but wind on the buffalo grass.

There was nothing but a marker, either, at what Caroline referred to as the "river place," the farm on the Niobrara under Indian Hill where much of the action of *Old Jules* takes place. All six children had been born there, but I supposed Caroline had been out so many times with tours and stray visitors that there was no longer the least tug when she leaned

over the barbed-wire fencing and pointed down to the marker, telling me
as she did so that some years ago an archeologist had been able to identify the arrangement of the farm's various buildings. Down among the grass
stems and gray, sandy soil I could see a few bits of sun-blackened tin and
some twinkles of glass, but that was all. Turning around from where we
now stood we could see to the west the three aluminum storage bins on
the dugout's site: not that far, really, but in the context Mari Sandoz created
in her book Old Jules's move to the river had about it the air of an ancestral
procession over the unsung land, each stop a sort of sacred site. Whether
or not it had ever seemed so to the daughter and sister who stood beside
me in the absolute flood of the panhandle's midsummer sun, I couldn't
have said. At the very least she had made it her business to learn her family's
Songlines and by repeatedly traversing them keep them current. Two grazing saddle horses, interested in human company but leery of contact, crop-
ped and snuffled the grass beside us as we stood there, Caroline talking
of the river below, of Crazy Horse, who "Marie was convinced had been
on this place."

We went down the grass-grown road to the river and its traditional
tribal ford and along through a grove of old cottonwoods to a much
dilapidated ruin of a house that Caroline said was part of the original river
place in which they had all been born. Subsequent owners had transported
part of the old house down to this site. One good stiff wind would bring
down what was still upright, as though the house had harbored such a
weight of life's struggle and unhappiness that at last it could bear it no
longer and had begun to sag and settle into the patient and forgiving dust.
One wall was still upright, and on its gray slabs were nailed rungs leading
to an attic window. I chose to regard these as the very steps Marie Sandoz
had mounted to her shared room, concealing beneath her baggy blouse
some contraband Conrad or Hardy. Beyond the ruin lay the twisted, gnarled
remains of Old Jules's orchard and beyond this the slope of Indian Hill
atop which the old man had rested his crippled foot after his hunts, the
secretly attentive daughter listening there to his stories of Switzerland and
the Sioux. We climbed the hill, slipping occasionally in the chalky, sandy
soil. On the summit Caroline pointed out the sites associated with *Old
Jules* and Crazy Horse, shading her eyes with her hand even under her
straw hat, the dragon's breath wind whipping her dress against her legs.
"Right below where we're standing," she said, "is the sand spot Papa used
to use for target practice. He'd sight guns in using this spot, and he could

tell from the shape of the bullet when he dug it out whether the gun was shooting high or low."

Caroline thought I ought to see the old Swiss burial ground several miles east of Indian Hill but warned me the going would be rough because in places the road had almost disappeared through disuse. She was right. After crossing Pine Creek we plowed steadily upward through the high grasses with the roadbed becoming fainter by the mile. The pickup pitched and swayed as it climbed, and ten feet in front of its hood ornament I could see nothing but grass and sky. I had to keep telling myself that I could not be in better hands. Caroline kept guiding me through the treacherous spots, telling me when I must suddenly switch off the "main" channel and onto a yet fainter one in order to avoid high-centering the truck. Under sullen, oranged tinged thunderheads we crested the long rise and came down the other side to a junction of two sections. "There!" shouted Caroline suddenly, "Straight ahead! Give 'er the gas!" I had been lazily imagining that I now knew how to negotiate these unfeatured upland stretches when there yawned before us a quaggy gumbo ready to mire the truck to its axle. But spurred by Caroline's shout, I gunned the truck, and it foundered through the trough and out the other side.

At the burial ground Caroline took me down the little rows, pointing out the graves. Here lay Mary's mother, Grandmother Fehr, whose bleak, cancerous death Mari Sandoz had described in *Old Jules.* "This is Uncle Emil's, the one murdered by the cattlemen. You can see from here, too—right over there—where they shot him. This was a suicide. This one was killed by a runaway team." The short, harsh litany was accompanied by meadowlarks and the long, rushing whisper of the grasses. I thought of Masters and his *Spoon River Anthology.*

Caroline Pifer was quiet through much of our return to her place. Perhaps the tour had tired her, or perhaps she was tired of the tour. But on the twelve-mile cut through to Route 27 she spoke up again, remarking that "Papa located him . . . and him . . . and him," referring to the sections of settlers of the long ago. I could see no sign of habitation and said so. She grunted and said that was because the settlers had almost all moved away, leaving the land to those who grazed cattle on it. "This used to be full of settlers," she averred with a short, emphatic shake of her head. "When I show this country to tourists they think it just looks all the same. And, really, there isn't anything here." She did not give me so much as a half-glance, only that voice gesture that had in it something like a challenge: what did *I* make of it, she seemed to want to know?

"Well, that depends," I said slowly. "That depends on what you can see in any place. Here I have you to tell me what I'm seeing, and I have Marie's books to tell me. But your guide can only take you so far, and the rest depends on you. I see what's obvious to see here: fencing, cows, grass, sky. And I see what you help me to see. And I also see all that history that Marie gave me in her books. But what I finally make of all this—that remains to be seen."

She looked sharply at me then, her eyes clear, young behind the glasses, and neither friendly nor hostile but simply appraising. In that instant I saw her as a Sandoz; I saw Mari, and beyond, I saw the looming, limping shadow of the old man himself. The look had a long, tough history to it. In part it said to me, "Well if you can make something of it, go ahead. But you'll have to show us."

Over a lunch of fried ham steak, baked potatoes, and green beans back at Caroline's, I wanted to ask about her sister's working methods in general and about the composition of *Old Jules* in particular. During her Lincoln years Sandoz had worked hard to make herself into a professional historian, taking several university courses in historical methodology and historiography and developing her own techniques for historical research. Yet when it came to writing her books of history—*Old Jules, The Buffalo Hunters, The Beaver Men, Crazy Horse, Cheyenne Autumn*— she characteristically eschewed scholarly trappings and employed fictional techniques, including the use of dialogue, and most of this was surely invented. Except when attacked on the score of strict historical accuracy, Sandoz seems never to have been much bothered by the question of genre, often referring to her biographies as novels and her novels as essentially history. When she was attacked, as she was about the accuracy of *Old Jules,* she was quick to cite the chapter and verse of her researches, even sending some doubters copies of historical documents she said supported her literary constructions. Everything in *Old Jules,* she was at particular pains to point out, was in strictest accord with the truth, and even its conversations she claimed were based on reports of them from witnesses or participants. How accurate, I now asked Caroline, did she find *Old Jules?*

"It's the truth as far as I'm concerned," she said immediately, "and it was the truth as far as Marie was concerned. On the other hand, I suppose if each of the six children had written the book, you'd have six different *Old Jules*es. I know there are some things I saw differently from Marie, but that's what you have to expect, isn't it? The man Marie described—he was gone by the time I came along. He was really an old man by that time. But I'll tell you one thing: that book was true enough that when it came out Mama began it but never finished it. She said it just brought back too many painful memories."

"Well," I persisted, "what about some of the specific anecdotes, such as the time Marie says Old Jules brought your mother into Chadron, then forgot about her until the end of the day?"

"It's true. That's the way it was then. I've seen wives in town just waiting and waiting for their husbands to come back from wherever they'd gone and pick them up. And that's the way he [Old Jules] was. He never knew how many children we were, never knew our ages. That was the way he was, and you just got used to it." She seemed willing that I go on in this vein, and for the rest of our lunch I did, asking her about other specifics of the book. When she could confirm them, she said so; when she could not, she admitted it. Finally, I asked about her father's death-bed request that Marie write his story, observing that it seemed so much out of character given his long-held loathing of writers and artists.

"Well," Caroline said carefully, "she [Marie] was the only one ever heard that. I'm not telling you Papa didn't say it..But we only have her word that he did say it. And, of course," here glancing up at me briefly over the rims of her glasses, "it would have been very convenient for her to have had that . . . go-ahead."

Caroline had told me there might possibly be a Sandoz family get-together on the Fourth, but it didn't materialize and I found I had the holiday to myself. Since I had earlier remarked to her that it would be a hundred years since Red Cloud had brought his Sioux into Chadron and her father had gone visiting among his old friends, it seemed fitting somehow for me to spend the day on the Pine Ridge Reservation just over the Nebraska–South Dakota border. A big Sioux I had met the day before in Gordon had told me there was to be a powwow near Batesland on the Fourth, and so early on this day I headed north, passing through the still silent town of Gordon, the flags along its main street hanging limply in wait for the oven of heat to come as the sun mounted (it would reach 102 degrees in Gordon by midafternoon).

East of Batesland, South Dakota, the blacktop took a swing north-eastward and right where it did so a dirt road led east. This, I had been told, was the road to the powwow, and following it I soon overtook other cars wallowing through the soft dirt. Immediately we were all enveloped in a huge billow of smoke-colored dust and traveled the rest of the way within it. When we arrived at the parking area by a small marsh we could hear the crackle of the loudspeaker from a ball field where a round-robin softball tournament was underway.

Behind us the sound of heavy drumming came from a pine-bough arbor. The arbor was arranged in a circle in the midst of which the American flag snapped straight east in the high wind. Under the arbor a small huddle of young men in ball caps and sunglasses sang and drummed with what seemed to me a real enthusiasm for their traditions, and it was soothing to sit in the needle-spangled shade, heavy with the resinous odors of pine, and have this profoundly American sound all about. Far to the east was the hectic show of patriotism in New York's harbor. The immigrants who had come through that portal had been assimilated, taken in; the natives around me had not. Old Jules himself had become an American despite his often lawless, violent ways, his publicly pro-German pronouncements during World War I, but his friends on the reservation and their descendants had remained as curiously alien as they had always been, unassimilated and, except when in trouble, forgotten. Yet here again, over the new marvels of mikes and wires, were authentic Songlines laid down of old by the ancestors, and it was good, even necessary, to hear them and so be reminded of other, deeper layers of history.

Around the edges of all this there were, of course, other things if you wanted to look. Buses and vans selling poisonous-looking junk food were invitingly snugged up to the arbor. Yesterday's *Sheridan County Star* had carried a listing of the DWI convictions of Two Crow, Charging Thunder, Red Wing, Sun Bear, Big Boy, Warrior, etc., etc., but as yet there was little evidence of that hopeless drunkenness I had seen on this and other reservations in years past. Instead there was a serenity that sat under the arbor, and it was symbolized by the row of ancient, shriveled women, blankets over their legs even in this heat, parked in wheelchairs, their brown, big-veined hands quiet in their laps, listening attentively to the young singers/drummers.

Not many miles west of the powwow grounds lay Wounded Knee, and to complete my Fourth of July I went over there, following the sinking sun along the reservation's dusty, empty roads. After the massacre of December 29, 1890, a blizzard had poured in, and when Jules Sandoz arrived at the scene a couple of days later he found the grotesquely frozen bodies of the dead Sioux still strewn about, half-covered with snow and blackened blood. Fully two miles from the spot of the Indians' original encampment he came across the bodies of the women and children who had tried to escape on foot and had been ridden down and executed by the troopers. The scene so shocked and sickened him that he could never again bear to hear of the "American tradition of the frontier." The only tradition of the frontier, he would say, was blood and plunder.

I had not been there since before the troubles of Wounded Knee II and found it much changed. The little settlement had disintegrated into a jumble of twisted, burned-out ruins, and the Catholic church that sat overlooking the site of the massacre had also been burned to its foundation. The one new thing was the large sign in place at the approximate site of Chief Big Foot's encampment. One side took the reader up to the morning of the event. Then he had to go around to the other side to find out how it had come out, knowing how it had come out, how it had to come out.

Up at what had once been the church was the mass grave of the Sioux dead, flung into a narrow ditch after the whites had first stripped them of their Ghost Dance souvenirs. The trench had a weedy, uncared-for look, and at its entrance a few faded bits of bunting tied to posts flapped and fluttered like orphans in the wind. A scrawled red spray-paint graffito on a piece of busted concrete read "Fuck A.I.M."

My final stop in Sandoz country was Old Jules's last homestead, the one in the sandhills to which he had sent Marie and James in 1910 and on which Mari Sandoz was buried. Flora Sandoz, whom Marie had been allowed to name, lived there now, in her eightieth year. She couldn't see me in the morning, she told me over the phone, because she had haying to do. "There's lots of determinants when you're cutting and stacking," she'd told me. "Stack it too wet, and it molds. Stack it too dry, and it won't stack." But now she was there in her living room, her big round table and bookshelves stacked and overflowing with Mari Sandoz materials, and I asked her, too, about the truth(s) of *Old Jules*.

"It's the truth, all right," she said definitely. "And one reason it is, is because she [Marie] was the one most like him." She gazed out the living-room window, the hot, hard, sandhills light reflecting on her face and changing its weathered tan to ash blond. "She thought like he did. That's why she could understand him well enough to write about him.

"Oh, there's some fiction mixed in there. When I was a child going to school history was just names and dates, and so naturally I hated it. There weren't any, oh, *associations* with those names and dates. They were all you were given, and you were supposed to memorize them. Well, for the class of people like myself, you have to have those things to associate with the names and dates. You've got to embellish it some. And to me, that's what Marie did. If you don't have all the facts, then you've got to add what you think was true, what you think must have happened."

And the portrait of Old Jules: was it essentially an accurate one?

"Well, he'd changed a good deal, I suppose, by the time I can remember him. He was an old man by the time I was born, you see. But was he mean? Well, I guess he was. Marie got blamed a lot for what the younger kids did. If they did something wrong, Marie got the whipping. One time I set the house on fire, but it was Marie that got the whipping. The two things I got beaten for most were crying when I saw Fritz getting whipped, and for gossiping, repeating things that I'd heard somewhere else. Dad was death on gossip, probably because he'd been hurt by it so much himself."

I asked her then what difference it had made to the family, to the community, to her own life that Sandoz had written *Old Jules*. "Well, of course, I'm glad she wrote it," Flora Sandoz said. "It's a good story, and it would have been lost if Marie hadn't done it. I supported her all along in her desire to be a writer, and I'm proud of what she accomplished, as you can see. But, what if she hadn't written *Old Jules?*" She looked mildly surprised, having taken the book so long for granted as a solid fact of her life, like the waiting hayfields beyond the house. "Oh, if you mean how has that book changed this community—why, I don't think it's changed it at all. Now, if Dad hadn't located all those people—and some of their descendants are living here today—that might have made a difference." She paused, looking out over Old Jules's fields, now hers. "I guess I've never really thought about how life might be different than it is. Oh, there's people here now who hate the Sandozes' guts because of what Marie put in that book. I know a man here—I won't say his name. When that book came out he'd read a few pages, then throw it across the room. Mad, you know. Then he'd pick it up again and start reading. And he did this all the way through, because there were things in there he thought oughtn't to have been printed."

Behind Flora Sandoz's house lay the graceful curve of Old Jules's last orchard and above it on a long hill the grave of Mari Sandoz. The orchard was untended now, though its rows were still intact. The branches of the unpruned trees interlaced to form a blanket of shade over the walker, and the ground underneath was rank with cactus, poison ivy, and other woody wild growths. Deer tracks abounded. Old Jules lay down in Alliance where

Marie buried him, perhaps a sort of permanent exile. But she was here, as she meant to be. "I should like to be buried," she had written her family in her last days, "on the Old Jules place in the Sand Hills where anyone can come and sit awhile, as the old Sioux used to come to the river place to throw their minds back, as they said."

If the pilgrim comes up from the orchard to the grave at sunset, he will see reflected in the granite stone's simple face the outline of a long, undulating sandhill to the south. The stone has a high finish, but the hill is far enough away so that the reflection is vague, making the land look still untouched as it was when old Jules first looked on it, unsung as it was until Mari Sandoz wrote. And if the pilgrim were to wait there beside the stone until darkness, falling slowly, effaced the last faint trace of the hill's reflection, leaving nothing but a name and a date, then indeed the mind would be thrown back and caused to muse on matters of mortality. And this, it seems—beyond the strife and sadness, the uncompromised spareness of the plains song she sang—was what Mari Sandoz had aimed for all along.

SEVEN

PLACE SPIRITS: WILLIAM FAULKNER'S <u>ABSALOM, ABSALOM!</u> *and* "THE BEAR"

I

AS SUMMER 1927 BEGAN TO drape its heavy folds over the Mississippi landscape, young Bill Faulkner made another of his apparently abrupt and aimless decisions: he would go south to spend the season on the Gulf at Pascagoula where his friend and mentor Phil Stone had a family camp and where Faulkner had made a number of other friends in summers past. Faulkner had been home in Oxford for several months now, knocking about in his characteristically shiftless way, seen at odd hours walking the roads around town, working occasionally at the local golf course where he dispensed soft drinks and snacks and would occasionally volunteer to someone who engaged him in conversation that he was in truth a genius. Few outside the family knew that the scruffy-looking little man, now nearly thirty, arose at first light each morning and worked steadily till noon on the manuscript of his latest novel. If the townspeople knew anything about him, they knew he had started, then stopped in about a dozen different directions and that his current "career" as a self-styled writer would likely come to little enough. Oh, he had published an obscure book of poems and two novels, it was true, and he also had a clutch of newspaper and magazine articles he had done in New Orleans. But he had said it himself in his second novel, *Mosquitoes,* when he had one of his characters remark quizzically of a strange little man who professed to be a writer, "'Faulkner? Never heard of him.'" More than likely Faulkner himself would soon tire of the current masquerade and drift on to another one. Already he had dropped out of high school, worked briefly as a bank clerk, been rejected by the U.S. Army as too small for service in the World War, had some sort of stint with the RAF in Canada, bummed around the state, tried school again at Ole Miss and failed again; he had worked in a New York City bookstore, then returned to Oxford, had led a local scout troop until dismissed because of his drinking, had gone to New Orleans and then Europe. Most recently he had been living at home in a state of arrested adolescence. Now he was going back down to Pascagoula to resume his shadowy and feckless existence as a beachcomber.

But the man himself knew Pascagoula was a lucky spot for him: he could work well there, away from the measuring glances of Oxford. The previous September he had finished *Mosquitoes* there, having worked hard on it through the summer. Now he felt he could do the same with a novel he was calling *Flags in the Dust*. He was right. On September 29, four days after his thirtieth birthday, he wrote the final pages and sent off an exultant wire to a relative in Memphis. A few days later he returned to Oxford and in a sustained exultant mood wrote his publisher in New York, claiming that this was "THE book, of which those other things were but foals." Then while he awaited what he confidently expected would be acceptance and high praise from Horace Liveright, he made one of the party that annually went to another Stone family camp, this one old General Stone's hunting and fishing camp thirty miles west of Oxford on the edge of the Delta. The bears that had once roamed the thick woods and somber swamps along the Tallahatchie were gone now, the wilderness rapidly contracting as land sales, timbering, and agricultural development squeezed its life out. There were stil deer though, and coon and squirrel, and there was still the seasonal ritual of the hunt with its veteran hunters, its dogs, guns, whiskey, the sharp, pungent odors of wet leather and wool, the foods fried too fast at too high a temperature. Faulker enjoyed the ritual dimensions of it all more than the actual hunting itself.

As it happened, Liveright did not like the new novel and frankly told an amazed and outraged Faulkner so. It was, he said, a lesser novel than Faulkner's first, *Soldier's Pay,* and lesser even than his second, *Mosquitoes,* which Liveright had not thought highly of. Faulkner's reaction was the more severe because his apprentice's road to the writing of *Flags* had been so long and circuitous and because in its pages he had for the first time written of his home ground of northern Mississippi. "I was shocked," Faulkner recalled a couple of years after. His first emotion, he said, was "blind protest," quickly changed to a strange, hard-edged objectivity, "like a parent who is told that its child is a thief or an idiot or a leper; for a dreadful moment I contemplated it with consternation and despair, then like the parent I hid my own eyes in the fury of denial." Following this and a period of numbed stupefaction during which he seemed incapable of any work, he plunged into extensive revisions and sent the manuscript off again to his New York agent. Ten rejections later Harcourt, Brace and Company agreed to publish a version of it under the title *Sartoris.*

The novel's beginnings can be traced to Faulkner's first New Orleans visit in the fall of 1924 and his impluse while there to call on Elizabeth Prall, for whom he had worked at a New York bookstore three years before. Now Elizabeth Prall was married to Sherwood Anderson, one of the voices of that bitter protest that had been an unexpected harvest of America's relentless westering. In 1924 Anderson was one of the most famous writers in America, at the acme of his critical and popular fame (though soon enough to slide down its other side). Hamilton Basso, himself then a fledgling writer in New Orleans, called him "Our Royal Personage" and said that all the young men clustered around Anderson there "owed him much." All of them, Basso recalled, "were young enough to profit by example, and Anderson's example, leaving aside the example of the dedicated artist, was basically that of benevolence. What he had, he shared. What was his to give, he gave—his time, his patience, his attention—the hospitality of his house." Many years later, easy at long last in his own literary fame, Faulkner himself looked back at New Orleans and Sherwood Anderson and recalled their leisurely times together, their afternoon walks and talks, the sharing of a bottle, the easy comraderie of those days. "I decided," he told an interviewer, "that if that was the life of a writer, then becoming a writer was the thing for me. So I began to write my first book. At once I found that writing was fun. I even forgot that I hadn't seen Mr. Anderson for three weeks until he walked in my door . . . and said, 'What's wrong? Are you mad at me?' I told him I was writing a book. He said, 'My God,' and walked out. When I finished the book—it was 'Soldier's Pay'—I met Mrs. Anderson on the street. She asked how the book was going, and I said I'd finished it. She said, 'Sherwood says that he will make a trade with you. If he doesn't have to read your manuscript he will tell his publisher to accept it.' I said, 'Done.'"

As Anderson was to learn it was dangerous to literally believe anything his younger friend said about himself. Even at the obscure beginning of his career Faulkner found personal disclosure distasteful and so was liable to say almost anything when asked about himself. When drinking his self-mythologizing was apt to become utterly unbridled. But we can at least believe this much of the Anderson/Faulkner anecdote: that the former by his example and precept was an inspiration for the latter and that his help was decisive in the publication of Faulkner's first novel.

Anderson's example was, first, as Hamilton Basso said, that of a dedicated artist who put in his hours at the desk and turned out work rather than mere talk about work. And there was too the specific example of his book, *Winesburg, Ohio* (1919), in which he deftly laid bare the secret lives and desperate hopes of residents of a small, undistinguished heartland town, much as had Edgar Lee Masters a few years before in *Spoon River Anthology*. Anderson (born in 1876) had missed the World War, and perhaps that accounted in part for the fact that at the war's end he wrote not of that immense tragedy, of the wreckage of young lives, the sudden impact of the Old World on a yet provincial America, but instead about the life of a landlocked town in the middle of the country. But more than any mere accident of his age, Anderson's decision to train his talents on his hometown of Clyde, Ohio, represented a triumph of artistic insight and courage. It was easy enough to say, as many were then saying, that an artist uses what he is given. Faulkner had said it himself in a book review he had written in 1922. But still the great question remained: how to know what it was in fact that you had been given by the evidently modest circumstances of your birthright. Somehow Anderson had for himself answered that question, and this above all was what he had to give Faulkner. In their afternoon strolls through the faded exoticism of what by then had become known as the French Quarter, Anderson would talk and Faulkner would listen. Anderson was a great talker and storyteller and Faulkner was still young enough to attend to what was being so freely given. One day, as Faulkner was to remember it, Anderson said to him, "You're a country boy; all you know is that little patch up there in Mississippi where you started from. But that's all right too. It's America too. . . ." If these were not the precise words, the import was the same, and this is what stayed with Faulkner even if at the time the imputation of his own unrelieved rusticity may have rankled; even if he was some years from being capable of acting on the advice.

Before he could turn his own talents homeward to northern Mississippi he had to write about his generation's great experience, the war, though, he too, in a way, had missed it, having not gotten out of flight training in Canada before the Armistice. So the book he was at work on when Anderson allegedly barged in to see him was a war novel, published with Anderson's strong support as *Soldier's Pay* in 1926. In it a doomed pilot comes home to his Georgia town to die, accompanied by a soldier he has met on the train and a war widow. It is an often self-conscious work, and there is in it more than a trace of its author's personal regret at having missed the big show. But in its depiction of war's psychic dislocations *Soldier's Pay* bears comparison with works written about the same time by Faulkner's soon-to-be distinguished contemporaries: Dos Passos's *Three Soldiers* (1921); e.e. cummings's *The Enormous Room* (1922); Hemingway's *A Farewell to Arms* (1929). And though there are few hints here of the kind of writer Faulkner would become, there is like a radiant lodestone this description of the dying pilot's hometown square:

> Charlestown, like numberless other towns throughout the south had been built around a circle of tethered horses and mules. In the middle of the square was the courthouse—a simple utilitarian edifice of brick and sixteen beautiful Ionic columns stained with generations of casual tobacco. Elms surrounded the courthouse and beneath these trees, on scarred and carved wood benches and chairs the city fathers, progenitors of solid laws and solid citizens who believed in Tom Watson and feared only God and drouth, in black string ties or the faded brushed gray and bronze meaningless medals of the Confederate States of America, no longer having to make any pretense toward labor, slept or whittled away the long drowsy days while their juniors of all ages, not yet old enough to frankly slumber in public, played checkers or chewed tobacco and talked. A lawyer, a drug clerk, and two nondescripts tossed iron discs back and forth between two holes in the ground. And above all brooded early April sweetly pregnant with noon.

There are even fewer portents in *Mosquitoes,* though for close students of Faulkner's biography there are intriguing glimpses into the life he lived in New Orleans in the mid-1920s. The framework of the novel comes from a hapless, drunken boating excursion Faulkner was party to on Lake Pontchartrain in 1925, and on this somewhat flimsy construction the writer hangs a lot of airy talk about Art and Life and Death. Horace Liveright was right in feeling that it marked no progression from Faulkner's first novel, and Hamilton Basso regarded it as Faulkner's "one truly negligible book."

Regardless of its modest intrinsic merits, *Mosquitoes* had at least one invaluable personal feature for its author in that it showed him that he was not yet being personal enough in his fiction. In showing him this the book served as a muted but insistent warning, a flashing red light in the night, that his talent might be dissipated in the creation of bright and hectic books that drew from no deep wells. True, he had here, as he had not previously, drawn on personal experience. But the experience lacked the heft, port, sheer displacement that could only derive from a particular kind of personalism where the individual life is subsumed by place and history, the weight of things undergone. Faulkner was still bemused by the topical and had yet to muse on what he knew in his blood, fiber, and bone, that little patch, as Anderson had called it, from which he had come. And just as Sarah Orne Jewett had once warned Willa Cather that unless she could come to her own unassailable territory and write out of that, so Sherwood Anderson warned Faulkner that he was too talented and that his perculiar, inmost gift might be squandered in Roman candle displays that would leave no least trace. "You've got too much talent," he once told Faulkner. "You can do it too easy, in too many different ways. If you're not careful, you'll never write anything."

Perhaps, without ever having seen it himself, Anderson understood the difficulty his protégé would have in laying claim to his little patch of Mississippi. For he had seen Clyde and Fremont and West Lodi, little northern Ohio towns no one had thought of writing about before he had claimed Clyde as his "Winesburg." Certainly Basso early sensed something of that potential difficulty whenever he and Faulkner happened to talk about their home places. Basso was himself a Southerner, but he was from cosmopolitan New Orleans with its rich, thickly-layered history and its own steadily accumulating literary tradition that included Cable, Charles Gayarré, Kate Chopin, Grace King, and a slew of contemporary writers like Lyle Saxon. "My South," he recalled, ". . . was not Faulkner's. Mine was Mediterranean, Catholic, and, in such families as my own, still essentially European. Faulkner's South was much less diluted, *sui generis,* Anglo-Saxon, Protestant, and, as it were, more land-locked, turned inward upon itself." Such a South was not only largely alien to the New Orleans man; it also seemed so intensely rustic that it might be unavailable for art. Years later Faulkner was to make use of this felt distinction between his place and New Orleans in his greatest novel, *Absalom, Absalom!,* when he created the tragic half-brothers, the cosmopolite from New Orleans, Charles Bon, and the rough-hewn Mississippian, Henry Sutpen. It was not, Hamilton Basso said, until he had himself moved away from New Orleans and had seen something of that inward-looking, insular South that he understood precisely the nature of Faulkner's challenge, what it was that Faulkner had been trying to get at.

As 1927 began perhaps Faulkner did not quite know what it was he was trying to get at. But a mighty and volatile pitchblende had been brewed within him: his New Orleans experiences—talks with Anderson and Basso and the others struggling to make art; his increasingly professional, knowing journalism; the writing of his stylishly sad and curiously outdated book of poems, *The Marble Faun;* and his two novels. And behind and beneath all this were his vivid memories of Mississippi, of Lafayette County and its seat, Oxford, of family lines and ties, of the rich humus of history and tradition that was his by birthright—the teeming world of his childhood and youth that at thirty he was apparently about to leave behind forever. While *Mosquitoes* went through the press in the winter of 1927, he mused on the loss of youth, on time's wingèd chariot, and on art. He mused also on that Mississippi world he had left to come to what Basso called Paris on the Mississippi. Now he felt upon him the urgent and immediate necessity of recreating "between the covers of a book the world . . . I was already preparing to lose and regret . . . and desiring, if not the capture of that world and the feeling of it as you'd preserve a kernel or a leaf to indicate the lost forest, at least to keep the evocative skeleton of the desicated [sic] leaf." He began to write on two projects at the same time, tentatively sketching in the lives of two Mississippi families, the Snopeses and the Sartorises. It was in the very act of writing about them that Faulkner came to understand fully and forever that his own memories could provide him with all the imaginative nutrients he would ever need because these same memories were really roots that connected him to the spirit of his place— and not only his place but to some significant extent a whole region, the South, as well. His diffidence about the autobiographical had been misplaced and had to some extent prevented him from writing out of his own background. Now he realized that to make his writing "truly evocative it must be personal," not in the narrowly autobiographical sense but in the wider, more authentic sense of birthright. "So," he recalled two years after he had begun on the family sketches,

> I got some people, some I invented, others I created out of tales I learned of nigger cooks and stable boys of all ages between one-armed Joby, 18, who taught me to write my name in red ink on the linen duster he wore for some reason we have both forgotten, to old Louvinia who remarked when the stars 'fell' and who called my grandfather and my father by their Christian names until she died, in the

long drowsy afternoons. Created, I say, because they were composed
partly from what they were in actual life and partly from what they
should have been and were not: thus I improved on God, who, dramatic though He be, has no sense, no feeling for, theatre.

In this passage in which Faulkner recalls the moment of his inspiration it is significant that his memories of his childhood and youth inevitably
lead him back to the elders of that stage of his life, those who surrounded
him, peopled his world, and who provided him with an introduction to it.
More particularly, his memories are not only memories of them as people,
they are memories of *their* memories as well, for it is their stories, their
recollected experiences of all that had transpired in that place that really
formed the child's understanding of the place. And as for the elders themselves, they too in their turns had been receivers of the memories of the
elders of their own childhoods, and so on back through the dimming yet
still vibrant chronicle of the generations. Here that very isolation and in-
wardness of Faulkner's place acted as a conservator of history and tradition—which is a version of history told and passed along. For you can
only be surrounded by the elders if you live where they have lived, and
they can only tell you the old stories if things have remained stable so
that the stories make living sense. Born and resident in his place, Faulkner
by that fact became one in a line of listeners and narrators, hearing the
stories since before he could remember, hearing them so often he could
not but remember them. A later Mississippi writer, Eudora Welty, would
one day try to explain to an outsider this special nature of Southern stories
that grow up in a place. We in the South, she told him, "have grown up
being narrators. We have lived in a place—that's the word, Place—where
storytelling is a way of life. When we were children we listened a lot. We
heard stories told by relatives and friends. A great many were family tales.
We naturally absorbed not only the fairy tales, but the sheer nature of telling them." She continued,

> it is true that a lot of history has happened in many of our front yards,
> but also, the person was there to see it happen, watching and being
> part of it over long periods of time, passing it carefully to the next
> generation, telling the stories which always grew, never got smaller.
>   We've known a community of life, as I say, through the years. If
> we weren't around when something happened, way back, at least we
> think we know what it was like, simply because we've heard it so long.
> We tend to understand what's tragic or comic, or both, because we

know the whole story and have been a part of The Place. Our concept of Place isn't just history or philosophy; it's a sensory thing of sights and smells and seasons and earth and water and sky as well.

For reasons he never explained Faulkner put aside for the time being the sketch of the Snopes family and concentrated his efforts on the Sartorises. If in the novel he finished in Pascagoula in the summer of 1927 he did not get the full spirit of his place between the covers of that book, he made a richly promising beginning and one he could build on. The war was still much on his mind, and in the mind and behavior of young Bayard Sartoris the war's deaths remain so current they finally reach through into the present and claim his own life. But the business of the war did not finally distract Faulkner from what he had learned was his purpose. For in these pages he staked his claim to his place, creating for the first time his county, Yoknapatawpha, its small seat of Jefferson, and some of those characters whose fortunes he would follow for more than three decades. Here emerged a world, intact, solid, complex with its landmarks, seasons, smells, dialects, and customs. If he was not its sole creator—since it was founded in the soil of a real place—he was its presiding and designing genius, and it was his to work with as he would. With the writing of *Soldier's Pay,* he said many years later,

> I found out writing was fun. But I found out afterward that not only each book had to have a design but the whole output or sum of an artist's work had to have a design. With "Soldier's Pay" and "Mosquitoes" I wrote for the sake of writing because it was fun. Beginning with "Sartoris" I discovered that my own little postage stamp of native soil was worth writing about and that I would never live long enough to exhaust it, and that by sublimating the actual into the apocryphal I would have complete liberty to use whatever talent I might have to its absolute top.

The great meditations on and evocations of the place spirits of his world are *Absalom, Absalom!* (1936) and "The Bear" (1942). *Absalom, Absalom!* has its immediate genesis in a story Faulkner wrote in the summer of 1931, "Evangeline." In it the writer/narrator gets a wire from a friend announcing the discovery of a ghost in an old Mississippi house once the home of a Colonel Sutpen. On subsequent investigation of the house the writer finds the ghost real enough.

Like Hawthorne before him Faulkner had a significant interest in ghosts. Part of that interest was playful, and when he moved with his bride into their ante-bellum Oxford home in 1930, he invented one to tenant the old place, as if the total design of Southern living required a resident haunt. And yet like Hawthorne ghosts were more than playthings to him, more even than literary devices. They were emblematic of the true nature of time, which was not linear, as commonly supposed, but rather a kind of suspended continuum in which nothing that had ever existed corporeally or even in the mind could ever truly die, in which there could be no such thing as *was,* only *is,* on and on as long as there would be an earth and people on it. Everything that once was in some form still is, preserved somewhere for use yet again. The earth, the land itself, he had come to feel, must be the great conservator, repository of all artifacts, of the bones and trinkets, and even the dreams and deeds of the ancestors, for the long, tangled, tragic, and sometimes wildly comic chronicle of the ages could not have been invented for nothing and then thrown into cosmic discard. The spirits of the departed were substantive.

As a son of Southern soil Faulkner was the more aware of the spirits of the Old Ones, those who had walked in this place in other times. In New Orleans they had said the white Creoles were like the Chinese: they ate rice and worshipped the ancestors. And if all Southerners were not rice-eaters, there was a shared veneration of past generations that caused their spirits to walk, to rustle in old houses, old families, to intrude in the shapes of repeated anecdotes, even casual allusions. As in "Evangeline" the spirits were there in old houses, too. Faulkner knew of a number of those sagging, settling ruins in and adjacent to Lafayette County, the seats once of prosperous planter families before the Civil War, and in his New Orleans days he would have had ample opportunity to see some of the magnificent ruins of plantation manors in that area, such as Seven Oaks at Westwego, which Cable had used as the model for his "Belles Demoiselles Plantation." Faulkner felt the spirits in these houses, and he saw in the houses themselves the vainglorious dreams of the elders, dreams which like the houses themselves continued their spectral existence into the present day. Here were gaunt reminders of an old order for which young blood had been shed, on which lives had been staked and lost.

One such house with which he was especially familiar was the Shipp place located south of the Oxford airfield and beyond the settlement of Water Valley. In 1833 Felix Shipp and his family had come here and bought Chickasaw lands, and subsequently Shipp had raised an imposing cypress-wood mansion framed by a long avenue of cedars. But the war destroyed the family fortunes, and by the time young Bill Faulkner knew the Shipp grandsons in Oxford the plantation had become a desolate scene of ruined fields and tumbledown outbuildings presided over by the grim old mansion. Evidently it was this place he had in mind as he wrote "Evangeline."

Faulkner had no luck with "Evangeline": the story was rejected by both the *Saturday Evening Post* and *Woman's Home Companion.* He had a better time with the same story told from a different angle when two years later *Harper's* bought a story he called "Wash" after a poor white, Wash Jones, who is sucked along in the wake of Colonel Sutpen's fortunes. But Faulkner too was haunted now by the spirits he'd raised in the old colonel and his ruined house, and the next year he returned yet again to this story, this time trying to lash together "Evangeline" and "Wash" into a novel he called *A Dark House.* "It is," he wrote his agent, "the more or less violent breakup of a household or family from about 1860 to about 1910. It is not as heavy as it sounds. The story is an anecdote which occurred during and right after the civil war; the climax is another anecdote which happened about 1910 and which explains the story. Roughly, the theme is a man who outraged the land, and the land then turned and destroyed the man's family. Quentin Compson, of the Sound & Fury, tells it, or ties it together. . . ."

The letter suggests the bent of the writer's thinking: that the story of the colonel and his failed dreams was fundamentally an ancient one, relived here in the Civil War South, relived yet again in the narration of Quentin Compson. There was something Greek about Sutpen's story as Faulkner created it now, and something older than Greek, older even than the ancient Israelites who had told the story in the books of *Samuel* and in *I Kings* as the story of David and his rebellious, fatal son, Absalom. At the home he called Rowan Oak he worked on through the spring of 1935 with a mounting sense of excitement as the ramifications of his narrative continued to unfold beneath his furiously scrawling pen.

Ever since he began on "Evangeline" Faulkner had had the essentials of his story, and he kept them as he worked through to his shattering conclusion. Thomas Sutpen appears out of nowhere one Sunday in 1833, in Jefferson's town square, carrying on his spattered and spent horse the only belongings he apparently owns. Then while the little town watches in amazement, the man acquires title to a hundred square miles of Chickasaw land, directs the building on it of a vast plantation house, and sets his fields into production. With an inscrutable deliberateness Sutpen lays siege to the daughter of a Jefferson merchant distinguished only by that single quality which Sutpen must acquire in the furtherance of what he calls his "design": respectability. He carries the siege, marries the daughter, and gets by her the son to carry on his name and the daughter to contribute her progeny to the Sutpen dynasty.

In the normal course of matters the son, Henry, goes off to college at Oxford, Mississippi, where he meets a handsome, older man from New Orleans, Charles Bon. Both Henry and Charles are unaware that the latter is Sutpen's son, too, by a Haitian woman whom Sutpen had deserted years earlier after learning that she had a trace of black blood. But the men are drawn to each other by an affinity neither can explain, and when Christmas comes Henry brings Charles Bon to Sutpen's Hundred for the holidays. Charles there begins a curiously languid suit of Henry's (and his own) sister, Judith. But the following Christmas, when all assume the courtship will come to a formal betrothal, Sutpen calls Henry to him and tells him Charles is his half-brother and cannot marry Judith. Henry renounces his father and patrimony, and without telling Judith why, the two young men ride off "side by side through the iron dark of that Christmas morning. . . ."

They go to New Orleans where Bon introduces Henry to the lush, half-hidden complexities of local custom, including, finally, a visit to one of those battened and shuttered cottages on the edge of the Vieux Carré in which resides Bon's octaroon *placée* and his infant son. Now the Civil War breaks out, and the brothers enlist in the army as, in another place, does Sutpen. Through the next four years of battle, bloodshed, and steady defeat Henry wrestles with demons not of his own making but now surely his: his love of his half-brother; his abhorrence of the potential incest Bon's and Judith's marriage would constitute; and the fact that Bon is already a husband of sorts and a father.

Meanwhile Bon himself bides his time, awaiting patiently some least sign from Sutpen—some glance, note, even a signature—that will acknowledge the blood tie, in return for which Bon would gladly renounce all his claims to Judith. It never comes, and at war's end it is Henry instead who is summoned before the colonel, his father, and told the final, trump truth: that Bon must not marry Judith because his veins carry a trace of his mother's impure blood. Henry has struggled to forgive all the rest, even the potential incest, but this cannot be forgiven, and so, the war over, the two men ride with a steady fatality up to the gates of the Sutpen mansion where Bon refuses to relinquish Judith. Henry shoots and kills Bon, then disappears.

Thus when he returns to Sutpen's Hundred, what the colonel comes back to is his wife dead, his unacknowledged son dead, too, and his son and heir vanished. His daughter is a widow without ever having been a bride, his domain is in tatters, his grand design a failure. His furious efforts to recoup are doomed, and he dies as a paunchy crossroads storekeeper, cut down at the hand of his long-time worshipper, Wash Jones, who at last sees through to the utter ruthlessness of this man who would sacrifice all, even himself, to accomplish a vain and inhumane goal.

Through all the permutations the novel took to its final form, Faulkner kept the essential story he had begun with in "Evangeline" because he recognized how good a story, *qua* story, it was. But in the final form of the novel he sacrificed both narrative pace and clarity because he, too, like Sutpen, had a grand design and would give up much to accomplish it. What he wanted was to tell this story in such a way that it would become evident it was much more than the story of a single man's vanity and overweening ambition. It was also and more significantly the story of a whole regional culture, the American South. And in its distinct Biblical echoes and overtones Faulkner was also asking readers to see in it something as ancient and recurrent as the story of human culture itself, of great opportunities missed, of folly, and heroism misplaced, and of the tragic consequences of both the folly and the heroism.

To achieve such a dimension Faulkner tells the story through a group of narrators, each of whom has his or her own surmises, blindnesses, and vested points of view, and all but one of whom is a tribal member of that culture haunted by its spirits. At the outset the narrator is Rosa Coldfield, sister to Ellen, Sutpen's Mississippi wife, and whom in the ruinous wake of the war Sutpen had outraged by suggesting they breed together as a test, and if the issue was a male heir, they would marry. For forty-three years Rosa Coldfield has lived on her implacable hatred of the man she calls "the demon," his shade so haunting her that it has warped her life to its own shape, so that she, too, has become a ghost. Now in September 1909, she suddenly chooses to tell her version of the Sutpen saga to Quentin Compson, just then on the verge of his departure for college at Harvard. Telling the story to him, hour upon hour in her hot, airless house, she conjures up for the youth the old demon with a "faint sulphur-reek still in hair clothes beard," astride his galloping stallion, wrenching his dream's domain out of the apparently passive and supine Southern earth.

The young man listening is her captive audience. But he is also like her a captive of his place, "the deep South dead since 1865 and peopled with garrulous outraged baffled ghosts." He listens to the old woman out of a Southern male's deference and politeness, but he listens also because as a Southerner he must listen to the tales of the spirits that come down the generations, teller to teller in an unbroken succession, a custom, a birthright, a curse. "Quentin," Faulkner writes, "had grown up with that; the mere names were interchangeable and almost myriad. His childhood was full of them; his very body was an empty hall echoing with sonorous defeated names; he was not a being, an entity, he was a commonwealth." So the question he poses to his father after Rosa has let him go home for awhile is in some sense rhetorical, though it is felt, too: "But why tell me about it? What is it to me that the land of the earth or whatever it was got tired of him at last and turned and destroyed him? What if it did destroy her family too?" The two of them are waiting for full dark on the Compson porch, at which point Miss Rosa will take the secretly resistant Quentin with her out to the dark ruin of the Sutpen mansion. She is convinced there is something out there, something hiding in that house, and she means to find it out.

"Ah," is what Mr. Compson answers him in mild but sure reproof, seeing in his son's impatience a thoughtless repudiation of his birthright. For Mr. Compson, too, is a teller of tales, and this one he has gotten from his father who was perhaps Sutpen's single confidant. So he proceeds to tell his son what he knows from his father, General Compson, and what further evidence has accumulated through the intervening years, adding in his own surmises where there is not so much as rumor. When he is finished it is full dark and time for Quentin to keep his appointment with Rosa Coldfield, to find at the dark house what Faulkner does not at this point tell.

Nor is this the end of the telling. In Chapter VI Quentin is seen telling the story again, this time to his Canadian roommate at Harvard, Shreve McCannon. McCannon is the only non-Southerner in the novel to hear the tale, and his difficulties in understanding the place spirits who stalk through it and who are evoked repeatedly in its successive retellings are instructive. *Tell about the South,* he says to Quentin in one of the novel's many italicized passages. *What's it like there.* Hours later, as Quentin has been trying to explain the South to Shreve through the Sutpen story, the Canadian interrupts him. "Listen," he says.

> I'm not trying to be funny, smart. I just want to understand it if I can
> and I don't know how to say it better. Because it's something my
> people haven't got. Or if we have got it, it all happened long ago across

the water and so now there aint anything to look at every day to remind us of it. We dont live among defeated grandfathers and freed
slaves (or have I got it backward and was it your folks that are free
and the niggers that lost?) and bullets in the dining room table and
such, to be always reminding us to never forget. What is it? something
you live and breathe in like air? a kind of vacuum filled with wraithlike
and indomitable anger and pride and glory at and in happenings that
occurred and ceased fifty years ago? a kind of entailed birthright father
and son and father and son of never forgiving General Sherman, so
that forevermore as long as your children's children produce children
you wont be anything but a descendant of a long line of colonels killed
in Pickett's charge at Manassas?

"Gettysburg," comes Quentin's curt correction. "You cant understand
it. You would have to be born there." But through Quentin's extended,
strangely incantatory telling Shreve does begin to understand something
of the South; the spirits visit him also, and so he begins to take his part
in the telling so that here once again Quentin must listen. *Maybe,* Quentin
thinks under Shreve's voice, *nothing ever happens once and is finished.
Maybe happen is never once but like ripples maybe on the water after
the the pebble sinks, the ripples moving on, spreading.* As Shreve warms
to their joint work, Quentin thinks, *Am I going to have to have to hear
it all again . . . I am going to have to hear it all over again I am already
hearing it all over again I am listening to it all over again I shall never
have to listen to anything else but this again forever so apparently not
only a man never outlives his father but not even his friends and acquain-
tances do. . . .*

Midnight comes, and the night rolls toward light, and still the young
men talk on in the freezing room until out of their reconstructions and
their renditions of all the earlier retellings something of the meaning of
the Sutpen story emerges, though Faulkner is careful not to say it is The
meaning.

It is the story of a man raised in the impoverished, intensely isolated
democracy of the West Virginia mountains who learns through shocking
and bitter experience in the tidewater lowlands that the South, his world,
is a hierarchical culture based on racial and class distinctions. Since this
is so, Sutpen reasons, it is better to be the one who is in the position to
say to others "Go here" or "Fetch me that" rather than to be an under-
ling, a hind, a Wash Jones who has to do the going and the fetching. The
boy who finds out the true nature of his world becomes the young man

determined to rise out of his circumstances so that he can justify himself to himself and so that he can justify himself to all his unknown, nameless ancestors who in their days did the going and fetching. For this he would need money to buy land and money to buy the slaves to work it and to raise his castle in the midst of his fields. Then he would need a wife and children to perpetuate his name in the land.

Sutpen achieves his spectacular rise, though none ever knew quite how. The story of that rise and of the fall that followed, told and retold in the intergenerational ritual of the South, now in a Harvard dormitory room yields up a dark and tragic truth: no order founded so substantially on the distinction of skin color can ultimately produce good, just as a thorn can never produce a myrtle. Evil, says the Bible, must beget evil, as King David's sinful congress with Bathsheba brought down upon him the Lord's curse: "Behold, I will raise up evil against thee out of thine own house"— the rebellious Absalom. When its details are shifted and pondered in the successive retellings the thread of race is seen running faintly but surely through the Sutpen story, just as it is there in the history humans have made in the American South. It was Sutpen's youthful rejection at the hands of a haughty, big house slave that sparked his consuming desire to own the big house and the slave, too. Again, in Haiti, where the very earth cried out in drumbeats against the crime of slavery there, it is Sutpen's discovery of his wife's tainted blood that impells him to forsake her and their son. But the dragon's teeth sowed there later spring up in Mississippi in the handsome form of the castaway son. Sutpen's final attempt to forbid the marriage of Charles and Judith is his disclosure to Henry that Charles is partly black, and it is this that Henry cannot forgive as he contemplates the union of his brother and sister—as Sutpen had correctly foreseen. When Charles flings this truth in Henry's face, the latter cannot answer, for there is no answer to it. But he does say, *You are my brother.* And Charles flings the dark truth again in Henry's face: *No I'm not. I'm the nigger that's going to sleep with your sister. Unless you stop me, Henry.* At the gates to the Sutpen mansion Henry does just that.

But that, of course, is not the ending, cannot be, since in Faulkner's view nothing is ever just ended. The Sutpen story does not end with the death of Charles Bon; not even with that of Sutpen himself, as Quentin's and Shreve's retelling forty-one years after that fact makes clear. The place spirits still haunt the South, the town of Jefferson, and all its residents. They linger too out at Sutpen's Hundred, and Rosa Coldfield is feeling them that hot September afternoon when she makes Quentin agree to accompany her that night out to the ruined old house. At the end of their long night of telling, Quentin supplies Shreve with the details of that night-time intrusion into the ghost-stalked past.

When he and Rosa Coldfield drove in the surrey out to Sutpen's Hundred what they found were the ragged, twisted ends of the vanished colonel's grand design: the skeleton of the huge house, and within it its ghostlike tenants: Clytie, half-caste daughter of Sutpen's casual union with a slave; Jim Bond, idiot grandson of the murdered Bon; and, in an upstairs bedroom, the wasted form of the fratricide, Henry, who had come home to the demon's castle to die. When subsequently Miss Rosa sent an ambulance out to the house to carry Henry to a doctor, Clytie mistook it for a police wagon and fired the house, so that the last physical vestiges of Sutpen's design vanished as smoke curling, then drifting off into the thin air.

When Quentin finishes the telling of this final episode Shreve says to him: "'Now I want you to tell me just one thing more. Why do you hate the South?'" and Quentin replies, too quickly, "'I don't hate it.

> *I dont hate it* he thought panting in the cold air, the iron New England dark; *I dont. I dont! I dont hate it! I dont hate it!*

II

❧ THERE ARE SIGNIFICANT parallels between *Absalom, Absalom!* and the novella Faulkner published six years later in *Go Down, Moses,* for "The Bear" is a continuation and extension of Faulkner's concerns in the earlier work. Once more the writer conjures up the spirits of his native place, summoning them to show themselves, to show how they shadow the present and so cause it to recapitulate the past. And once again Faulkner shows the reader a young man coming of age in Yoknapatawpha County who must in the process come to terms with his place spirits. Like Henry Sutpen and perhaps Quentin Compson, too, Isaac (Ike) McCaslin finds certain of these spirits so repulsive and morally contaminating that he attempts a repudiation of his inheritance and his heritage, hoping to exonerate himself from the sins of the fathers. Once again the specific, immediate sin is chattel slavery and its avatars, and because of it Isaac comes to feel that his place and all of the South lie under a curse. Faulkner again employs Biblical references and overtones to suggest this, working here with the Abraham/ Isaac story as earlier he had worked with that of David/Absalom.

For "The Bear" he drew on his memories of the Stone family hunting camp in the Big Bottom of the Tallahatchie, thirty miles west of Oxford on the edge of the Delta. About 1915 Faulkner had begun to accompany the hunters out there and to hear the tales they told around the fire after a day in the woods. Many of the tales were regional descendants of those the circuit-riding lawyers of the old Southwest had heard a century before, that Twain had heard in his days on the river at the end of the 1850s. The men said that in the days of creation, when America was still an unscarred wilderness, the trees of the Big Bottom had been so monstrous that two hundred-pound bears were apt to climb into woodpecker holes when the dogs treed them. All the game was huge and in astonishing abundance then, and there was one particularly grand bear known as Reel Foot whose left front paw had been mutilated by a trap and who left his distinctive track through a wide range of impenetrable bottomland for nearly a quarter century. No dogs had been able to bring this hoary monarch to bay, and maybe, the hunters suggested, old Reel Foot was still at large in what remained of his kingdom. Here Faulkner also heard tales of dogs famed for their courage, or their stamina, or their noses, or distinctive voices. He may even have seen a great, deep-chested mongrel, thirty inches tall at the shoulder and with a voice like a hunting horn. When he came to compose the first version of "The Bear" he would start with this memory of the great dog and call him "Lion."

All of this talking and the big woods themselves would make a good story, he knew at the time, and during those autumns out on the Tallahatchie he began to store up the experiences, putting them away, as he once said, like lumber in the attic for future use. A man who hunted with Faulkner in those years said he never understood much of what Faulkner wrote, but he understood a great deal about the writer's sources because some of those sources were in the talk of the Big Bottom hunting camp. Later, in a book of his own, John B. Cullen was able to identify many of the originals from which Faulkner had drawn his situations and characters.

By the time Faulkner got around to using this particular stack of "lumber" the woods on the Tallahatchie were mostly memory. As early as 1916 General Stone, Phil's father, had begun selling off timber tracts, and by 1935 the old Stone camp was part of something called the Okatoba Hunting and Fishing Club, and William Faulkner was one of the incorporators, hoping thus, as he said, to preserve some remnant of the old wilderness there. As it turned out, he was to preserve it with his art, beginning with the story "Lion," which he published that same year. What apparently started him was the sad, dramatic contrast between the Big Bottom he had known in young manhood and that pitifully little patch of woods now circumscribed by the drained and leveled fields of cotton and soybeans. The dominant note of "Lion" is threnody: the passing of the wilderness. In it a boy witnesses the hunt for a legendary bear called Old Ben and is forever changed by the killing of the great beast.

As with *Absalom, Absalom!* Faulkner realized from the first that he had here a deeply resonant story, and through all the subsequent versions of the work that became the centerpiece of *Go Down, Moses* he kept it. The longer he thought about it, the more he could see there was more to it than a great hunting story, more even than that classic of the national culture, the male initiation story. "The Bear" is both of these, of course. It is a great hunting story, and as an initiation narrative it takes its rightful place with Hawthorne's "My Kinsman, Major Molineux," Melville's *Typee*, *Huckleberry Finn, The Red Badge of Courage,* Hemingway's "Big Two-Hearted River," Steinbeck's *The Red Pony,* and Bellow's *The Adventures of Augie March.* "The Bear" is more even than a symbolic story of the American South, though it is this, too. In the five-section form Faulkner gave it in *Go Down, Moses,* "The Bear" is a meditation on the history of the New World, from its divine creation, through its discovery, to its spoliation at the hands of men. Here in the speculations of Isaac McCaslin, Faulkner takes the voice and the part of the American earth itself, telling of the savage disregard of its guardian spirits and of the still-unfolding consequences of this.

From the outset, though the setting could not be more particularized, Faulkner's language places the hunt for Old Ben in a very spacious frame. This, he writes, was the story of

> the wilderness, the big woods, bigger and older than any recorded document—of white man fatuous enough to believe he had bought any fragment of it, of Indian ruthless enough to pretend that any fragment of it had been his to convey; bigger than Major de Spain and the scrap he pretended to, knowing better; older than old Thomas Sutpen of whom Major de Spain had had it and who knew better; older even than old Ikkemotubbe, the Chickasaw chief, of whom old Sutpen had had it and who knew better in his turn. It was of the men, not white nor black nor red, but men, hunters, with the will and hardihood to endure and the humility and skill to survive, and the dogs and the

> bear and deer juxtaposed and reliefed against it, ordered and compelled
> by and within the wilderness in the ancient and unremitting contest
> according to the ancient and immitigable rules which voided all regrets
> and brooked no quarter---. . . .

Faulkner's choice of the bear as the embodiment of this wilderness
is a master stroke that at once comprehends and binds together the wilderness of northern Mississippi with that of the New World as a whole. And
beyond this, the New World becomes one with the entirety of the earth
that in the Biblical myth man was granted as God's steward: for everywhere
bears have been, there men have venerated them as standing in some
special relationship between the unseen powers on the one hand and the
world of humankind on the other. In "The Bear" Old Ben is one of the
spirits the boy, Ike, must confront in the process of discovering and coming into his birthright. Old Ben, Faulkner writes, was

> not even a mortal beast but an anachronism indomitable and invin-
> cible out of an old, dead time, a phantom, epitome and apotheosis of
> the old, wild life which the little puny humans swarmed and hacked
> at in a fury of abhorrence and fear, like pygmies about the ankles of
> a drowsing elephant;—the old bear, solitary, indomitable, and alone;
> widowered, childless, and absolved of mortality—old Priam reft of
> his old wife and outlived all his sons.

As Sam Fathers, the half-black, half Chickasaw master of the hunt, tells
Ike, Old Ben is the "head bear. He's the man."

It is Sam who instructs the boy in the lore of the woods, how to
track, how to shoot, how to find your way again when you have lost it.
Sam also instructs him in the deeper lore of the woods: that attitude of
reverence and humility that was all the fee the Creator asked of man in
the original paradise. Finally, through Sam Ike learns that the hunt for Old
Ben is in truth the hunt for that very spirit of wilderness that was all of
the New World. Learning these things, Ike knows he can never kill Old
Ben even if he should miraculously have the chance. Thus, on the day
he sets out to track the bear it is not with any desire to kill him but rather
simply to behold him and to confront that wildness, naked and unarmed,
as perhaps the first humans did here. Without a gun and bearing but a
watch and compass (two crucial tools white civilization used in conquering the New World) the boy sets out for the bear, but he is unsuccessful
until he discards everything alien to the woods and accepts this world

on its own terms. "He stood for a moment," Faulkner writes, "—a child, alien and lost in the green and soaring gloom of the markless wilderness. Then he relinquished completely to it. It was the watch and compass. He was still tainted. He removed the linked chain of the one and the looped thong of the other from his overalls and hung them on a bush . . . and entered it." Now he is vouchsafed a vision of the bear:

> It did not emerge, appear: it was just there, immobile, fixed in the green and windless noon's hot dappling, not as big as he had dreamed it but as big as he had expected, bigger, dimensionless against the dappled obscurity, looking at him. Then it moved. It crossed the glade without haste, walking for an instant into the sun's full glare and out of it, and stopped again and looked back at him across one shoulder. Then it was gone. It didn't walk into the woods. It faded, sank back into the wilderness without motion as he had watched a fish, a huge old bass, sink back into the dark depths of its pool and vanish without even any movement of its fins.

Later, both Ike and Sam are armed when they surprise Old Ben, but neither pulls the trigger because neither wishes to. Yet both know that soon enough someone will kill the "head bear," and with the coming of Lion, the huge mongrel dog, both the old man and his apprentice know that the hunt is over:

> It seemed to [Ike] that there was a fatality in it. It seemed to him that something, he didn't know what, was beginning; had already begun. It was like the last act on a set stage. It was the beginning of the end of something, he didn't know what except that he would not grieve. He would be humble and proud that he had been found worthy to be a part of it too or even just to see it too.

The premonition is accurate. In 1883 when Ike is sixteen the men with their guns, horses, mules, dogs, and the fierce Lion to lead the pack finally run the bear to earth. Near the river they bring him to bay, and Lion drives straight in on the bear who rises on his hind legs and clasps Lion to him with a loverlike embrace. Both creatures go down, and then Boon Hogganbeck, one of the hunters, races forward and, leaping astride Old Ben, finds his heart with a knife. The bear, Faulkner writes, "surged erect, raising with it the man and the dog too, and turned and still carrying the man and the dog it took two or three steps toward the woods on its hind feet as a man would have walked and crashed down.''

It was over then, the guardian spirit of this wilderness a lifeless trophy and the doom of the Big Bottom sealed and struck. Shortly thereafter Sam Fathers dies, and within a year Major de Spain will lease the timber rights of his camp to a Memphis lumber company.

The same year Old Ben is killed Ike embarks on quite another phase of his rite of passage. Here Faulkner turns his attention from the wilderness to what has replaced it, the agarian civilization of the South, and causes Ike to encounter the tangled, tragic truths of that civilization as these are chronicled in the ledger books of the McCaslin family farm. In the commissary of the farm Ike is to learn that his inheritance is blighted by the sin of slavery's racial distinctions: if Sutpen is the brooding spirit of Sutpen's Hundred, so old Carothers McCaslin, Ike's grandfather, is the equally forbidding spirit of the McCaslin family farm. Like Sutpen, Carothers has a design in mind for these lands—productive fields, slaves to work them, big house, family—though evidently he is less singleminded than Sutpen since he never finishes the building of his castle and in general leaves more matters to chance. Ike grows up with the spirit of the departed ancestor hovering over everything: there in the shape of the huge, unfinished house; there in the conversations of Carothers's twin sons, Uncle Buddy and Buck, Ike's father; there in the productive fields that will one day belong to Ike. Occasionally in the commissary Ike would idly glance up at the row of old ledger books on the shelf above the counter, intending to take them down and read in them some day, knowing that old Carothers would be in them, too, and knowing that in the accounts of the farm, kept down the years, he would learn much of his family's history. At sixteen he does read them, the first entries in Carothers's hand and recording how he had come into the country from Carolina in 1787, purchased land from the Chickasaw chieftan, Ikkemotubbe, purchased slaves, and set his fields into production.

Gradually the entries become those made by the twins, Buddy and Buck, who assume the farm's management in the 1830s, and it is the sight of his own father's semi-literate scrawl on the yellowing pages that quickens the boy's interest. The brothers drift into the habit of communicating with each other through the ledger books, and it is this habit that allows Ike to piece together the history of the farm and the family. That history is for him one of fundamental racial injustice, continued and compounded through the long years of the black laborers' ceaseless toil and tragically meagre recompense, an injustice which not even the Civil War and still less Reconstruction have been able to rectify or ameliorate. The more Ike reads—the entries drawing nearer to his own time, then entering it—the more he comes to see that he is not reading history as it is given in school textbooks—something past, fixed, and having nothing to do with him. Instead, he is reading the book of his people, and so he is reading about himself, the more particularly so here since the farm will be his when he reaches his majority. And there is also for him to read here "not only the general and condoned injustice and its slow amortization but the specific tragedy which had not been condoned and could never be amortized. . . ." Old Carothers, Ike learns in these pages, once fathered a child by his slave Eunice. Then, twenty-two years after, that child, Tomasina, becomes pregnant, and Eunice learns the father is Tomasina's own father, the master of the lands, old Carothers. In her grief and horror Eunice drowns herself in the creek on Christmas day. Subsequently Tomasina dies in childbirth. In 1837, when Carothers dies, his will cynically provides a thousand dollars for his son by Tomasina, Tomey's Terrel. *So I reckon,* Ike thinks, *that was cheaper than saying My son to a nigger. . . . Even if My son wasn't but just two words.* Like Thomas Sutpen, Carothers McCaslin cannot bring himself to say these two words to a mixed-blood son, and so the crime goes apparently undetected and unpunished. Except that now it falls to the grandson and heir to acknowledge it, not only for what it specifically is, but also as a symbol of the corruption at the heart of the whole system, that design of which Carothers McCaslin and Sutpen, their slaves and heirs, are parts.

Thus on his twenty-first birthday Isaac, refusing to be a sacrifice on this foul altar (and so troubled by his legacy that he has tried to make restitutions to living descendants of old Carothers's slaves) confronts his kinsman, Cass Edmonds, and repudiates his inheritance. He has himself by this time come to write in the books with a penmanship that curiously resembles that of old Carothers, and now he wishes to write himself out of them, to put a stop to this history. Appropriately the confrontation takes place in the commissary, the place where the black laborers must come for the goods necessary to work the white men's land, and where the ledger books reside on their shelf as witnesses. In a lengthy and in places heated discussion Ike and Cass range over the history of this place; its legal and moral foundations; the history of those who raised its crops in the sweat of their brows; Reconstruction; and the character and destiny of the black race. The longer the discussion goes on, the wider and deeper its implications until finally the events of the novella's first three sections, the hunt for the bear, are included. For his experiences in the woods have taught Ike that the farm now legally his cannot morally be his because ownership of land or the people on it constitutes a fundamental violation of the divine plan for the human habitation of the earth. "'It was never mine to repudiate,'" Ike tells Cass.

> It was never Father's and Uncle Buddy's to bequeath me to repudiate, because it was never Grandfather's to bequeath them to bequeath me to repudiate, because it was never old Ikkemotubbe's to sell to Grandfather for bequeathment and repudiation. Because it was never Ikkemotubbe's fathers' fathers' to bequeath Ikkemotubbe to sell to Grandfather or any man because on the instant when Ikkemotubbe discovered, realized, that he could sell it for money, on that instant it ceased ever to have been his forever, father to father to father, and the man who bought it bought nothing.

The spiritual logic of his Big Bottom experiences now leads him beyond the matter of old Carothers's patch of Mississippi earth, back to the origins of the New World, and beyond even these to the Biblical Genesis. Like water ripples, which Quentin Compson in *Absalom, Absalom!* uses as a metaphor for the continuing effect of any event, so the McCaslin farm now becomes in Ike's mind merely the widest evident ripple out from the original mischance that was the Europeans' discovery of the New World. Having failed the first, aboriginal opportunity in the Garden of Eden, man had struggled on, had fought, carved up, and enslaved much of the Old World until God, having seen the opportunities there utterly exhausted, had granted humans yet another chance in the New World. Here was a chance to begin anew, to serve as worthy stewards of God's creation, holding the earth and all its bounty in commonwealth. Instead of which men instantly repeated on these shores the self-same mistakes already ancient in the known world: fighting, carving out national and personal domains, enslaving. Nowhere, thinks Ike, was the sorry procession more obvious and more painful to contemplate than here in the American South, a place blessed with rich soil, a mild, beneficent climate, abundant game—another Eden ripe for the hands of the newcomers. Yet here again they fell into the old ways, corrupting all—land, the aboriginals who had lived innocently enough in it, and those future generations who would in time inherit the full consequences of their misdeeds—and this despite the efforts of the American founding fathers to create a new order based on the brotherhood of men.

As with Quentin Compson in *Absalom, Absalom!* Faulkner does not say that Ike's argument contains the entire truth. Indeed, in the character of Cass, Faulkner has created a skillful and intelligent opponent. Cass's defense of the South and of their grandfather, Carothers, sometimes has the effect of making Ike seem shrill and callow. Cass has no interest in spirits, though like the others he has grown up around them, and to him an inheritance is not something you can make choices about. It is yours, whatever its genesis. For him the ledger books tell nothing but the steady outgo of supplies balanced against the income from the crops harvested. Yet the manner in which Faulkner ends the novella suggests he wishes the reader to ponder the thrust of Ike's version of New World history more than Cass's.

In Section V Faulkner flashes back to Ike's eighteenth year and his last trip into the Big Bottom with the Memphis lumber company poised to make its assault on the ancient and unresisting woods. Riding horseback out from Jefferson to meet Boon Hogganbeck on a squirrel hunt, Ike comes to the log junction of Hoke's and sees the stacked and assembled materials of the woodcutters—the half-completed planing mill, the piles of cross-ties and rails, the mule corrals, the employees' tents—and feels a "shocked and grieved amazement." Riding the log train into the doomed woods he thinks back on the quite recent days when the same train, running between the towering walls of the wilderness, had seemed harmless and almost toylike in its puniness. Now he sees it for what it is and always has been, the advance agent of destruction, its once amusing whistle now the triumphant scream of conquest.

He swings down from the slowing cars at the old hunters' stop and is met there by Ash, the camp cook, who tells him Boon is already hunting on ahead and will meet Ike at the big gum tree. Though this is ostensibly a hunting expedition, it is in reality an excursion into the past, and Ike is not watching for game. His thoughts run to other hunts of other days, to the men of those hunts, and, most significantly, to the unchanging background of those days and hunts, the natural world itself in all its timeless and silent strength:

> . . . summer, and fall, and snow, and wet and sap-rife spring in their ordered immortal sequence, the deathless and immemorial phases of

the mother who had shaped him if any had toward the man he almost
was, mother and father both to the old man born of a Negro slave and
a Chickasaw chief who had been his spirit's father if any had, whom
he had revered and harkened to and loved and lost and grieved. . . .

Unconsciously his steps take him away from the gum tree and toward
the knoll atop which lie the graves of Sam Fathers, his spiritual father; Old
Ben, the guardian spirit of the woods; and the mongrel dog who fought
Old Ben to the death. And it is here on the knoll—the small plot reserved
by Major de Spain out of the sale to the Memphis lumber company, the
last retreat of the unspoiled wilderness—that Faulkner reminds the reader
of Ike's version of New World history. For the musing, almost inattentive
hunter nearly steps on a six-foot rattler, "the old one, the ancient and
accursed about the earth, fatal and solitary and [Ike] could smell it now:
the thin, sick smell of rotting cucumbers and something else which had
no name, evocative of all knowledge and an old weariness and of pariah-
hood and of death." Here is the snake in the garden.

Once when hunting with Sam the boy had seen a tremendous, heavily antlered buck pass right by their stand, its eyes wild but unafraid, and
had heard Sam salute it, saying in Chickasaw, "Grandfather," with palm
lifted and outstretched. Now in ironic salute to this symbol of death, fatality, and the destruction of paradise, a salute that comprehends everything
from Genesis through the coming of white civilization to the New World
to his own family history, Ike utters the same word, "Grandfather."

III

OXFORD, MISSISSIPPI, DOESN'T MAKE either an industry or a community
fetish out of William Faulkner. Coming down from Memphis on Route
7, the pilgrim looks vainly for any signs proclaiming the town as Faulkner's
home. Once in Oxford itself there are almost no signs of any sort that the
man was born here, lived here much of his life, is buried here with his
many kin. His name has not been used in any commercial ventures, nor
have the titles of any of his books been similarly used. No characters of
his have streets named for them. In fact, copies of his books are not con-
spicuously displayed anywhere, though they are in good supply at Square
Books. The Chamber of Commerce map and guide to the town devotes

almost as many lines to Faulkner's brother, John, as to Faulkner and does not list his home, Rowan Oak, among its "Points of Interest." In recent years a plaque bearing a quotation from *Requiem for a Nun* has been added to the Lafayette County Courthouse in the town square, but, after all, it is not really much of a show of pride in the native son who is arguably the greatest American writer of the twentieth century.

When Faulkner died on July 6, 1962, at a sanitarium in nearby Byhalia it was a major story in the international press, and if there was a certain ceremoniousness to the town when he was buried in St. Peter's Cemetery the day following, it was a far cry from the homage paid Hugo in Paris when great throngs gathered at his bier under the Arc de Triomphe to chant his poetry. A decade after Faulkner's death there was hardly an official trace here of his existence. Beginning in 1973 signs of local recognition began to appear, largely through the efforts of the University of Mississippi (located in Oxford), which purchased the writer's falling-down home and inaugurated annual Faulkner conferences. The university could not boast it held the bulk of Faulkner's papers. It had itself been so tardy in its recognition of its one-time dropout that these had gone elsewhere, principally to the University of Virginia, which had recognized Faulkner's stature and made the aged lion welcome in its groves. The literary pilgrim who comes to Oxford now can learn a good bit about the man, but the serious literary scholar will have to go to Charlottesville to examine the texts, precisely as Faulkner wished.

The peculiar modesty of the town in regard to its most famous son is a muted and continuing announcement that folks here are not altogether pleased to be residents of Faulkner's Yoknapatawpha County. Groping toward the greatness he felt was within him, Faulkner came to understand how much he needed his native place. His native place, though, seems still not so sure it has needed Faulkner. His lack of visibility, so to say, does not mean the place has forgotten him or is not aware of him. It means that it is very much aware of him and wishes to keep him at a special, reserved distance.

Faulkner himself fairly early on understood that however artistically fruitful his relationship to his native place might be, it would never be a happy one, could not be. As he entered his major phase in the 1930s, he wrote that for a Southerner to try to write of his place was equivalent of taking a clawing, spitting cat and forcing it into a croker sack. "We need to talk," he wrote in 1933, "to tell, since oratory is our heritage.

> We seem to try in the single furious breathing (or writing) span of the individual to draw a savage indictment of the contemporary scene or to escape from it into the make-believe region of swords and magnolias and mockingbirds which perhaps never existed anywhere. . . . I do not believe there lives the Southern writer who can say without lying that writing is any fun to him. Perhaps we do not want it to be.

If it was not fun for Faulkner to write of the South, dead since '65 and still struggling to remake itself, it was even less fun for Southerners in general and those of his native place in particular to read what he wrote. Many of them, observed his fellow Southern writer, Robert Penn Warren, regarded Faulkner as "somehow a family scandal, a skeleton in the closet, a libeller who will never do much good for the Chamber of Commerce." John B. Cullen, who hunted with Faulkner in the Big Bottom, put it more pungently when he said that to Faulkner's neighbors some of his work was about as popular "as a dead skunk would be in a sleeping bag." Faulkner, Cullen claimed, "knows as well as the rest of us that these sordid stories are read as insults to the people of this community." Cullen's guess was that the books must have been written to sell to "the Yankees." Thinking over the subjects Faulkner treated—the pervasive violence, the rapaciousness of the Snopes clan, idiocy, sodomy, grave robbing, fratricide, incest, putrefaction, an Ole Miss coed raped by a gangster using a corncob—it is easy enough to see what bothered Cullen. And it is easy enough to understand why, despite the fact that Faulkner was often very specific about the sites of his fictions, there could never be a literary tour of his Jefferson or his Yoknapatawpha County. ("Here's the famous square, folks, that Faulkner wrote about in so many of the books. It was here, you'll recall, that the townspeople gathered to lynch Lucas Beauchamp in *Intruder in the Dust* because they thought he'd shot a white man in the back. And it was here the idiot Benjy got so upset when the Negro drove him the wrong way around the monument. This house here is the one Faulkner had in mind in 'A Rose for Emily' where the woman poisons her Yankee lover, then keeps his putrefying corpse upstairs in her bedroom. Out here is where Joe Christmas in *Light in August* almost severed the head of his white mistress, an event Faulkner borrowed from real life. And this house here is where Percy Grimm at the head of a lynch mob paid him back by castrating him." Et cetera.) Plainly, such a tour will be forever impossible. No theme park, either.

But it is possible even this might have been forgiven had not Faulkner written so compulsively of the race question. Without in the least assuming any sort of moral superiority here, the non-Southerner must conclude that in all the years since the Civil War Southerners have remained profoundly sensitive on this question. And here was Faulkner, the cocky little strutting ne'er-do-well some Oxonians sneeringly referred to as "Count No 'Count," writing again and again and again of the dark, twisted relations between whites and blacks. Writing of a white woman who loved a black man; of masters who went to bed with their slaves and of the half-caste children of these illicit, unmentionable affairs; of lynchings; even (in *Absalom, Absalom!*) daring to suggest that the whole hallowed institution of the Southern white virgin of ante-bellum days was only made possible by the easy availability of slave women to their white masters. In his book about Faulkner, John Cullen devotes an entire chapter to proving how wrong his neighbor was on this crucial question, what a great lot of trouble he had stirred up when, in fact, the only trouble between the races in Mississippi was caused either by those who wrote or spoke without thinking or by outside meddlers.

I asked Richard Howarth if this was in fact the basis of what I perceived as the town's peculiar official regard for Faulkner. Howarth was a native Oxonian whose family roots were deep in the town's soil and who now ran Square Books. "I think there's something of that," he replied cautiously. "Yes, there's that. But there are a lot of other things, too. He was a very private individual, and even more so after he became famous, and I think people here wanted to respect that privacy. There's a sort of residue of that, I think. And I guess there's also a kind of reluctance to admit they were wrong about him. For years, you know, people here were sure he wouldn't amount to anything. Then they found out the rest of the world considered him a literary genius. It's kind of hard, sometimes, to have to admit you didn't have good judgment about one of your own. I don't think it [the racial matter] is the whole story, thought it's certainly a factor."

Thomas Freeland III, who knew Faulkner, was somewhat more emphatic on the subject. "I don't think that has anything much to do with local attitudes at all," he said. Freeland and his son, Thomas IV, now occupied the old Stone law offices just off the square where the letters "STON" are just legible in the walkway to the little, turreted building. The Freelands were lawyers, too, and we talked in a room beneath a large framed photo captioned "The Bench and Bar of Mississippi, 1902–3," on either side of which were photos of General Stone and his son Phil. In glass cases were ranked volumes of the *Southwestern Reporter* and the *Pacific Reporter* in their brown leather bindings and red and black spines. "I doubt you'll get very far with that," Freeland senior said affably enough, brushing cigarette ashes from the ample bosom of his white shirt. "It's more a mixture of lack of awareness of what he actually wrote and a kind of diffidence. They felt funny about the fact that he didn't have any qualms about using characters and incidents that really existed and happened here. When his books started to come out people were really disturbed about his use of things and people they personally knew. When the book club got together to discuss *The Mansion,* Maude Brown just clucked her tongue over and over: 'How did William get hold of *that?* That happened so long ago he couldn't have known about it himself.' 'How did William find that out? Nobody was supposed to know about that!'" Freeland senior laughed and tilted back in his swivel chair. "I think you'll find that's the root of it."

Whether Freeland senior was right about the root of the town's attitude toward its famous son, he was certainly correct in predicting I wouldn't get very far in pursuing the racial angle. It was simply not a matter area residents wanted to discuss—at least not with a visitor with a non-Southern accent—and after a while I had to give it up or risk losing what contacts I had established, meanwhile privately consigning it to the crowded realm of hunch. By the time I visited Faulkner's handsome restored home on Old Taylor Road I had given up asking about it. There the young, obliging caretaker told me William Faulkner was very much honored here, thank you, as witness this museum in which we now stood talking. And, no, there weren't any ghosts here except the one Faulkner himself invented. I didn't believe this but said nothing as he briskly showed me through the rooms, delivering intelligent and uninspiring responses to questions I put to him. Midway through we were joined by two retired women schoolteachers from Wilkes-Barre, Pennsylvania, down south for a literary tour. The deep, tired lines in their martial faces brought to mind the Kafka story where the inmates of a penal colony must have the nature of their offenses carved into their flesh with a huge mechnical needle. Neither, they told me, had read Faulkner, but they understood he was a significant author, and so they wanted to see his home. Could I perhaps tell them something of his subjects?

Together we inspected Faulkner's smudged handwriting on the walls of a study where he'd outlined *A Fable,* then gazed at the silent little portable typewriter whose keys the vanished man had once made flash and clatter in his furious need to tell stories. Oh, there were ghosts here, all right. They were upstairs, too, in and about the bedrooms where the real life of any family unit is enacted. Standing at a west-facing window in the hall, I thought of the years of bleak misery and pain Faulkner had created and shared here with his wife, of the stories their daughter Jill had told and those many more she might have wanted to tell but didn't. One, perhaps, would have been enough to stand for an unspoken host: how, once, when she'd seen her father getting ready for one of his colossal benders, she'd begged him not to spoil her birthday, and he had replied that nobody remembered Shakespeare's children.

Behind the house are various outbuildings, including the old kitchen and cook's quarters from slavery days and the little cottage in which Mammy Caroline Barr had lived until her death in 1940. Born a slave, she had looked after Faulkner children for forty years, and Faulkner had immortalized her in the character of Dilsey in *The Sound and the Fury*. On a wall of the now barren and dirty room was a crucifix:

$$
\begin{array}{c}
\text{H} \\
\text{I} \\
\text{MY} \;\; \text{S} \;\; \text{SINS} \\
\text{L} \\
\text{O} \\
\text{V} \\
\text{E}
\end{array}
$$

From here at any hour you can easily pick out the deliberate strokes of the courthouse clock, reminding you of the Reverend Hightower of *Light in August* who measured his own life by those strokes and the passage of time they told.

Out in the countryside he liked to roam there were any number of unofficial but significant signs telling the literary pilgrim that whether folks here liked it or not this was Yoknapatawpha County, "William Faulkner, Sole Owner & Proprietor," as he had put it in a map he once drew of the place. In the pine hills east of town you passed mailboxes bearing names that startled you, like the one that said, "Bundren," the family name in *As I Lay Dying*. The hamlets of Toccopola and Tula were said to have been inspirations for Frenchman's Bend from which place the Snopes clan began its crawl toward Oxford, and at both places there were old crossroads stores like the one Will Varner runs at Frenchman's Bend. At Taylor weed-shrouded railroad tracks reminded you that it was here Temple Drake jumped from the train and so began to move toward her rendezvous with the vicious gangster, Popeye. The Old Jones place west of Taylor was a congeries of Faulknerian associations. There sat the splintered, sagging ruin of the big house surrounded by sentinel cedars that muttered in winds sweeping over the cotton fields. In the briary little cemetery plot, among the Jones family graves, was a tombstone for Lucy H. Bond who died in 1890 and whose name may have suggested to Faulkner that telling orthographical corruption in *Absalom, Absalom!* where Charles Bon's idiot grandson is known as Jim Bond. Two Ole Miss Faulkner specialists told me the Jones place might have been part of what went into the making of Sutpen's big house, though Professor James Webb said he agreed with the general opinion that if any one house could be said to be the model for Sutpen's place, it would be the Shipp place past Water Valley.

Directions to it were vague, though, and I went up and down Route 32 several times hunting a certain turnoff into the woods before I inquired of Mike McGregor, a jewelsmith and leather worker who lives close to where I imagined the turnoff must be. I waited patiently in his shop while he worked through a ticklish piece of business on a small gem, but it was worth it since when he turned his attention to me he gave such explicit directions I couldn't miss the dirt lane that led through a stand of woods, then across cotton fields, and ended at the Shipp family cemetery. Here were the graves of Felix Shipp, his wife, and their descendants, ranked under big spreading pines. Published sources on the Shipp place suggested the family's slaves were buied here, too, but if they were, I couldn't find their markers within the small chain-link enclosure.

In the days when Shipp settled here a stage road ran hard by the house, and in the lowering spring light I followed old ruts into the dry woods. The twigs and leaves of yesteryear exploded underfoot as I hunted farther and farther in the gloom for a row of cedars big enough to have been set out a century and a half before. McGregor had told me they once had formed a stately *allée* leading to the house, and that now they would be my only clue, for the house itself was gone, burned to the ground several years ago by Ole Miss students—an odd reprise, I thought, of what Faulkner had imagined as the end of the Sutpen mansion. By the time I had come into a clearing it was almost twilight, and if nothing else was very clear, it was evident I'd come too far for the house site. Seeing a stack of stove wood it was also evident I'd been following someone's seasonal wood-cutting road. The stack had been there long enough to grow a girdle of green mold around it, and at some point the woodcutter had brought out an office chair missing its pedestal and had screwed it deep into the earth. There it was beneath a pine where the man might sit back and gaze upon the neat product of his industry, perhaps to muse upon the warmth it would one day bring.

Twilight was Faulkner's favorite time of day, and he uses it often in his work. For him personally it was the end of another day of trying to make life jump off the page, time to cease for a time the telling of the stories, time for a drink. Twilight in this place his words had made so alive for me brought back the twilight time in *Absalom, Absalom!*: the father and son sitting on the Compson family porch, talking of Sutpen, waiting for full dark when Quentin would unwillingly accompany Rosa Coldfield to Sutpen's Hundred to discover what secrets still lurked in the gaunt, ruined house; and then in the hot September night riding the surrey out there, the young man's eyes straining forward but the mind's eye of both Rosa and Quentin expecting at any moment to see the fine, bearded figure on the black stallion burst again into their sight out of the past—Sutpen, destroyer of his domain. Groping back toward the Shipp cemetery, my own mind's eye was working, imagining a meeting here with Sutpen's shade, imagining pushing aside some vines and seeing in the gray, thin light of the woods the spirit of the old man rise up from its unquiet grave, swag-bellied and grizzled as Faulkner portrays him in his final phase, his Confederate uniform rotting about him, but his eyes burning and intact, saying to the hapless intruder, "What do you want of me?" And what could have been the answer?

As it turned out, Mike McGregor told me the next day that I hadn't gone in far enough. He offered to take me there himself and did so. While he talked to me over his shoulder I followed the twinkle of the big nickle-plated revolver that swung in its holster at his hip. Of the revolver, he'd said simply when I watched him strap it on, "In the woods you never know. I've been around enough to learn that, anyway." Now as we made our way through a hip-high tangle of blackberry brambles he was saying he'd lived here a quarter of a century ago, then had gone up to Memphis where he'd become a general factotum for "The King," Elvis. "In those days," he said, "I used to come out here with my wife for an afternoon. Of course, the road wasn't choked up like it is now. The kids from the college would come in here with their cars, and that kept it down. But often it was peaceful here, and my wife and I would spend the day under those cedars and dream about buying the old place and restoring it. I guess even then, though, it would have been too late. Here we are. See the cedars?''

I saw them, rising tall, columnar, and still stately in this wilderness of brambles and creepers, their bark a scaly silver. As McGregor had warned, they now led the way to nothing, for the only relics of the house were a couple of crumbled chimneys and a bit of the brick foundation buried in the rioting green. "It would be hard for you to imagine," McGregor said as we stood in the warm, dappled spring sunlight, "just how big this place was." But I was beginning to get an idea, both from the *allée* of cedars, which had obviously been planted to scale, and from the scatters of rusted tin that poked brown edges into the sunlight over a wide expanse. The house's timbers, McGregor was saying as I poked about the ruins, had been notched and pegged in the old-time plantation style.

"All that work must have been slave labor, I imagine," I said.

"Oh, for sure," he said glancing briefly at me. "What else?"

Back in the Freeland law offices I scanned a large wall map with father and son, trying to figure out the location of the old Stone hunting camp on the edge of the Delta that Faulkner had used as the setting for "The Bear." The younger Freeland felt the directions I'd gotten a few years ago from the late Professor Webb couldn't be right since they seemed to place the camp considerably west and south of the town of Batesville, whereas the map we now looked at plainly showed the Illinois Central railroad line running right through Batesville. "There used to be a stop on the line called 'Stone Stop,' and it was used for the hunters," he told me. "So the camp has to be pretty close to the line, I'd think. Possibly if you go out to Batesville, then turn south on 51, you'll run across somebody who remembers where Stone Stop was." He laughed briefly. "I have a copy of 'The Bear.' That might get you there." All of this seemed a slender lead, but at the least I could see for myself how the Big Bottom had changed, and then, as Richard Howarth said when I told him I wanted to find the camp's site, "The Delta's a good place to get lost in. If you're hunting Faulkner, you'll do better to go out there and get a little lost. They say that's what he used to do himself."

It was bright and windy on the way to the Delta. Where there were tractors in the fields I could see brownish-red clouds of soil funneling up and away in the wind, and the metal of the machines was dull with dust. Batesville was a child's version of geography where the divisions between

terrains and states is finite and absolute: east of town the country was rolling; west of it was the utterly flat Delta. At the intersection of 6 and 51
I followed the younger Freeland's suggestion and turned south, the I.C.
tracks on my right. But with each mile I came to feel a little more hopeless:
how to know where to stop and ask? Finally I saw two men at work on
a house and pulled over to ask them about Stone Stop. Neither, as I had
foreknown, had ever heard of it. But, said one of them, his mouth and
chin stained with chew, a claw hammer hanging from a thick hand, "See
that white house yonder? That old man's been here forever. If there ever
was such a place as Stone Stop, he'll be the one to tell ye. Say," he added,
"ain't a carpenter are ye? We sure could use an extry hand today." He
shook his head and smiled.

E.G. Walker, the man in the white house yonder, had not been in
it quite forever, but he had been in this area all his eighty years, and though
so crippled with arthritis it took him several minutes to answer my knock,
his mind was supple, and he instantly knew what I was talking about when
I asked about Stone Stop. It had never been, he said, a stop on the I.C.
line, only on a logging line that had run some miles west of Batesville.
"Many and many a time my daddy drove a wagon with supplies from Oxford clear out there for them fellows," he said. "I seen that camp myself,
too. You have to go through Batesville, then take the Crowder road. When
you come to the white grocery store, look off west about a mile, and that's
where it was."

"Was?"

"Yes. Was. Ain't there now. There's nothing to show for that now.
It was tore down many a year ago."

I had to see for myself, having come this far. I turned south from
Batesville on the road to Crowder, out into the flat, cleared country, unex-
ceptional now in its agricultural productivity, reforesting it in my mind
so that the sun on the broad fields was eclipsed by the gloom of that
mighty, soaring wilderness of which Faulkner had written: that primeval
New World wilderness Isaac McCaslin says had been granted by God as
a chance for the Europeans to recover the loss of the paradisal garden and
begin anew in humility and brotherhood. Whether Isaac is right—and
whether or not Faulkner meant him to seem right—there is no doubt that
in "The Bear" Faulkner touched the profoundest theme in the American
experience. Thus, looking out the window of a speeding automobile in
the common light of a late twentieth-century spring day, I could see a
stranded row of old cypresses in a cleared field and feel loss, the great

chance missed. Nineteenth-century Indian speechmakers used to warn white men that the spirits of the departed tribes, game, and trees were not without their own peculiar powers and might haunt the places of their former habitations. Faulkner was too good a craftsman to risk a speech of this sort for his Ikkemotubbe, and still less for the corrupted Issetibbeha. But he does let this ancient woods wisdom, this voice of the American earth, issue from Ike's mouth as the bequest of the half-Chickasaw Sam Fathers. In these opened, drained lands with "The Bear" to guide you, you could for a moment re-encounter a vanished land, and the stranded cypresses could stand for what all the New World once had been, as fact, as metaphor.

Looking for my marker, "the white grocery store," I found I had misunderstood E.G. Walker and that what he had instructed me to look for was Ira White's store, one of those now rare country places that carries everything from canned meats to tack. It was closed this Sunday morning, but a young man in the next yard knew what I was looking for when I asked. "You mean about that man from Oxford?"

"I guess you've been asked this before," I said. "But I'm told it's been torn down." He shook his head and pointed across the fields, and there it was, a small, weathered cabin with a low-slung tin roof, standing in the midst of a large soybean field. It was watched over by one pine and one grand old hardwood that stretched benedictory arms over the ruin. Up close I could see it was covered to its eaves with briars and vines. The little slough that ran behind it was vine-snarled, too, and alive with red-winged blackbirds. In the high wind skitters of plowed soil ran up the furrows, and hundreds of feet above two red-tailed hawks swung and sailed on the updrafts. Inside: a cheerless ruin of crumbled chimney, shards of glass, old stove, curling hunting boots, and a rotten mess of women's undergarments. Just a busted old building in a soybean field of the Delta but as alive with associations as the slough's bushes were with blackbirds. Through the 1920s Faulkner had seen the changes here and brooding on them had created a narrative capturing something so essential about the whole of the Americas that even readers who race through "The Bear" for the sheer thrill of the chase finish it feeling an indefinable loss. Major de Spain, like Faulkner, sees the future with the death of Old Ben and so begins to sell off his land. Subsequently he refuses all invitations to come out again on hunts, regretting, like the rest of us, the changes he has been a party to. Like Buffalo Bill Cody, Daniel Boone, the mountain men, and a nameless, unrecorded host of pioneers, hunters, and trailblazers, de Spain has killed what he loved, perhaps for no other reason than that his love for the wild was hopeless.

"But above all," the plaque at the courthouse front read, "the courthouse: the center, the focus, the hub; sitting looming in the center of the county's circumference like a single cloud. . . ." On another side of the building was another plaque, this one to the local soldiers of World War II: two columns of the names of those who served and beneath these, in a segregated section of the plaque, those "Of the Negro Race" who also served. Around both plaques and beneath the four-faced clock tower were green benches, and on this day as on many another they were filled with the old boys in synthetic fedoras, flannel shirts with suspenders over them, some with canes, their knees crossed, craning their creased and sun-mottled necks at the traffic swishing slowly about them. They were talking of just what you'd expect: weather, crops, aging, death.

Weather and crops: "I don't believe I've ever seen it so dry this time of year." General nods and murmurs of assent. Talk of soybean prices and of so-and-so who had rented out his farm. "He don't keep but a few cows now." "I don't believe he was ever able to do more'n that, do you?" Chuckles deep in throats.

Aging and death: "I'm seventy-two now," said one, leaning forward to scan the bench for the effect of this disclosure. "Hell, boy, you ought still to be in school then." A discussion of the death yesterday of Buddy whose funeral will be the day after tomorrow. "What kin was you to Buddy?" "He was my wife's first cousin."

Meanwhile the big clock aloft in spring's soft blue solemnly noted the passage of time. I was waiting for the arrival of Clew Moore whom I had been stalking for several days now. I had seen an interview with Moore in a Faulkner documentary at Ole Miss's Center for the Study of Southern Culture, and he had interested me with his blunt, salty observations on Faulkner and his community, yet when I contacted him by phone he had been evasive, claiming others in town knew much more about these things than he. Finally, he said he was going to a pig auction the next day in Pontotoc, and maybe we could talk there.

It was a thirty-mile drive east along Highway 6 to Pontotoc, and along it you had to pass Thaxton where Dean, Faulkner's younger brother, had died in a plane crash; then through a stretch of poor-soil pine hills; and then Pontotoc, a small agricultural center where in the long-ago a Chickasaw chief sold his birthright for what he could get. But if I was disappointed by Clew Moore's nonappearance at the auction, the trip was far from a failure, for at the auction barn, at the holding pens, and in the parking lot I saw descendants of those stolid yeoman farmers of whom Faulkner had written: men with names like Grissom and Lovett and Robinson and Jernigan, in ball caps, jumpers, and boots, their jaws slowly working over plugs of tobacco, staining the splintered boards of the auction barn with great brown splotches of spit. They said little to one another while the auctioneer filled the air with a steady mixture of vital information and filler ("*Ain't* them good-lookin' rascals, now!"), occasionally raising blunt forefingers in bids. Up in the topmost cheap seats was a small knot of young black men, same uniform, though through the whole afternoon they took no part in the bidding.

Out in the Delta the yeoman farmer was gone with the game and the big woods, replaced by agribusiness. There all you saw of the older way were occasional forlorn, abandoned sharecropper cabins, their tin roofs scrolling slowly upward in neglect. But here in the pine hills the yeoman farmer had managed to hang on, with what margin of success I couldn't guess. But I did notice a number of their kids helping out at the holding pens, one of them wearing a denim jacket with an F.F.A. emblem stitched on its back. He was a big, freckled kid in his early teens, and I asked him if in fact he saw himself as a Future Farmer of America. "Sure do," he said, keeping his eye on the moiling shoats in the pen. "I like working with animals."

All this was to the good, but there was still the Clew Moore problem, and in the way of such things, when I missed him again the next day at Smitty's Diner in Oxford I became more than ever determined to corner him, convinced he was the one person here I had to talk Faulkner with. The proprietress at Smitty's (just off the square), stirring her plate of hash browns, said I had just missed him, that he'd popped in only long enough to say he couldn't make our appointment, that something had come up. When I called his home again his wife said he was out inspecting some property but might possibly be down at the square later in the afternoon. Which is why I now found myself listening to the old boys and the clock as the afternoon waned. When I ventured to ask one of them if he had seen Moore he pointed to a big, late-model Cadillac illegally parked not twenty feet from our bench. "That's his car settin' right there," he told me, pointing with the rubber tip of a cane. I felt now I had my man.

Soon enough he appeared, a tallish fellow in a blue-and-white checkered flannel shirt and polo cap, bearing a striking resemblance to the old-time Dodgers pitching star, Dazzy Vance. Moore had been a bootlegger in the 1940s, and the look of his great nose suggested he'd done more than vend his product.

Now that he had almost blundered into me he was affable enough, seating himself next to me on the bench from which he saluted every pickup and car that circled past so that our conversation was peppered with "Hey, boy!" "Hey!" and so on. I told him what I'd learned about him from the film interview, and he sniffed. He'd run a "joint," he corroborated, "out on the old Highway 6 where we had poker games, dice, slot machines, any goddamn thing you wanted, we had it." But, he said, "I never once sold a bottle directly to Faulkner, like you may have heard on that interview. But I sold plenty of 'em to Mrs. Faulkner. She was a worse alkyholic than him. Sure 'nough! Hey, boy! Yeah. She used to come out there regular and buy it. So I know that many a bottle I sold her was drunk by him. Yes sir."

Faulkner's drinking is pretty well documented, I said, but that wouldn't in my opinion quite account for what I saw as the somewhat peculiar way his hometown regarded him. How, I wondered, did the town think of Faulkner during his lifetime or after? Moore shot right back: "They thought he was a damn fool, that's what! Oh, not all of 'em. There was a few that thought he hung the moon, like Stone and the fella ran the drugstore over there [pointing]. But most folks didn't have any use for him. I know I didn't. Liked his brother, though, the one got killed. Dean. Shit! You could be sitting right here [slapping the bench slab lightly], and if you didn't get outa the way when he come along with that cocky little walk of his, why, he'd liable to walk right over you. Half the damn time wouldn't even speak to you."

"Well," I said, "it must have changed local opinion of him some when the film crew came here to make *Intruder in the Dust,* didn't it? I mean, that's fame. Who else here has had a book of his made into a Hollywood film?"

"Shit!" Moore spat. "That wasn't made by people that came from here." He gave me a small sideways glance. "That was made by folks like you that don't know no better. Shit! If it was up to us, he'd a starved to death out there in that rotten old barn he called 'Rowan Oak.'"

I asked whether Faulkner had taken much from the town and region even if the town itself seemed not to take much pride in him. Was he, for instance, a good listener, someone who heard and remembered many of the town's stories?

"Hell no, he wasn't a good listener. Hey, boy! I said, *Hey!* Speak to me, you old bastard! Where were we?" Stories. "Oh, yeah. No, he never listened to you. Half the time wouldn't speak to you, like I said. Cocky little sum bitch. Now, I don't know this for a fact, but I heard he couldn't write less'n he was drunk. And I do know for a fact that he drunk many a bottle that come from my place. And once, after I'd quit that business and gone legit—worked for Schenley's then—I seen him out to the Oxford airport and offered him a good deal on a case of whiskey. 'Mr. Faulkner,' I say to him, 'I'll get you a whole case at a real discount.' And he said, 'Sorry, Clew, I done just bought myself a case yestiddy.'

"Anyway, they tell me he used to sit out there at Rowan Oak propped up in bed, drunk, and write. And after he died they found where he'd wore a damn hole clear in the the wall where he rested his head. Shit! I s'pose you know he died in Doc Wright's clinic up there at Byhalia. Died right in the arms of a nurse up there. Know her well."

How obviously puzzling it was to Clew Moore and to others that little Count No 'Count could have by some forever mysterious process distilled something of the quality of local life in such a way that years after the man's death he, Clew, should be sitting here in the lengthening shadow of the courthouse, enduring again the foolish probings of another stranger— a stranger whose only interest in Clew was in the connection he once had had with a man for whom he had had no use. Among the mortal mysteries Clew Moore had to live with there was this recurrent one, and clearly it galled as much as it puzzled. That was why Moore continued into a tirade against the cruel tricks of fate, and while I half-listened to it, I saw in the person of this red-faced, bulbous-nosed, ex-bootlegger the personification of that local attitude against which so many American writers have had to struggle. As if it were not already sufficiently difficult for writers here to find their voices in their native places, to divine in the sad, stolid facts of the daily round—the noon whistle, June bugs popping against street lights—the songs, the stories these things really were. For in addition to this there was the bafflement and even hostility of that very place whose life the writer would evoke. To be sure, the phenomenon is not limited to America: as Ronald Blythe deftly observes of Thomas Hardy, "The more he used his background, the less he was of it." The process of artistic creation seems necessarily to involve estrangement, in other words. But what was true for Hardy in his country with its long history of belles lettres, its ancient traditions, was even truer for American writers in a rude, democratic republic with little inherent respect for the arts and still less for those solitaries who were seeking to draw creative nurture from their common places.

Moore was still talking. "No, sir. I come up the hard way. Raised in a house without electric lights nor runnin' water. Had to shuck corn many a night by farlight. And do you think we could waste them kernels we spilt on the hearth? No, sir! My Daddy would'a beat the dog shit outa us if we'd a swept them into the far. No, sir! We parched them spilt kernels and ate 'em for supper. What do you think of that?" He fixed the stranger with a hard, defiant look. "I made two fortunes," Moore concluded, his hands rising briefly from his lap, then settling again in a gesture of existential resignation. "Made two fortunes and don't have nobody to leave 'em to. No kids. Me and my wife, we don't have a happy home. All that shit. I've had hell with her."

Trying to raise him out of this autobiographical slough, I asked again what things of community life Faulkner had drawn on.

"Names. Places. Those for sure. Names of people. Names of places. Hey, boy! He got aplenty of them from right around here. Dutch Bend. We called it Springdale. We got some hellacious names around here. He got Yoknapatawpha from here."

EIGHT

THE VALLEY OF THE WORLD: JOHN STEINBECK'S EAST OF EDEN

〰 "A LOT OF PEOPLE seem to have a special feeling for that book [*East of Eden*]," I said to Pauline Pearson. "Steinbeck told someone he'd had many letters from people who said it was their book, that it seemed to have been written especially for them."

"That is the way people feel about it," she rejoined. "It does have a special quality to it that speaks right to you, whatever your experiences may have been."

We were in the Corral de Tierra country that lies between Monterey and Salinas, and Mrs. Pearson was giving me a guided tour of Steinbeck country, of which she had made a long and careful personal study. Her husband Merle was at the wheel on this mizzling March morning, the breathtakingly soft and green slopes of the hills veiled in the mist with shreds of fog hanging motionless against them. For more than thirty years Mrs. Pearson had worked at the Salinas Public Library and had taken a special interest in the library's Steinbeck collection. She sat in the back seat with a small sheaf of notes in her lap, pointing out the sites associated with Steinbeck's *The Pastures of Heaven* and *The Long Valley*, but now our talk had veered away from this landscape to *East of Eden* for which Steinbeck had used the Salinas Valley from below King City to the town of Salinas at the valley's head. Mrs. Pearson said she had always regarded *East of Eden* as Steinbeck's finest work and the one for which he will finally be remembered. I told her it had long been a favorite of mine.

"When I was a kid," I told her, "there was an old black woman who worked as a maid and cook for the family of a friend. Now that I look back on her I see she was a pretty unusual person. I don't know anything about her background, but anyway, she read a lot, and she must have been pretty observant in that way some 'domestics' can be, you know: they can take in an awful lot without seeming to be nosy."

"Like Lee in the book," Pauline Pearson added.

"Exactly. That's one of the odd parallels I didn't see at the time. It just never occurred to me, and neither did the fact that she gave me that particular book to read. I thought she was giving it to me because she knew I liked Steinbeck. Now it seems clear she saw some things in it that pertained to me. I was kind of a 'troubled' kid, as we say, and I had a brother who did well so that by comparison I looked even worse. And I had a father who seemed sort of God-like in his power and Old Testament temper. Anyway, Virgie gave me *East of Eden,* and I read it with tremendous absorption. I can't say at the time I was aware of the parallels between Cal's situation and mine, but looking back, I'm absolutely certain Virgie wanted me to see those parallels and to get Steinbeck's point: that a man has a choice, that he isn't fated to be bad. I guess that makes me like those readers who wrote Steinbeck to say *East of Eden* was theirs."

Pauline Pearson smiled at me from the back seat. "That story," she said, "sounds like so many others I've heard [about *East of Eden*], and I'll tell you: I just want somebody to write about *East of Eden* and about how much of the valley's story Steinbeck got into that book. That book, to my mind, has never gotten its due, but someday it's going to be recognized as the great book it is. Maybe not greater than *Grapes of Wrath* but as great. And when that happens, wherever I am, I'll know it."

If you count from Steinbeck's first published book, *Cup of Gold* (1929), *East of Eden* (1952) was more than twenty years in the making, much longer if you accept Steinbeck's own valuation of it as the book for which he had all his life been in training. The formative years were spent in Salinas where he was born in 1902 and on the Monterey Peninsula at Pacific Grove. Like other middle-class Salinas families, the Steinbecks had a small house in Pacific Grove they used as a summer retreat, and Jackson J. Benson, Steinbeck's biographer, says the early exposure to the sea was crucial for Steinbeck as a writer. Indeed, *Cup of Gold* had as its hero the seventeenth-century buccaneer, Henry Morgan, and books connected with the sea were interspersed throughout Steinbeck's career: *Sea of Cortez* (1941), *The Pearl* (1947), *The Log from the Sea of Cortez* (1951). When he died in 1968 he had long been in exile from the Pacific, but he was still living by the sea at Sag Harbor, Long Island. But for all this, Salinas was home and the valley below was his chosen literary terrain. It was the place he imaginatively returned to again and again, the locale of his best work, and the locale also of his most ambitious novel, *East of Eden,* his "last stand."

When he was born to John and Olive Steinbeck, Salinas was a town of 2500. It was less than a half-century old and owed its origins to nearby cattle ranching and to the Southern Pacific that used it as a railhead for its southern expansion down the valley. It had its goodly sprinkling of German citizens, Italians, and French, as well as its native Hispanics. On the wrong side of the Southern Pacific tracks was a Chinatown and adjacent to it the Row, a strip of whorehouses that catered to visiting farmers, ranch hands, and railroad men as well as to the local trade. The Salinas River ran just west of town on its way to its debouchment in Monterey Bay at Moss Landing. By the point at which it reached town it was a sluggish river, and in summer and during dry years it disappeared for stretches beneath the sand or else stood in green, stagnant pools in the shade of its banks. In the beginning of *East of Eden* Steinbeck disparages it as "only a part-time river," yet he knew well that the Salinas, small and sluggish though it might be, had over the course of geologic ages made the northern portion of its valley one of the most fertile spots on earth. The valley floor from Salinas south through Gonzales, through Soledad and Greenfield and King City was incredibly rich farm land with deep, black soil. Here were extensive fields of lettuce, beets, beans, and sweet peas. Below King City and down through Paso Robles at the valley's bottom the great soil began to thin and give out, flinty rocks sticking up through sparse grass cover. In summer the heat down there was fierce.

The valley was closely bounded by mountain ranges, the Gabilans on the east and the Santa Lucias on the west. The foothills of the Gabilans were pocked with potreros and the ranchos of the old Hispanic settlers. The Santa Lucias had another character altogether: dry, stony, uninhabited, they rose up from the valley floor to naked, scoured heights of nearly 5000 feet at the Ventana Cones before hurtling into the Pacific. The Monterey Peninsula looked out on the glittering expanse of the Pacific with "Asia at your doors and South America," as the aged Emerson had put it when he beheld the ocean in 1871. But the Salinas Valley was another world, though but a few miles away, an enclosed, bounded world, and at certain seasons it seemed more unto itself when thick white fogs enveloped it and sealed it off even from the sky. Though he mulled throughout his career the ocean tide pool as a metaphor for life in all its various in-rushing vicissitudes, Steinbeck came to see in the valley's extremes of lush plentitude and dry poverty, its black flatlands and craggy heights, its mingled nationalities, its settled families and migrant workers a naturally defined arena in which a writer, any good craftsman (all he ever professed to be) could find everything he would ever need.

He had relatives and friends scattered over the entire region, from the peninsula east to the Gabilan foothills, from Salinas south to King City. In the Corral de Tierra valley that lay between the Salinas Valley and the peninsula he had an uncle and aunt. The latter, Molly Martin, was a story-catcher who had learned much about the history of that velvet pocket in the hills. East of King City was the bare-bones ranch that was the seat of the Hamilton family, his mother's people, a populous Irish clan that eventually sent its offspring abroad in the valley from Salinas to Paso Robles. East of Salinas in the Gabilan foothills was the Hebert ranch where as a boy Steinbeck often visited his friend Max Wagner. From the ranch's hilltop buildings you could look east to where the Alisal Creek flowed through thick beds of wild azaleas, and if you climbed the slope behind the bunkhouse, you could look out west on Salinas lying beneath the gentle inclines of the intervening pasturelands.

From childhood Steinbeck was unusually sensitive to the spirits of the places of this valley world. Max Wagner told Steinbeck's biographer that as a boy Steinbeck developed a cluster of secret places—wooded nooks, secluded trails, stream banks—where he would go to be alone and to watch the small, special life that went on otherwise unobserved. In *East of Eden* Steinbeck recalls this childhood trait. "I remember," he writes in the novel's second paragraph, "my childhood names for grasses and secret flowers. I remember where a toad may live and what time the birds awaken in summer—and what trees and seasons smelled like. . . ." Early, there were opportunities to honor the expression of such feelings, for the Steinbecks were a family of readers. A photo taken when Steinbeck was in his teens shows the family poring through magazines and books in the living room of the house at the corner of Central and Stone, and though the shot is clearly staged, it speaks to a reality. Everyone read, and Steinbeck and his sisters were actively encouraged in this. His aunt, Molly Martin, aided too: she gave the boy a copy of Malory's *Morte d'Arthur,* and he never got over the wonders it disclosed. His mother considered the reading of good literature a necessary part of good breeding and personal culture.

But no one in the family imagined that the strapping, somewhat indolent kid would eventually seek to express his own inner feelings, his kinship with the soil and spirit of the valley in literature of his own making. Yet, unlike Twain or Cable or Sandoz who came to writing having tried other things, Steinbeck apparently never seriously wanted to be anything other than a writer. When this singular ambition became evidently more than an adolescent dream—early in his desultory undergraduate years at Stanford—there was some family consternation, especially as it was already then evident from young John's fledgling literary efforts that he was not interested in creating "respectable" literature. At this point it was his father, stolid, silent keeper of ledgers, who encouraged the youth. In a letter to his friend and publisher, Pascal Covici, Steinbeck said it had been his father who had provided what family support there was during his earliest artistic struggles. "I often wonder about him," he wrote Covici. "In my struggle to be a writer, it was he who supported and backed me and explained me—not my mother. She wanted me desperately to be something decent like a banker. She would have liked me to be a successful writer like Tarkington but this she didn't believe I could do. But my father wanted me to be myself. Isn't that odd."

In 1926 he began in earnest trying to be himself. He was twenty-four then. He had finally admitted to himself and his family that he was never going to finish at Stanford. Feeling curiously isolated from what he imagined were the main literary currents of his time, he shipped by freighter for New York and there worked as a manual laborer while seeking some ingress to the literary world of the big city. He never found it and came back home in apparent defeat. But if it was a defeat, it was a salutary one, for back in California Steinbeck was forced to begin the pioneering process of rediscovering for himself his native turf.

As the young man began on this he was unaware of what it was he had begun; he was aware rather of the fact that he felt driven to write and that as yet no truly compelling subjects had come to him—a familiar dilemma of the younger artist. He was aware, however, that others in California had found plenty to write about, for the state had been discovered for literature with that same quality of instantaneous vengeance that it had been discovered for commerce in the wake of James Marshall's gold strike at Sutter's Mill. Then Bayard Taylor, Bret Harte, Twain, Joaquin Miller, and Ambrose Bierce had mined the crude ore of the new culture. Subsequent to them had come another generation of writers—Clarence King, John Muir, Jack London, Frank Norris, Helen Hunt Jackson, Ina

Coolbrith, and Mary Austin. In Steinbeck's own time there was yet another generation breaking into national notice, the most significant voice of which was probably Robinson Jeffers. Thus, however much he felt himself to be on his own in his search for a voice and a vision, Steinbeck was hardly in the situation of Hamlin Garland, raised in a region then almost utterly unsung; or that of Sandoz whose family and local culture were actively hostile to belles lettres; or even that of Cable whose native city had its lively arts but which before him had not produced a single distinguished work of fiction. Still, as Steinbeck discovered, every writer must be his own Boone if he wishes to stake claim to a place. If more than seventy years after *Walden* there was the incalculable value of precedent, Steinbeck had to find for himself the worth of that world given him by birth.

Cup of Gold (1929), Steinbeck's Henry Morgan novel, was a false start, as he recognized even before its publication. Nor did he appear to make much progress with a novel-length manuscript he was calling "The Green Lady" while he worked at it through the last years of the decade. But he was learning about himself as a writer, learning that he had three sovereign strengths: he knew good stories when he heard them and could effectively retell them; he was a gifted observer of local life and customs, one whose eye took in significant details; and he was a hard worker.

In his knockabout days as a construction worker, day laborer, ranch hand, and bench chemist at the big Spreckels sugar mill outside Salinas, Steinbeck heard the stories the men on the jobs told one another. Some of the stories were coarse, some frankly obscene, many without real point. Still, there were occasional nuggets buried in them that he thought might one day prove usable, and he developed the habit of scrawling little notes on what he'd heard. At least once he offered to pay men for any stories they might have, and on that occasion a young hobo told him about a man saved from starvation by a nursing mother who gave him her breast. A decade later Steinbeck was to use this for the controversial last scene of *The Grapes of Wrath*.

From childhood Steinbeck had been a close observer of those small pieces of the natural world he privately sought out. The habit carried over in his apprentice years. He became an inveterate hiker of the Salinas Valley roads, of the streets of Salinas and Monterey. Several times he hitchhiked up to the Bay area. Later, when he had a succession of battered autos he used these to get out into the land, absorbing the look, the feel, the smell of places and seasons. If he thought he was hunting for subject matter, he was in fact storing up a fund of observation that he would effectively draw on all through the 1930s. And he kept writing. He kept working steadily, doggedly through the last years of the twenties, on through the discouragement of a mounting pile of rejection slips, the failure of one large project after another, the false start of *Cup of Gold.* He had made a promise to himself that he would write at least something every day, and usually he was able to keep that promise.

In the late winter of 1931 he heard more stories, and these especially intrigued him because they all came from the same restricted region, the Corral de Tierra valley. The tellers were Elizabeth Ingels, a young woman who worked briefly with Steinbeck's wife in Monterey, and Edith Gilfilin Wagner, related by marriage to the Wagners who owned the Hebert ranch in the Gabilan foothills. Through his aunt Molly, Steinbeck already had heard some of the local lore of that place, and a year before he had toyed with the idea of trying to write a collection of loosely connected short stories based on a specific place, the structure to be modelled on that of the *Decameron.* Now he put together what he had heard from his aunt with what Elizabeth Ingels and Edith Wagner were telling him. And he began to haunt the Corral de Tierra roads himself, hunting for more stories but really for those details of light and seasonal mood that might lift a slight narrative into a work of art. He renamed the valley Las Pasturas del Cielo, the Pastures of Heaven, and began his projected collection with a quick, deft sketch of a Spanish corporal in pursuit of some Indians who had escaped salvation at the Carmelo Mission and had run eastward over the mountains. The corporal eventually catches up with them, ties them into a coffle, and starts back to the mission. On their way a deer starts up before them out of the tall grass and lopes over a hill, the corporal in pursuit. At the hill's ridge the soldier looks down into a secluded valley "floored with green pasturage on which a herd of deer browsed. Perfect live oaks grew in the meadow of the lovely place, and the hills hugged it jealously against the fog and the wind." Faced with such soft splendor the lean man of violence whispers, "Holy Mother. . . . Here are the green pastures of Heaven to which our Lord leadeth us."

Steinbeck doesn't let his corporal get back to this paradise, though he tells us the man always intended to get back some day. Instead, the corporal dies of syphilis, his face rotting away, while still he dreams of the happy valley he accidentally discovered. But the real discovery here is Steinbeck's, for in *The Pastures of Heaven* he saw as if for the first time the literary possibilities of what he had known all his life. His, too, is that whisper, tinged with awe, as he sees his familiar landscape as the garden he has been given, Adam-like, to work. *The Pastures of Heaven* was published in 1932 to no great acclaim, but the book is an achievement nevertheless, and it marks the essential beginning of a career.

Like so many others, Steinbeck has a debt here to Sherwood Anderson, though it would be too much to call *Pastures* derivative. These are small, slight stories, sparely told. Here are the small tragedies, the blunders, and self-deceptions of those living in a small place under the merciless scrutiny of their neighbors, much as in *Winesburg, Ohio*. Here too is that ironic loneliness of those who live in a community yet suffer like castaways on a desert island. There is John Battle trapped within a religious mania that at last proves fatal. Shark Wicks pretends so successfully that he is a sharp investor with a fat bank balance that he succeeds in walling himself away from everyone. Tularecito is an idiot savant whose efforts to rejoin the elf people under the earth get him a one-way ticket to the insane asylum at Napa. Helen Van Deventer becomes the prisoner of her insane daughter until she liberates herself by murdering the girl. The fat, jolly Lopez sisters wish only to make the valley's best tortillas and "entertain" their male customers on the side until a witless practical joke gets them run out of the valley up to San Francisco where they are forced to set up as sure-enough prostitutes. Raymond Banks is a successful chicken farmer whose only exceptional trait is his pleasure in attending executions at San Quentin.

Prompted by what Elizabeth Ingels told him about a meddling family that sowed discord like seed through the valley, Steinbeck introduces a family he calls the Munroes whose arrival in the valley sets off a series of small disasters in which the Munroes marginally figure. This is patently a unifying device and not a very convincing one at that. It is as if the writer did not quite trust his own vision to hold the stories together. What in fact does hold them together is Steinbeck's grasp of the characters' shared landscape and the fact that the lives they lead on it are not equal in richness or contentment to the bounty and beauty of the landscape itself. In contrast to that deep green browse, those beguilingly rounded hills, and the perfect live oaks the old Spaniard had seen, the lives of the current inhabitants of the Pastures of Heaven seem mean, even twisted. This place may be paradise, Steinbeck is saying, but we are in a curious way exiled from it even as we live in it.

Steinbeck finally broke through to fame and some money with *Tortilla Flat* in 1935. In between *The Pastures of Heaven* and his novel of paisano life in Monterey he had published *To a God Unknown,* a novel begun years earlier as "The Green Lady" and badly marred by his undigested reading in Frazer's *The Golden Bough* and Jung. But he had also been writing the kind of stories of his home place that made up *Pastures,* and he knew these were good, that they derived a real power from his increasingly sure grasp of his place in all its moods and variety. In the spring of 1933 he wrote his old friend, George Albee, expressing his excitement over the world he was creating in these stories. "I think," he said, looking ahead, "I would like to write the story of this whole valley, of all the little towns and all the farms and ranches in the wilder hills. I can see how I would like to do it so that it would be the valley of the world." But before he made that attempt there was the life of the Monterey Peninsula, and he wanted to stake his claim to that portion of his heritage too.

He had hung out on the Monterey waterfront, its canneries and bars long enough to have developed a sympathetic knowledge of the gamey underside of the old town's life. He was especially intrigued by the town's paisano population. These were people of mixed Anglo, Hispanic, Italian, and Portuguese blood. Some ran small shops, some worked in the canneries or on fishing vessels. Others—and these he saw as the best potential material—lived hand-to-mouth, working when they had to, cadging drinks and meals, going on colossal drunks and ending up in jail or drying out in the pine woods up behind town. He supplemented his own observations of them through talking with Susan Gregory, a Monterey Spanish teacher and a gifted collector of tales, and Harriet Gragg, a longtime resident deeply informed about local history and folkways. Monty Hellam, the town's chief of police, told him other paisano stories. Two stories he had heard when he worked at the Spreckels sugar mill. As with *Pastures* he worked close to his oral materials and drew his characters on the models of living members of the paisano community. Again, as in the stories he had been writing about life on the other side of the Santa Lucias, there were his characteristically precise descriptions of the way a place looked and felt and smelled. Here is the town preparing for a big paisano party:

> All Monterey began to make gradual instinctive preparations against the night. Mrs. Guttierez cut little chiles into her enchilada sauce. Rupert Hogan, the seller of spirits, added water to his gin and put it away to be served after midnight. And he shook a little pepper into his early evening whisky. At El Paseo dancing pavilion, Bullet Rosendale opened

a carton of pretzels and arranged them like coarse brown lace on the big courtesy plates. The Palace Drug Company wound up its awning. A little group of men who had spent the afternoon in front of the post office, greeting their friends, moved toward the station to see the Del Monte Express from San Francisco come in. The sea gulls arose glutted from the fish cannery beaches and flew toward the sea rocks. Lines of pelicans pounded doggedly over the water wherever they go to spend the night. On the purse-seine fishing boats the Italian men folded their nets over the big rollers. Little Miss Alma Alvarez, who was ninety years old, took her daily bouquet of pink geraniums to the Virgin on the outer wall of the church of San Carlos. In the neighboring Methodist village of Pacific Grove the W.C.T.U. met for tea and discussion, listened while a little lady described the vice and prostitution of Monterey with energy and color. She thought a committee should visit these resorts to see exactly how terrible conditions really were. They had gone over the situation so often, and they needed new facts.

Steinbeck described Tortilla Flat so vividly and precisely that soon after the book had become a success out-of-towners were asking the locals to point the way to it.* Actually, it was to be found nowhere in specific but was instead an artful combination of several places drawn from the area. Eventually, so Steinbeck told a friend, the town's chamber of commerce began telling inquiring visitors that there was no such place and that Steinbeck had invented the entire book out of whole cloth—a minor irritation to Steinbeck but one symptomatic of a new problem for the writer. For *Tortilla Flat* had in fact brought Steinbeck the kind of fame he did not cherish. After a decade of struggle for recognition and financial stability he now had both but found that fame of this sort was a "pain in the ass." When a stranger recognized him on the street Steinbeck said he felt sick to his stomach. What he had wanted, and what he still hungered after, was another sort of fame, the kind that can only come from having written so well that it was your books' indisputable literary merits that made your name known.

More than a decade after *Tortilla Flat* he was still looking for this more genuine sort of fame and feeling more and more aggrieved that it continued to elude him. Somehow the high panjandrums of literary criticism, led by Edmund Wilson, refused to take him seriously. Too often he found himself written off as a merely popular writer or a sentimental

*Charlie Chaplin, a great Steinbeck fan, was so taken with the writer's descriptions of the places in his fictions that he used to come up from Hollywood to identify the locales.

one or as a muddled thinker without a serious, consistent philosophy. Alfred Kazin, in what remains one of the most influential studies of American prose literature of the first half of the twentieth century, *On Native Grounds* (1940), articulated what many critics felt was the truth about John Steinbeck. Kazin admired the depth and sincerity of Steinbeck's understanding of his valley landscape. There were, Kazin felt, certainly things to admire in *The Long Valley* and also in *In Dubious Battle,* Steinbeck's novel about a strike of fruit pickers. Finally, however, Kazin could not grant Steinbeck status as a first-rank writer, judging him to be crude, slyly sentimental, and consistently unable to breathe life into his characters.

Then—and always—Steinbeck made a public presence of indifference to his critical notices. What do I care? he as much as said: my books sell. And he knew also that at least some people spoke of him as ranking with Hemingway and Dos Passos among his contemporaries. But the disdain of the critical establishment hurt, and especially since it was becoming evident by the war years that he had written three works that looked very much as if they might be classics of the national literature.

The first of these was *Of Mice and Men* (1937), the story of two bindle stiffs chasing their paradisal dream of a homestead as they work the ranches up and down the Salinas Valley. Steinbeck's portrait of George and the huge, shambling half-wit he cares for, Lennie, was especially affecting in the California of the late 1930s when there were thousands of homeless men like them wandering the state and dreaming of somehow carving out a small piece of it for themselves. But the power of the portrait is not wholly topical, and in the post-Depression years *Of Mice and Men* remained just as powerful and poignant as it had been in the thirties. This was because its author had seen in the migrants and hobos of the Salinas Valley a truth about America, as it was in that terrible decade and as it had been from the very first. From the eighteenth century on there had always been the have-nots who had come here in hopes of establishing themselves on the land, and the older the republic became, the more of them there were until in the Depression years the gap between the haves and the landless was truly tragic. Nowhere was this gap more heartbreakingly obvious than in California, that "Garden of Eden" as the Okie folksinger Woody Guthrie put it, and nowhere in California was it more obvious than in the fertile northern reaches of Steinbeck's valley. At Soledad, where *Of Mice and Men* is set, there were miles of broad, burgeoning fields, but those who worked them had shockingly little to show for their backbreaking efforts. " 'Guys like us,' " George says, " 'that work on ranches, are the loneliest guys in the world. They got no family. They don't belong no place. They come to a ranch an' work up a stake and then they go into town and blow their stake, and the first thing you know they're poundin' their tail on some other ranch. They ain't got nothing to look ahead to.' " Later, Crooks, the black stable buck, supplies the antistrophe to George's and Lennie's dream of a place of their own: " 'I seen hunderds of men come by on the road,' " he says bitterly, " 'an' on the ranches, with their bindles on their back an' that same damn thing in their heads. Hunderds of them. They come, an' they quit an' go on; an' every damn one of 'em's got a little piece of land in his head. An' never a God damn one of 'em ever gets it. Just like heaven. Ever'body wants a little piece of lan'. I read plenty of books out here. Nobody never gets to heaven, and nobody gets no land.' "

The second of Steinbeck's major works of the thirties was *The Red Pony,* which began as a memory of his childhood when his parents had bought him a red pony named Jill. In 1933, back in the house in which he'd been born to care for his dying mother and deteriorating father, Steinbeck thought back on those days of primitive wonder when he'd had Jill, and he set his story out on the Hebert ranch where he had so often visited Max Wagner. Writing under the pressures of his parents' illnesses, Steinbeck was somehow able to reach back and recreate the boyhood world of Jody Tiflin in all its rich details: the mossy spring up behind the hilltop ranch house; the way clouds look when a boy lies in the tall grass and dreams up at them; a lucky hit with a slingshot; the look of puddles in a spring rain; the forbidding strangeness of the Santa Lucias to the west. And in dramatic tension with all this were the narratives of the red pony that dies of the strangles and of the mysterious old paisano who rides off into the mountains carrying only the Spanish rapier his father had given him long ago. In 1937 Steinbeck added a third story, "The Promise," and the next year as part of his collection of stories, *The Long Valley,* he completed the novella with the addition of "The Leader of the People."

This last is one of the best stories Steinbeck ever wrote and one of the best written by an American in the twentieth century. Like many of his stories it is a simple thing, telling merely of a visit to the Tiflin ranch by Mrs. Tiflin's father, a garrulous old gentleman whose life was effectively ended many years before when he led a wagon train across the great plains to California. But Steinbeck is able to impart to the visit and to the old man's musing reminiscences the whole restless yearning of a nation of people forever dissatisfied with wherever they were—that unappeasable hunger for something more, something further that for the nineteenth century was the very engine of Progress. When the boy Jody suggests to Grandfather that maybe some day he too could be the leader of the people Grandfather smiles down at him sadly. " 'There's no place to go,' " he tells the boy. " 'There's the ocean to stop you. There's a line of old men along the shore hating the ocean because it stopped them.' " Suddenly, in the light of the old man's forlorn words to a wondering boy the whole history of the settlement of California looks different, and even the great ocean is seen anew.

At the end of the thirties there had been *The Grapes of Wrath* (1939), which Steinbeck had written in a single year. Because of its depiction of the tragic plight of the Okies, the brutality of the California growers and law enforcement officials, and because of its last scene the book was the subject of explosive controversy from the first month it was published, and there are still places in California where you can get into a pretty serious argument about it. Though it won that year's Pulitzer Prize, from the first the literary merits of the novel went all but unremarked in the furor. Steinbeck was either a hero of conscience or a Communist rat, depending on the point of view of the debater, but he was rarely a *writer* who had created a fictional work of art. As the controversy continued to build through the next year and as his mail continued to mount—praise, vicious hate mail, death threats, pathetic requests for money from migrants—Steinbeck grew to loathe the whole subject and especially so since he now saw, or at least suspected, that *The Grapes of Wrath* had irrevocably remade him into a celebrity whose name would be forever linked to this one book. His chance at that more genuine literary fame he had quietly and honestly worked for now seemed blasted, not by failure but by a strange sort of success. Always a quiet man in a crowd, almost shy at times, and modest in his assessment of his talents, he had wanted only to be judged on the intrinsic merits of his works. Now in the wake of *The Grapes of Wrath* he appeared doomed to permanent public misapprehension.

This is the context in which Steinbeck began to make plans for his big book about his valley, the long-deferred novel about all the valley's ranches and farms, even about the quality of its soil. This would be the book of which he had written George Albee in '33, saying he could see a way to write about the Salinas Valley so that it would seem the valley of the world. For it was evident to him now that he needed a big book, one of sufficient range and depth to explain once and finally who he was and what all along he had meant. In these post-war years he knew that *The Grapes of Wrath* was already being dismissed by some as a period piece just as earlier it had been dismissed as a piece of propaganda. Similarly, the critics could dismiss *Tortilla Flat* and its later companion, *Cannery Row* (1945), as pleasing, slight books of local color. True, some of his short stories were becoming staples of anthologies, but there was a clear need for a magnum opus, and it would have to be about his home place and about his own history in that place.

He had for some years been in exile from California, living much of the time in New York, and he had been gone even longer from Salinas and the valley. In his home state opinions about him were still sharply divided, and on his part he harbored grudges against his home town where he felt he had never really been understood. In 1944 when he had tried to move back to California with his second wife he had discovered quickly that Thomas Wolfe had been right about one thing at least: you couldn't go home again because home was different than it was when you left it. And indeed, the very success with which he had described Monterey in *Tortilla Flat* and *Cannery Row* had played a significant part in so changing the town that now it felt strange to him. Monterey, Salinas, and the valley below had nurtured what was best in him as a writer, and now he would have to find a way of reattaching himself to this place if he was to write his book.

So in the first months of 1948 he went back to begin his researches. In a rented car he traveled the old roads just as he had in his apprentice years, "going around the country," as he wrote a friend, "getting reacquainted with trees and bushes." He drove down through the locale he had made famous with *Of Mice and Men* to the old Hamilton ranch east of King City and poked about in the Gabilan foothills where in childhood his uncle had taken him on fishing expeditions. He talked again in Monterey with Harriet Gragg who had given him material for *Tortilla Flat* and combed the files of the Salinas newspaper. The more immersed he became in the landscape of his past, the more he convinced himself that this was the time for him to make his major statement. In a characteristic mood of mingled confidence and anxiety he wrote Pascal Covici that he had

> been into the river beds now and on the mountains and I've walked through the fields and picked the little plants. In other words, I have done just exactly what I came out here to do. What will come out of it I don't know but I do know that it will be long. There is so much,

so very much. I've got to make it good, hell, I've got to make it unique. I'm afraid I will have to build a whole new kind of expression for it. And maybe go nuts doing it. . . . But it's the whole nasty bloody lovely history of the world, that's what it is with no boundaries except my own inabilities.

To his wife he wrote that the book might "be my swan song, but it certainly will be the largest and most important work I have or maybe will do."

But before he would actually begin the writing there would be a three-year delay. In large measure this was the consequence of the messy ending of his second marriage, the death of his old and close friend from Cannery Row days, Ed Ricketts, and the beginning of his third and last marriage. But in addition to these events there was the natural procrastination of a writer who believes he has staked everything on a new book. To so much as begin it was to begin to fail a little of that perfection at which he was aiming.

His working title was "The Salinas Valley," and writing his friend George Albee about it he told him the book would be fundamentally different from any of his previous works because in those he had always had what he now called a "personal out. I could say—it is just practice for 'the book.' If I can't do this one, the practice was not worth it. So you see I feel at once stimulated and scared." Then in January 1951, newly remarried and resettled in Manhattan, he confronted the inevitability of the actual beginning of the writing, but not before commencing a habit he would continue throughout the writing of the book. He began each working day with an unsent letter to Pascal Covici, telling him of his struggles with the manuscript, of his hopes for it, of the decisions he daily faced in writing about matters of such intense personal concern to him. Posthumously published as *The Journal of A Novel,* the book, while it is not quite the private communication of author to publisher it seems, is nevertheless an intimate, revealing look at a writer who saw himself scaling with bare hands toward the summit of his life and career. "We come now to the book," he wrote Covici on January 29. "Maybe I can finally write this book. All the experiment is over now. I either write the book or I do not. . . . In a sense it will be two books—the story of my country and the story of me."

February 12: "This is the book I have always wanted and have worked and prayed to be able to write."

February 20: "I truly want it to be the best I have ever done."
February 26: "This is my big novel."
March 21: "This is my big book."
June 26: "I know it is the best book I have ever done."
July 9: "I still think this is The Book, as far as I am concerned. Always before I held something back for later. Nothing is held back here."

On November 16, he wrote his friend Bo Beskow: "I finished my book a week ago. Just short of a thousand pages—265,000 words." It was still for him "the book," and having written it he said he felt released to do anything he wanted. "Always I had this book waiting to be written."

The book Steinbeck had written is a daring combination of fiction and biography, realism and allegory. He had long been interested in stories of families down the generations, an interest that probably derived from the Old Testament of which he was an avid reader. Then, too, his home country researches of recent years had turned up a lot of his own family history, and now his own history had come to include two young sons by his second wife. In a way he had not yet worked out he knew he wanted to forge a connection between the generations of Adam and the generations of his own people. This would be no light-hearted look at the paisano subculture nor an affectionate portrait of the loafers of Cannery Row: this was to be the story of the family of humankind in the valley of the world, told through the fortunes of two generations of a fictional family whose fathers and sons recapitulate what Steinbeck saw as the primal situation, the story of Cain and Abel. He introduces into the narrative his own people as well, the Hamiltons, and there is even a small boy, John Steinbeck, then living in his parents' house on Central and Stone in Salinas.

In the story of the sons of Adam the writer found everything essential and recurrent in the human condition. Here was inexplicable favoritism and violent reaction; here were mingled inextricably love and hate; here was the inevitability of sin and also the potential to triumph over it. Adam and Charles are the sons of Adam Trask of Connecticut, the former favored by his tyrant father and for no apparent reason since Charles is his evident superior in most things. The favoritism, expressed through the years in countless ways, kindles in Charles a murderous rage, and one night he just misses killing his brother. In time, Adam Trask has his own sons and lives in Salinas. Aron is the favored one, Cal the dark and secretive night-walker who discovers that his mother, long vanished from the home, is in fact now a notorious madam on Salinas's whorehouse row. When Cal tries for and fails to get his father's love and approval he vengefully takes the trusting Aron to see their mother and so blasts his brother's innocence to bits. Aron runs off to the army and is subsequently killed in action in World War I, and this precipitates Adam Trask's fatal stroke. Thus Cal, like Cain before him, has in effect been both his brother's keeper and killer, driven to be the latter by his father's greater regard for Aron. He has also become his father's killer. He explains his behavior to himself as a consequence of his inheritance of his mother's evil nature, but Lee, the longtime, sagacious Chinese servant, won't let Cal off so easily. He wants Cal to see that everyone has a choice to make, that sin is not inevitable. He rouses the dying Adam from his stupor and begs him to give Cal his blessing. Adam does so, whispering the Hebrew word, *"Timshel,"* which can be translated, "thou mayest," as in the line from the Cain and Abel story where God speaks to Cain and says to him of sin, "and thou mayest rule over him.' "*

It is a powerful narrative, and in it Steinbeck has effectively merged the history of his valley and town. The first chapter, in which Steinbeck gives a panoramic view of the valley, is one of the most poetic things he ever wrote. Reading it, you can hardly fail to feel the urgency Steinbeck brought to the writing of it. This, you can hear him saying beneath the words of the page, is where I come from, these are my roots, and this is me. If the whole book is not as daring as Whitman's audacious effort of *Leaves of Grass,* where he proposes that to read the book, or even just to hold it, is to read a man and hold him, still, this is a bold, daring effort Steinbeck makes. Every detail of geography and incident and name is utterly faithful to the way it was, for Steinbeck had convinced himself that in this book he could make fiction, history, and family biography harmonious, could make them comment on each other. In the main, he is successful in doing this. His evocations of the valley and the town provide the novel with precisely that rich soil of reality he wanted to ground his fiction in, and the inclusion of his own family, rather than being distracting and obtrusive, gives an unusual sort of substantiality to the book. In the figure of his grandfather, Samuel Hamilton, blacksmith, dowser, well-driller, philosopher, Steinbeck wanted to create a grand,

*As Steinbeck discovered after the publication of the novel, the word is actually "timshol." Ernest Tedlock and C. V. Wicker in *Steinbeck and His Critics* suggest "timshawl" might be an even more accurate transliteration of the Hebrew characters. The Jewish Publication Society's *The Holy Scriptures,* does indeed give the translation, "thou mayest," though most other translations make the verb a command, as in the King James version's "thou shalt".

folkloric figure. If Sam Hamilton talks too much and too wisely for our current taste, there is no doubt such well read, speculative men were once to be found all along the moving frontier. He is a significant, plausible character, one of Steinbeck's best.

It is in attempting to combine mid-twentieth-century realism with allegory that Steinbeck crucially fails. Driven by his own hugely inflated expectations for this book and by those of others—including, apparently, those of his agents and publisher—he wanted desperately to sound the great voice in these pages, and for him that sort of greatness was available only to works that had their roots sunk deep in myth. From early in his career allegory had tempted him, and he seemed always to wish he had had the advantages of Malory or Bunyan whose works so naturally drew from the well of Christian mythology. Thus, in the opening lines of *Tortilla Flat* he says that his hero's hovel in the Flat was "not unlike the Round Table and Danny's friends were not unlike the knights of it." Throughout the book there is an effort to make that equation work. *To a God Unknown,* a novel about a terrible drought that is broken by the blood of ritual suicide, literally drowns in mythology. There is a preliminary effort in *Cannery Row* to get readers to see Mack and the boys as saints, martyrs, and holy men and their actions as the misunderstood acts of the blessed. It is significant that after having thought of the "big book" for some time as "The Salinas Valley" Steinbeck then went to "My Valley," "Valley of the Sea," and "Cain Sign," before settling on *East of Eden,* the place to which Cain is banished. But the procession of provisional titles probably reflects more the hardening of his resolve that this should be an allegorical/realistic novel rather than a belated arrival at such a decision, for from the outset he was certain that the book about the valley and the town had to have the largest possible resonance. So, in February 1951, shortly after he had begun its composition he wrote in one of his letters to Covici that it would not be wise to burden superficial readers with "profound philosophy." Such readers, he says, might well be contented with the surface narrative. But "your literate and understanding man will take joy of finding the secrets hidden in this book almost as though he searched for treasure, but we must never tell anyone they are here. Let them be found by accident. I have made the mistake of telling my readers before and I will never make that mistake again." But, as these lines sufficiently suggest, he did so and in spades.

He had always been a good storyteller, and in *East of Eden* he still was. The story of the two generations of the Trask family is often riveting and surely could have carried the burden its creator wished. But he could not trust it enough and continually intrudes, not only with heavy, direct philosophizings, but worse with all the creakily obvious machinery of his allegory. Charles Trask is already a recognizable Cain-like figure, and the scene in which he beats his brother senseless, then comes back for him with a hatchet is chillingly believable. Still, Steinbeck has to have Charles smash his forehead with a crowbar so as to give himself the Cain sign. Adam Trask's twin sons must be named Cal and Aron since Adam, Samuel Hamilton, and Lee collectively contemplate the Scriptural account of Cain and Abel. Lee has to be an all-wise observer/commentator on the vicissitudes of Trask family life ("The book needs his eye and his criticism which is more detached than mine," he wrote Covici). Lee even turns a group of San Francisco Chinatown elders into a Hebrew study group that debates whether God commanded Cain to rule over sin or rather said he had the potential to do so.

Worst, there is the character of Cathy Ames Trask, later in her Salinas whoredom a.k.a. Kate. She is the most glaringly obvious of those Steinbeck frankly spoke of as his "symbol people." He tells us she is a "monster," and like Cain and Charles she is marked with a scar. She is the embodiment of that sin, that evil which is inevitable in life but which humans have the capacity to resist and even defeat. Here is a being who while still prepubescent railroads two male peers to the reformatory, who, still a teenager, incinerates her parents in the family home, and then becomes a prostitute. Barely two weeks after giving birth to the twins she shoots and severely wounds Adam, then leaves the ranch for Salinas. There she murders the madam of a whorehouse and succeeds her, turning it into the most viciously depraved such place in the whole state. She is brilliant, inflexible, remorseless, and unbelievable—even in our own day of mass murderers like Richard Speck and John Wayne Gacey. And in our own time of Hitler and Stalin, of Mengele and My Lai we know too well there are monsters among us and within us (which was Steinbeck's point). But in fiction that makes some pretense to realism the portrayal of such monsters demands some psychological plausibility, some depth of being, and Cathy has none. Perhaps, stung by his second wife's infidelity and desertion of him, her persistent subterfuge (as he saw the case), he simply could not supply this. More likely, he felt he needed the figure of Sin, the beast that crouched in the Scripture, needed it even as it is there in *Pilgrim's Progress*.

And still *East of Eden* is almost a great book, almost the book Steinbeck so fiercely wanted it to be. It represents the near-triumph of a gifted chronicler of his home place over disastrously flawed expectations. It is almost what at one point he told Covici it would be, the autobiography of the long valley. In his deepest, best self Steinbeck had always trusted the land, what it in itself was and where as a writer it would lead him. His father, that silently howling man, as his son once characterized him, had missed his calling. He ought to have been a farmer his son thought, and always he liked to have his hands in the soil, even if it was only a prim little backyard garden. His son knew the impulse and respected it. In one of his best stories in *The Long Valley*, "The Chrysanthemums," a ranch wife tries to articulate to a traveling tinker her feel for the valley's soil:

> "Did you ever hear of planting hands?"
> "Can't say I have, ma'am."
> "Well, I can only tell you what it feels like. It's when you're picking off the buds you don't want. Everything goes right down into your fingertips. You watch your fingers work. They do it themselves. You can feel how it is. They pick and pick the buds. They never make a mistake. They're with the plant. Do you see? Your fingers and the plant. You can feel that, right up your arm. They know. They never make a mistake."

But in *East of Eden* Steinbeck lost the feel of his place—or maybe he had lost it when he moved away and then couldn't get it back when he most needed it. Something of this seems to lurk in the pages of *Travels with Charley* when he returns to the scenes of his youthful struggles and triumphs. He was sixty then and tired beyond his years. And, of course, he was a stranger, but he expected that. What he had not expected was the feeling that Monterey and the valley were like a graveyard to him. Too much was buried there. His last stop was Frémont's Peak in the Gabilans. From it he could look down into the landscape of his birth, childhood, and youth and remember all its bright promise for him as a writer. Perhaps he was at that moment not unlike the lean man of war, the Spanish corporal, when he looked into Las Pasturas del Cielo. But what Steinbeck says he thought about then was that once he had wished to be buried up there.

II

❧ STEINBECK LEFT HIS HOME PLACE disappointed that it had changed so much. But his deeper disappointment may have been that it had changed so little. Perhaps every writer feels misunderstood and undervalued in his or her home place. So many of them who wrote evocatively of their places left them in search of a community that might nurture them! Jackson J. Benson in his biography of Steinbeck quotes Arthur Miller, who knew Steinbeck in New York, on Steinbeck's search for community:

> He was trying to find a community in the United States that would feed him, toward which he could react in a feeling way, rather than merely as an observer or a commentator. And I don't know if there is such a place left in the world. Faulkner tried to keep it alive, in Mississippi, but him apart, I don't know if it is possible.

The problem of the relationship of the writer to his home community becomes considerably more tangled if what the writer has said of it brings to it an unwanted publicity (as was true of Faulkner and Oxford). There is on the Monterey Peninsula and in the Salinas area a certain amount of lingering resentment of Steinbeck on this score. A man I talked with in Monterey, a long-time resident who didn't want his name used, said Steinbeck had ruined the town: "You go down there [to Cannery Row], and take a look. Just take a look! Of course, you won't know what you're looking at, but, still, if you've got eyes, you'll see what I mean."

Indeed. Everything along and adjacent to what was once Ocean View Avenue is now Steinbeck something-or-other, all based, of course, on that one book. Even the drinks and sandwiches bear the heavy stamp of Steinbeckiana: "Flora's Fantasy. While her girls were entertaining upstairs . . . Flora relaxed downstairs sipping this fantastic blueberry cream froth. $2.75." "Mack and the Boys Special." "Rickett's [sic] Melt." Et cetera.

Well, that is one thing, and in Monterey at least some people must be happy with what use they can make of the man's work. Over the mountains in the Salinas area, though, there are far fewer public references to the writer. Chief among them is the John Steinbeck Library (a part of the Salinas Public Library) outside of which is a life-sized statue of the writer. There is an annual festival. The home on Central and Stone is a combination restaurant and museum, and behind it are apartments called Steinbeck Flats. But this is about all. There is too much to remember, too much to forgive, Pauline Pearson was saying as we drove through the Corral de Tierra hills. "You can't get on these ranches," she said, "and especially not if you mention John Steinbeck. Why? Because these people thought he'd betrayed them by writing up their stories."

"Jackson Benson quotes Steinbeck as confessing to someone whose stories he'd used that he was a 'shameless magpie' who just gathered up every narrative scrap he could," I put in.

"Oh, he was! These people out here had told those stories to each other for years. They really felt betrayed when they were printed in *Pastures of Heaven*. See that hill back in there? That's where the outlaw Vasquez had his cabin. Steinbeck used that in one of the stories. Well, *Pastures of Heaven*. Others hated *Grapes of Wrath*. People all through this country dreaded every book as it came along. Just when they'd gotten over the last one and more or less forgiven him, along came *East of Eden*. And what a scandal that caused with all its talk of the Row and its respectable customers. There was tremendous speculation in town about the originals of some of the characters. Who was Kate? Lots of people thought they knew. You see, here again, these stories were theirs, they thought, and they thought Steinbeck stole them from them. And then, too, you have to appreciate the fact that they all knew who Steinbeck was writing about, they knew the references."

As for Pauline Pearson herself, she had more than intellectual and esthetic reasons for her interest in Steinbeck and her spirited defense of his achievement. She had come to California out of Seminole, Oklahoma, in 1935, at the age of eight. She and her parents, she wanted me to understand, were not technically Okies "because we had family already here. My grandparents had been out here since the twenties. But my dad came out because of the climate. Back in Oklahoma he became crippled when he drank some 'Jake' in the back of somebody's drugstore. They all got terribly crippled. 'Jake'? Well, it was some kind of alcoholic drink that was supposed to be good for you.* Well, a lot of people got crippled drinking it. It was a big scandal then. My father always claimed that Roosevelt never had polio, that what he had was 'Jakeitis.'" She laughed. "Well, they told him at Mayo's that he might do better in a climate that was less extreme than Oklahoma's, so we made 'the crossing.'

*Woody Guthrie from Okemah, Oklahoma, dispensed "Jake" in the early 1930s as a bartender. It was, he recalled in his autobiography, *Bound for Glory,* a deadly mixture of ginger and high-proof alcohol, and many people became permanently paralyzed by it.

"But I was identified as an Okie, yes. In school they put down where you were from, and I guess we had that accent, too: kind of flat, you know. But I can tell you that we felt that discrimination even though we never asked for a handout, not a thing. My parents never wanted to be a burden on their adopted state, and they never were. My mother was raised to be a lady, and she actually was a teacher. But she wasn't certified to teach in California, and she had to take what work she could get. She worked in a hospital, and then she went to work in the lettuce sheds—you know, where they did the packing. Of course, the hours were irregular, and the conditions were terrible, but you could make good money at it, and so a lot of women worked there and were glad to get it."

Two Corral de Tierra residents I talked with later that day seemed to have nothing against John Steinbeck. One was Gene Rowlands. Bull-shouldered, red-faced beneath his Resistol cowboy hat, and with the beginnings of a serious belly, he was the manager of the Walter Markham ranch that sat right under the huge sandstone cliff castle that was a local landmark. I sought him out there to ask if I might walk up to the ranch house Steinbeck had used as the setting for "The Murder," one of the stories in *The Long Valley*. Sure, he'd heard of Steinbeck. Hadn't read anything, though. I told him the story was about a rancher who comes home one night to find his wife in bed with her cousin and blows the man's brains against the bedstead with a shot from his carbine. "Listen to this," I said to Gene Rowlands, pulling from my bag of Steinbeck books a copy of *The Long Valley*: "At the head of the canyon there stands a tremendous stone castle, buttressed and towered like those strongholds the Crusaders put up in the path of their conquests. Only a close visit to the castle shows it to be a strange accident of time and water and erosion working on soft, stratified sandstone. . . . Below the castle, on the nearly level floor of the canyon, stand the old ranch house, a weathered barn and a warped feeding-shed for cattle."

"Wait a minute," interrupted Rowands, then he stepped to his pickup where he brought out a large, wrinkled envelope. "I just dug these up the other day," he said, withdrawing from the envelope photos of the ranch house. He seemed surprised and mildly interested that someone would have used it for a fictional setting. He knew nothing about *The Grapes of Wrath*, though he had come out here from the Texas Dust Bowl in 1939.

"The first job I got here was cutting hay over there where the country club is now," gesturing northward with his thumb. "Fella says can you drive a team? Shit! I'd been driving a team since I could stand. So he says, 'Well, don't be standing around talking about it: harness those two horses and get to work.' In those days you didn't have to be told twice. You were glad to get what you could." Rowlands said he'd bought a piece of land up the valley a bit and also had a ranch in the Carmel Valley he'd work when his present job ran out: the Markham ranch had recently been subdivided, he told me, with lots going for $750,000 apiece. I dug in my bag again, pulling out *The Pastures of Heaven* this time, and told him Steinbeck had predicted the coming development of the valley in the book's last chapter. There a rich man looks down into the Pastures and says, "Some day there'll be big houses in that valley, stone houses and gardens, golf links and big gates and iron work. Rich men will live there. . . .' " This was in 1932, I told Rowlands.

"Shit!" he snorted. "I wish I'd read that back then. I'd be a wealthy man today, that's for sure."

Roy Diaz has lived under the valley's Mt. Toro for most of his seventy years. "Sometimes," he told me as we sat in his cheerful kitchen, looking down on his small sloping vineyard of Muscat and Zinfandel grapes, "I look at all this building, these fancy new houses, and I could just kick myself in the pants. I coulda bought this whole area around me now for $11,000. But I didn't have the $11,000 to pay, you know? Now. . . ." He shrugged his shoulders. "But I've sold most of my land. I used to run some stock up here, but mostly I raised tomatoes. Not now, though. I'm seventy years old. I'm the last of the old-timers.

"My grandfather, Francisco Ambrosio, he was a number-one harpooner on a whaler. Came out of Spain when he was fifteen. He went around the world I don't know how many times and ended up in Monterey. He came in here in the 1840s and planted 360 acres in potatoes. That's my beginning here. When I was a kid we had mostly number-one oat hay up here. Sold it to fellows over in Monterey that had polo ponies. Then one day a guy says to my dad, 'Let's try tomatoes.' " He got up and went into another room for a moment. His wife Lorraine refilled my coffee cup as he re-emerged with a scrapbook. Leafing through its bulging pages, he came to a browning snapshot of himself and another young man in a sun-drenched field, holding up for the camera huge specimen tomatoes. "Now you're talking tomatoes," he announced proudly. "We sold 'em to the canneries over in Monterey."

About Steinbeck? "Living back in here Steinbeck was nothing to us. We really didn't know who the hell he was. Blanco Diaz, my dad, used to run into him over in Monterey. Liked him, too. Then that book *The Pastures of Heaven* came out, and there's no question there was a debate around here. Some said he told the truth, and some said he exaggerated. But," waving somewhat scornfully at a couple of big new homes visible on distant slopes, "you won't find out anything about him from all these newcomers here. I'm surrounded by 'em. That's all there are now. They don't know how it was then. You ask 'em about Steinbeck, they'll say, 'Love him.' But they don't know what they love.

"But I can't really say I knew him. Met him once or twice in here. Then, way later, when he was working on *East of Eden,* he wrote me and asked me some questions. I researched the answers and wrote him back." He shrugged again. "He asked me to help him with something, and I did. So. . . ."

Roy and Lorraine Diaz led me out into the bright sunshine and to her brilliant flower beds. The fog had lifted, and the last tatters of it were now running away over the shoulders of the impossibly green hills. "There's lots of history here," Roy Diaz was saying, and Lorraine Diaz told me I ought to interview her aged mother, half-Swiss, half-Carmel Mission Indian, and all character. Roy Diaz wanted to tell me about the outlaw Vasquez: "Old Ambrosio fed him and his gang. He was kind of a Robin Hood type of fellow. There's lots of stories about him around here. Now, Three-Finger Jack, he had a horse-collar fringed with gringo ears. He was a mean one."

Among other services Pauline Pearson had rendered Steinbeck aficionados was the printing of a map of Salinas that shows the many sites in town associated with *East of Eden*. I told her I was going to take one of Cal's nighttime walks over the Southern Pacific tracks to Chinatown and the Row. She traced it for me on the map—up Main to East Market, then right across the tracks, and there you are. But she advised against it, day or night, telling me it was as rough over there now as it was during the early years of the century when the novel was set. Joyce Conrow, a lifetime Salinas resident agreed. "I wouldn't do that if I were you," she said as we talked by telephone. "When we were kids—oh, thirty-five years ago, something like that—we might take a drive through there, just having a thrill, I guess. And your date would tell you, 'Now, you get down (on the floor of the car)—they were afraid, you see, of what might happen or that we'd be shocked. And there they were, the 'ladies of the evening,' right out on the verandas, yelling down at you, all kinds of things. It was two complete streets—wide open: gambling, prostitution, drugs, what-have-you. There were Filipinos, Japs, Chinese—you name it." She said she hadn't been over there in years but knew it was still rough.

I decided I had to go anyway, though I would do the walk in daylight, beginning at the Steinbeck house, then walking over to West Market, which in the novel's time was Castroville Street. Here was immediate evidence that Hispanics had largely replaced the Orientals on the town's wild side. A disco bar was called Pancho's Village. There were El Aguila Bakery and El Charito Market and signs in most windows announcing that Spanish was spoken there.

Downtown Main Street had been renovated, but out here where it crossed West Market the renovation had run out of steam, and the intersection was a huddle of boarded-up stores, transients' hotels, and vacant lots. On this bright, hot spring day paper sacks and newspapers, caught in a high wind that swirled between buildings, spun and floated in hopeless circles. In the entryways of the dead stores greasy chop bones blackened. Old men with canes lurched along the sidewalk or paused beneath spindly curbside trees. On a wall, this legend:

> Nick the Mouth Spoke Here
> And Here!
> And Here!
> And Here!
> And Here!

Beneath which was a pile of human excrement and an empty bottle of Thunderbird.

On the wrong side of the tracks the streets of Pajaro, Soledad, California, and Lake were sparsely populated with drinkers, loungers, and a few streetwalkers. A mustachioed young Hispanic man in front of the Do Drop Inn on Soledad smiled at me and eyed the camera I had slung about my neck. "Just out taking in the sights and scenes, eh? I hear you," he said cheerfully enough. A sign in the Do Drop's window behind him read, "THIS IS NO PLACE TO MAKE CONNECTIONS FOR ANYTHING. CATCH YOUR PUSHER OUTSIDE AND ON CITY PROPERTY. NOT MINE!"

In *East of Eden,* Jenny's house, long a Salinas institution, stands at the corner of East Market and California, just as its real-life prototype did. The building was still there, though now it was a produce market. The Hispanic kid trundling crates from a truck to the store said he knew who Steinbeck was but couldn't remember the names of the stories. Just around the corner on California I met up with a black streetwalker. She might have been sixteen, though with her huge sunglasses it was hard to tell. After the usual offer had been tendered and declined she cased me quickly: if not a john, then a cop? Not likely, she decided, seeing my camera. Well, if I was neither, what was I doing around here? When I told her, she barked a harsh, short laugh. "That's a good one," she said. We walked on a few paces, then suddenly she stepped in front of me, her legs spread defiantly, and pulled up her blouse exposing her full breasts with small, dark nipples. "Here," she coaxed mockingly. "Come on, take a picture of my titties for your souvenir." I focused, clicked off the shot, then turned back toward East Market feeling her laughter jabbing me between the shoulder blades until I turned the corner.

I walked past the last small vestige of Chinatown where Cal looks in on the fan-tan games. From the Green Gold Inn's second-floor balcony two Chinese men regarded me casually as I passed along the opposite side of the street. Outside the Swinging Door a knot of men with too much time on their hands and too little in their pockets lounged against the building, sat on the curb or on the hoods of cars. I watched them a few minutes from a discreet remove and saw trouble in the making. A car pulled up to the curb, a man got out, and plunging into the knot of men, dragged one of them toward the car. He spun him around and slammed him backward over the hood, smacking his head sideways with one short, vicious chop. The crowd began to move then, perhaps taking sides, and I began to move, too, across the vacant lot toward the tracks and away from the action. Pausing to look back I saw immediately behind me a young woman, her heavy brunette hair blowing behind her. She was not quite fleeing the scene, but she was putting distance between it and her with a good, determined gait. "Keep going," she said to me with no more than a sideways glance. "Just keep going." I fell into step beside her as we made the tracks.

"What's going on back there?" I wondered. She shook her head.

"Oh, I don't know," she replied wearily. "Some deal, I guess." She asked me what I was doing over there, and I told her. "Do you know," she asked, "that Steinbeck's ghost is over in that house on Central? Well, it's true. Yes, I believe in ghosts. I believe in life after death. You don't just pass on. I think that some persons don't want to leave the places they've lived, so they hang around. Maybe Steinbeck has something he'd like to finish up here. Why knows?" I asked her if I could buy her a cup of coffee, and she took me into Tico's restaurant and told me to wait for her while she went across to the Hotel Columbo to "pay her rent."

"I'm a hooker," she casually told me when she'd returned and we sat drinking coffee. "But I'm a pretty smart one." She said she worked out of the Colombo and was getting "back to normal" after she'd been stabbed in the lobby over there last year in what she said was a case of mistaken identity. "I was just passing through the lobby when I felt this strange sensation through my whole body, and things started spinning around. I made it to the top floor, and I said to my friend, 'You know, I think I just got stabbed.' They took me out in a body bag. I had no blood pressure: D.O.A. But I can actually remember being in the body bag! It was weird! Anyway, he didn't just stab me: he went right through my lung."

"This must be a pretty rough place to work even without cases of mistaken identity," I suggested.

"Well, it can be. I don't work at night anymore, and I'm a lot more choosy now than I used to be. I'll take clean-cut men. No blacks." I asked her then if she did drugs.

"I'm not going to lie to you. I did heroin. I had a $400- to $500-dollar-a-day habit. Just to get out of bed in the morning was $200. I don't know how much you know about drugs, but that's a habit. But now I've been two months and eleven days on methadone—clean time. Yes, I find it effective. I take it once a day, drink it in the morning." Sweat was beginning to stand out on her broad but handsome face with its hazel eyes and windblown hair. Her name, she said, was Ryan, and she had lived here all her life. She remembered when they'd filmed some of the scenes for the Elia Kazan/James Dean version of *East of Eden* over where she now worked. "They used the Green Gold for that," she recalled. "I was just a kid, but we went over every day they were there."*

"How long will I do this? I don't know. But I'm getting very interested in paralegal work. You know, I think I could do that, especially the criminal side of it." She laughed. "I know it would be hard work—but so is this. Maybe a couple more years of this, and then I'll try that."

---

*She was probably recalling the television production of the novel in 1981. The Kazan/Dean film was made in 1955 and utilized only some local footage shot in the fields at Spreckels.

She stubbed another cigarette out in the ashtray. In the crook of her arm there was a small deep scab.

"Look," she said, glancing away from me and out toward the Colombo. "I'd really like to talk longer, but I have to hustle $25 for my rent, and this has been a real slow day [glancing again at the half-empty street]. And anyway I'm behind: I got scooped up last night." She explained what this meant: the casual, occasional mass arrests of streetwalkers in the area. "All night, sitting down there doing nothing."

"Are you going over to Monterey?" she asked me out in the street's sun. "There's lots of Steinbeck stuff over there. I'd love to take you there. Such peace and serenity. . . ."

Where you find the first gentle rolls that will become foothills and then the Gabilans you come to the former Hebert ranch Steinbeck used as the setting for *The Red Pony*. A long straight ranch road leads upward past big strawberry fields on either side. This day there were gangs of Hispanic workers in them—straw hats, ball caps, flannel shirts against the cool gray of early spring. There has been some debate among Steinbeck scholars over whether this was the place Steinbeck indeed had in mind in writing that novella, but once the careful reader sees the layout of the ranch all doubts are banished: this was the spot. There was the old ranch house and up the slope from it the big barn wearing scales of red lichens like medallions on its gray boards. A large double-trunked camphor tree in front of the house reminded you of the ranch dog, Doubletree Mutt. There was the bunkhouse, home to that consummately professional horseman, Billy Buck. It is Billy who feels responsible for the death of Jody's pony in the novella's first story, and it is he who tries to make amends by getting Jody another colt in the second part. The small building was untenanted now, sagging at the roof and with the weeds reaching up toward it. Climbing up the slope behind the bunkhouse, I could see the hills declining to the eastward creep of Salinas's subdivisions. This was still country, but I had to wonder how long it would be before the pressure of those subdivisions so inflated land prices that agriculture would no longer make sense here.

The current owners were Bruno and Terry Sala. Mrs. Sala, a lithe, straight young woman clad in distractingly tight jeans, obligingly showed me about. She knew the novella well and had no doubt herself that her place was the one referred to in those pages. She told me she and her husband raised beef and dairy cattle and leased portions of their 2,000 acres to individual farmers. We talked in the dim, high-pitched barn where the rafters were alive with swallows and sparrows, and I thought of the wind slamming the big doors in the night storm when Jody awakens to find his pony gone. Outside, the barn pigeons strutted its roof ridge making their guttural cooings, and behind it in a pasture a four-day-old colt with a red coat and white blaze stood stiffly by its dam who had her ears laid back in warning. As Terry Sala and I talked we could look down on the bent backs of the workers in the strawberry fields below.

The Salases rented stables to those from town who had horses, and wandering by I came upon two young women from nearby Hartnell College. While we chatted they curried their horses. "*The Red Pony*! Really? Right here?" Tessa Ward asked with what seemed genuine excitement. "That's neat! I'm going to have to read that book." Tiffany Nunes said she had read it. "I read all those books, but I didn't know this was the place."

"You grow up around here hearing about Steinbeck," Tessa Ward came back. "I grew up in Hollister, and, you know, when you're little someone's always saying, 'Halfway between Hollister and wherever Steinbeck's blank-blank was set.' And, you know, it's kinda exciting in a way. I mean, Hollister's Nowheresville, and here a famous author has written about it. It makes you think."

Terry Sala routed me back to Salinas through more sites associated with Steinbeck's life and work. I went up the San Juan Grade past the little schoolhouse Jody Tiflin attended; swung south past the Alisal Creek where once the wild azaleas grew and a narrow-gauge train brought picnickers out from town (the creek had dried up and with it the azaleas); then by way of East Alisal back into town, passing on the east side of the tracks the one-time neighborhood of the Okies.

From Salinas Route 101 cut right through the heart of Salinas Valley, and I wanted to see those towns—Chualar, Gonzales, Soledad, Greenfield, King City—and outside of King City, the old Hamilton ranch. On 101 southbound I was out in those broad, flat fields of lettuce that allowed the Salinas area to advertise itself as the "Salad Bowl of the Nation." (In *East of Eden* Adam Trask tries to refrigerate lettuce heads and ship them to the eastern markets, but a combination of untoward events literally sidetracks his train, and the lettuce rots in the broiling yards at Chicago. Kazan's film version does an effective job with this section of the novel, including a fine scene filmed at the Spreckels sugar mill where the town's battered brass band plays "Avalon" in salute to the train's departure.) I could see the Spreckels mill's huge smoke stacks and storage tanks as I went south toward Chualar. All along the way there were the workers in the fields, stooping amid the lush green or standing upright for a moment's relief, leaning on rakes and hoes, their faces shaded by hats. Huddles of pickups and outdated Detroit hunks were parked by the roadside or driven into the fields where they sat beside buses and portable toilets. Sprinklers threw blazes of water high up into the sun and out onto the plants and the deep black soil.

The farther south I went, the more the lettuce fields began to give way to fields of onions and oriental kale and to vineyards. Below Soledad the Arroyo Seco River, a tributary of the Salinas, cut in from the southwest, and here I swung off 101, following the river road through fields of barley and oats. There were long lines of eucalyptus trees set out as windbreaks across the broad expanses, and nearing them you could understand their special virtue, for each tree was a murmurous, full-leafed colony that interlaced with its neighbors to break the force of the earth-scouring winds. Long narrow lanes led away from the road into the fields, and here and there a big, lone oak dropped its blue pool of shade in the dust.

I followed the river road to where it swung up a hill, and there beneath me was the still, light green river spanned by a short bridge with two iron arches. The day had gotten hot, and I could not resist the urge to swim. The water was warm on its sunny surface, but there was a small current beneath, and there it was cool. Lying back you could hear the smaller stones roll along the shallow bed near the shore. Three Hispanic families shared the rocky river's edge with me. The parents sat on blankets while the kids, in shorts and t-shirts, laughed and splashed in the shallows. There were coolers and canned soft drinks, chips, and sandwiches, and, hovering in the air above the river's gurgle, the soft lilt of Spanish.

A few miles south of King City, Wild Horse Canyon Road veered off into the barren glare of the hills. Where there were fields in production they were already at this early point in the year a molten gold: oats and long-bearded barley for stock feed. The large wine-growers were trying the area as well. There were extensive vineyards tucked in here, and later I learned that many of them were leased by the Almaden company. In *East of Eden* Steinbeck makes much of the fact that Samuel Hamilton took up ranching in here only because the best valley lands had long since been bought up by earlier settlers. This, Steinbeck says, was desperately poor ranch land, and without a good source of water. In dry years, the half-starved Hamilton cattle moped over it with gaunt ribs showing through poor, dusty hides. And, to be sure, the hills still wore a ravaged, naked look, but the fields alongside the road were high in barley, and a deer browsing in one of them could hardly see over the grass heads to flirt its long ears at the sound of my car in the soft dust.

Pauline Pearson had warned me that the ranch's owners discouraged the visits of Steinbeck fans, but Mr. and Mrs. Ernest Grab said that wasn't exactly so. Standing in the ranch's dusty yard between the house and the barns and sheds, Ernest Grab adjusted his battered straw cowboy hat and smiled. "If you want to get yourself a map, and take the trouble to get in here," he said affably, "why, fine. Those that're serious, well, all right. But those that're merely curious, we discourage them." He told me he had been raised on the place, that his family had bought it from the Hamiltons somewhere around 1910, and that his family had moved onto the ranch from King City in 1916. Mrs. Grab was herself a native of that town.

I was curious to see the original house, which, Steinbeck writes in *East of Eden*, "was designed to be unfinished, so that lean-tos could jut out as they were needed. The original room and kitchen soon disappeared in a welter of these lean-tos." Indeed, when the Grabs gestured toward the gray-slabbed structure half hidden by bushes and the rise of subsequent buildings it looked to me like a playhouse—though immediately I thought of Frost's "Directive" and its admonitory lines about an abandoned New England farmhouse: "This was no playhouse but a house in earnest." Ernest Grab said it was single-wall redwood and that in refurbishing it he had discovered pages of the *Christian Science Monitor* from 1892 used for insulation.

The other building I wanted to see was the blacksmith shop where the historical as well as the fictional Samuel Hamilton had dispensed Irish folk wisdom while attending to the intermittent work his neighbors brought him. There in the sanctity of the shop the men had exchanged talk and gossip while a gift bottle of whiskey went around the small group, well out of the sight of the fiercely teetotalist Liza Hamilton. Steinbeck says she never took a drop until her late years when a doctor prescribed a spoonful of port wine as a digestive, after which she was never entirely sober again, though always careful to take her "medicine" in tablespoons. It was still there, Ernest Grab said, pointing across the yard to a shed on the same scale as the original house. Inside in the tenebrous light slashed with daggers of sun that cut through the holes in walls and roof were jumbles of ancient nails rusted into Gordian knots, harness hames, buckles, metal dippers, stone jugs, as well as a variety of mechanical castoffs: cams, spark plugs, a drive shaft. From the beams dripped four ragged sheep hides with hooves still attached. Standing in here for a few minutes I thought of Steinbeck and his effort to make this place come alive, but then the present filtered through the cracks on the sounds of a horse plodding heavily by in the brown-gray dust without, a goat bleating in a corral, the sighs of a dove.

Out in back of the old shop I discovered the present-day counterpart of Samuel Hamilton's clandestinely convivial gatherings of long ago. Ernest Grab and his hands were leaning against a pickup stacked with hay and enjoying a round of beers from a cooler perched on a bale. Wayne, Dave, and Jim waited until Ernest Grab invited me to join in. The boys were hot and thirsty, their lined faces red beneath their cowboy hats. The beer went down quickly, and as it did the talk picked up. They were too polite at first to ask what this slight stranger was doing way out here with notebook and camera, but when Ernest Grab interceded in my behalf and told them about the background of the ranch, they smiled at me and their eyes twinkled. *Have another beer. This here is hot work you're at, ain't it? Who did you say that fella was, the writer? Know how to tell a real cowboy? He's got to have chrome hubcaps on his pickup. Yeah, but they got to be muddy, don't forget that! Naw, that ain't how to tell a cowboy. This guy wants the straight stuff. How to tell a cowboy is this: ever notice how much room there is around the tops of his rubber boots? Well, that's so's he can fit the sheep's hind legs in there snug* (giving thrusting pelvic moves to a burst of laughter). *They ain't only Montana whores! There's California whores, too: b-a-a-a!!!*

The Grabs said I had the run of the spread, and I took it, climbing in the late afternoon up a steep cowpath etched in a chalky zigzag into one of the naked hills. Atop the hill there was an outcropping of white rock that served as a good seat. A dry wind waved the thin grasses, and among them crickets and grasshoppers went about their desiccated daily chores. A thousand feet below was the old, warping ranch with its gray clutch of buildings. Its corrals and outbuildings that at ground level had seemed especially random up here assumed a geometric pattern. In the center was the original building, its roof covered with an aqua blue plastic tarp. Right there, for Steinbeck, was where it had all begun.

Like every writer, his career had had a blunt and blundering beginning as he bumped and butted himself against potential themes. Joyce Conrow had told me on the phone that the best advice Steinbeck ever got was from a somewhat scornful Salinas neighbor who had told him, "If you're going to write, for God's sake, write about something you know." In *East of Eden* he was at last circling back to write about something he knew, all right, knew in the very blood that flowed into the hand that held the pencils he preferred as the tools of his craft. With the great, thunderous Biblical narrative on one hand and this pure personal knowledge on the other, it must have seemed to him he could hardly fail. When he turned over to Covici the last draft of the manuscript he gave him a wooden box on which he had carved the Hebrew characters which he believed could be translated as "thou mayest."

NINE

OF LOCAL INTEREST: WILLIAM CARLOS WILLIAMS'S <u>IN THE AMERICAN GRAIN</u> *and* <u>PATERSON</u>

I

SATURDAY, JANUARY 19, 1924. The doctor and his wife met their old friend from the States, novelist and publisher Robert McAlmon, on the Boulevard Raspail at the head of which stood Rodin's heroic statue of Balzac. Then the three of them walked through a corner of the Luxembourg Gardens to the rue de l'Odéon to call on Sylvia Beach at her bookshop, Shakespeare and Company, the clubhouse for the city's expatriate literary community. The doctor, William Carlos Williams, was forty then, older than most of the American expatriates he was just now among, and this trip abroad was his first real break from fourteen years of medical practice in his hometown of Rutherford, New Jersey.

Oh, he'd seen something of the world before this, had been a schoolboy outside Geneva in 1897, then had spent several months here in Paris with relatives. He had also studied pediatrics at Leipzig in 1909 and then had traveled extensively that year on the Continent. Still, as he stood there in the rue de l'Odéon looking at a few copies of his books in the window of Shakespeare and Company he would not have been human if his blood wasn't singing in his veins. He had long been an active, vocal member of the modernist movement in letters, and now he stood in the very capital of the avant-garde. "Paris!' exclaimed Malcolm Cowley, one of the American expatriates:

> You leaped into the first empty taxicab outside the station and ordered the driver to hurry. In Paris the subways were impossibly slow, and the taxis never drove fast enough as you raced from one appointment

285

to another, from an art gallery to a bookshop where you had no time to linger, and thence to a concert you could never quite sit through— faster, faster, there was always something waiting that might be forever missed unless you hammered on the glass and told the driver to go faster. Paris was a great machine for stimulating the nerves and sharpening the senses. Paintings and music, street noises, shops, flower markets, modes, fabrics, poems, ideas, everything seemed to lead toward a half-sensual, half-intellectual swoon. Inside the cafes, color, perfume, taste and delirium could be poured from one bottle or many bottles.

A couple of days later McAlmon staged a party for the Williamses at the Trianon. James and Nora Joyce were there. So was Ford Madox Ford. Harold Loeb who had been editing the avant-garde magazine *Broom* in Rome showed up to see his old friend from the States. George Antheil, the modernist composer (another New Jersey man), was there along with Louis Aragon, Marcel Duchamps, and Man Ray who the next day would take Williams's picture at his studio. Before he and Floss left for Italy Williams would be in contact with most of the leading figures of the modernist movement, including old friend Ezra Pound, Fernand Léger, Cocteau, Brancusi, Ernest Hemingway, and H.D., Hilda Doolittle, the Imagist poet.

For years Pound had been hounding Williams to bust out of that stodgy, stale New Jersey swamp he called a home and come to Europe where things were really happening, where modernism was not a hope but a daily reality, and where Williams could encounter the sort of creative stimulus he (Pound) knew Williams needed. To Pound, Williams's Rutherford rootedness was little less than a species of perversion. Why was it, Pound wanted to know, if Williams was really serious about writing modern poetry that he would not come to where the stuff was actually being produced, argued over, and read, instead of staying put and turning out neo-Keatsian, neo-Whitmanian botches?

Though Williams was a skilled, tough, verbal fencer and could answer Pound back all right, in truth he was hard pressed to offer up a good explanation of his persistence in Rutherford. Indeed, in recent years he seemed to have been digging in even deeper: marriage, children, a house at 9 Ridge Road in his hometown and only a few blocks from where he had been born. Paul Mariani, Williams's biographer, quotes a 1914 letter from Williams to an old girlfriend who had been jibing at him for his failure to be as bold as Pound and follow him to the freedom of Europe. "Do you think I like to practice medicine here in Rutherford?" Williams had written. "And see others skipping over the face of the globe from the two poles to London, New York, Hoboken and Pekin?" But, he added, in a significant if slightly elliptical fashion, he did not see how he could be "anywhere else and do what I have determined to do." In 1914 just what that was remained, really, to be seen, though it was clear that the doctor was already deeply committed to the writing of his poems.

And, of course, if it had been only Pound, it would have been another matter. Williams's old friend from college days at Penn had ever been sedulous in keeping ahead of the pack, in calling the new turn or creating it. But it was not just Pound. It was a lot of writers, artists, and musicians who had fled America in the years after the war. That had not been Williams's war—he had been too old for it—but like the younger ones he too had hoped peace would bring in more than mere material prosperity, that American culture would enter upon an era of authentic expressiveness, openness to inquiry, and tolerance of diversity, instead of which there had come the Volstead Act, the Palmer Raids, Harding, the all but unchallenged cultural suzerainty of the *Saturday Evening Post,* and yet another quantum leap in the nation's obsession with the making of money. At last even Greenwich Village, the retreat of embattled artists and intellectuals, seemed unbearable with its plainclothesmen hunting Reds and anarchists and arresting girls as streetwalkers because they smoked in public. In the pages of newspapers and the *Post* there was a relentless campaign of slander aimed at the Villagers: these were poor poseurs who ought to grow up and go back where they came from. Certainly for the many who had come to the Village to escape the hinterlands there was no going back. The landscape of childhood, both physically and spiritually, was gone. Williams himself knew that well enough looking about his own landscape of Rutherford, Paterson, and Passaic, which bore astonishingly scant resemblance to that of his childhood years.

So they left for Europe where, it was said, the ugly and repressive force of Puritanism could not reach, where "they knew how to do things right," where there was an easy tolerance bred of the centuries, where you could get a drink, loosen up, and let the creative juices flow. "New York is very delightful this fall," Williams had sarcastically written in 1921 to Harriet Monroe of *Poetry* magazine, "with all the neurosis in Paris. One walks the streets in quiet enjoyment knowing that there is no one of any importance to be met at the next corner." The next year Harold Stearns's anthology, *Civilization in the United States,* proclaimed America culturally dead, the cemetery of creative talents, and it was hard not to agree. Williams tried to remain calm in the face of mass defections to Europe, disdainfully writing a friend, the critic Kenneth Burke, that when on a morning walk he'd stepped in dog shit he was seized with an almost "irresistible desire to study French literature." But even Burke, who like Williams was staying put, was soon to sound the general alarm, writing in *Vanity Fair* that there was in America "not a trace of that really dignified richness which makes for peasants, household gods, traditions. America has become the wonder of the world simply because America is the purest concentration point for the vices and vulgarities of the world." Had anyone been charting the course of American writers' accommodation to their native land the terms of the current debate would have seemed an odd reprise of those of the previous century when writers routinely complained of the bitter thinness of the cultural soil and longed for the dense texture of the Old World. Here Kenneth Burke sounded like Henry James explaining why Hawthorne's way had been so hard.

Not Williams. He was angry at the loss of so many of his friends— his creative context—the draining off to Paris and London and Italy of the most progressive talents of his time and place. And he could hardly have escaped knowing that to some of these expatriates he seemed a sort of throwback. If he was, it was to Thoreau that he was thrown. Emerson, who could value the Old World for what it had to give him, had been peeved at Thoreau for pretending that he could find all the cultural nourishment he would ever need right there in Concord, and in fact as his brief years went on Thoreau became progressively ever more local so that at last it was hard to get him to Worcester, let alone New York or Europe. The red snow of the Andes or Himalayas? Thoreau said he'd seen snow with a reddish cast in the Concord area. There was so much material right under your feet that if truly seen could last any man a lifetime. And still it would remain unexhausted. So too for Dr. Williams in a far different landscape. Thus he could write Harold Loeb about the magazine *Broom,* wanting out of the generosity of his heart to send all best wishes but instead writing, "What in hell can I bless you for?—the good work you might just as well be instigating in some actual locality where you would be IN IT—up to the ass hole. . . . I can't for the life of me see any use in shooting off jism into the atmosphere of Italy."

And now here he was himself in Paris, having arranged at considerable expense a year's sabbatical from his practice and the children safely stashed at home with their familiar old nanny.

He had to admit it was exciting and stimulating among all this talent and the grand indifference of the great city. "*Everyone* was in Paris—if you wanted to see them," he was later to write in his autobiography. A thought raced through his head: maybe he should set up practice in some quiet suburb outside the city. It was said to be a simple matter to establish yourself. Why not? Now he too began to sound just a bit like one of the expatriate crowd as he wrote back to stay-at-home Marianne Moore in New York, telling her of the splendors of the south of France where he and Floss had gone for several weeks:

> Dear Marianne: Florence, Bob McAlmon and I are here at a quiet pension-with-a-garden-overlooking-the-sea for perhaps a month. We came down from Paris last week where the Europe virus was injected into Floss and me, and are now, each in his own way, working. . . . Oranges, mandarins, lemons and tasty little ground cherries in their flimsy shells are all in our yard, while a man that dresses and looks like Geo. B. Shaw throws feed to the chickens next door every morning.

But then in Rome in March he wrote in a much more native voice to Burke in protest against the multiple seduction of the Old World. He loved what he had been seeing—how could he not?—and yet that very love, that enchantment, so he told Burke, made him see clearly how crippling for an American the Old World was:

> I never so fully realized, as in the smell of these relics of the old battle, how maimed we are—and how needlessly we are crippling ourselves. Frascati in full "wildflower" yesterday won me again, just as I have been won over and over here by the bits of wisdom that I've seen in museums—the statues, the whole colossal record of their oldtime fullness and our unnecessary subservience to our crippledom. We love it. That's the hell of it. We eat it, lie in it. Sing about it and build our monuments on it. . . .
>
> Well, so here we are again, back where I started.

Indeed. As the trip neared its appointed end Williams realized it had not in fact moved him at all from where artistically and spiritually he felt he ought to be. And as if to emphasize, if only to himself and Floss, his abiding sense of priorities he worked during these Old World weeks at a project which, had others there known of it, could only have seemed a further manifestation of his perversity. This was a book of essays on the roots of American history, to be called *In the American Grain.* In what became the book's fulcrum, an essay on the late-sixteenth/early-seventeenth-

century French Jesuit missionary, Père Sebastian Rasles, Williams set down
a recollection of his kaleidoscopic Parisian days and nights:

> Picasso (turning to look back, with a smile), Braque (brown cotton),
> Gertrude Stein (opening the doors of a cabinet of MSS.), Tzara (grinning),
> André Germain (blocking the door), Van der Pyl (speaking of St. Cloud),
> Bob Chandler (prodding Marcel), Marcel (shouting), Salmon (in a cor-
> ner) and my good friends Philip and Madam Soupault; the Prince of
> Dahomi, Clive Bell (dressed), Nancy, Sylvia, Clotilde, Sally, Kitty, Mina
> and her two lovely daughters; James and Nora Joyce (in a taxi at the
> Place de l'Étoile), McAlmon, Antheil, Bryher, H.D. and dear Ezra who
> took me to talk with Léger; and finally Adrienne Monnier—these were
> my six weeks in Paris.

And yet. And yet, "nothing came of it save an awakened realization within
myself of that resistant core of nature upon which I had so long been driven
for support." He felt, he continues, not released in the alleged freedom
of Paris, the Old World, but "beaten back" upon his own resources, upon
America. Paris had been a relief from the daily and nightly pressures of
his medical practice, and even more a relief from that growing sense of
artistic isolation he had been feeling in the States at the beginning of the
decade. But after all, that relief was "like a fairy tale." He had had only
a few weeks in which to try to soak up all that heady atmosphere. "Could
I have shouted out in the midst of it, could I have loosed myself to embrace this turning, shouting, rustling, colored thing, my mind would have
been relieved. I could not do it." Going one day to talk of American things
with the French scholar Valéry Larbaud, Williams says he felt the sullen
lump in his breast harden and become "like the Aztec calendar of stone
which the priests buried because they couldn't smash it easily, but it was
dug up intact later." If the others felt they could not go back to America
because the places they had known—the whole country itself as once it
had been—no longer existed, he could because imaginatively he had never
left it. Writing Burke from Austria in April, Williams said that he would
be glad to get back home. Paris would be wonderful "if I could be French
and *Vieux;* it would be still more wonderful if I could only want to forget
everything on earth. Since I can't do that, only America remains where
at least I was born."

The tone of this last sounds more resigned than its writer really felt as
he and Floss took ship for home in June. He was *not,* he insisted, doomed

to America, even if it was his historical fate to have been born there. There was much to be discovered about that place, and he meant to discover it. On the ship for home he worked hard on his essay on Daniel Boone.

First and last he thought of himself as a poet, though in his long career he experimented with about every form there was—novels, short stories, improvisations, drama, autobiography—and in returning to Rutherford he looked forward to urging on with his own poetry the creation of a modern American poetic voice. Yet there were good reasons why he wanted first to complete his book of historical essays. They were for him an opportunity to read widely in the primary documents relating to the discovery of the New World and the development of the American republic. So much, he never tired of claiming, was taken for granted here that the real character of the country remained unknown. He wished, then, to make his own voyage of discovery. There were obscurities lurking in the New World's past, and he wished by his reading and writing to root them out. Then, having found whatever it was he should find in the old chronicles, he would with *In the American Grain* clear the field for his poetry—and perhaps he might at the same time perform a like service for others. Then the poetry would rest on something solid, the actual rock of the country itself. As Thoreau had written of his own earthward drive for origins, "This is, and no mistake."

He had begun *In the American Grain* in 1923, sparked no doubt by the departure of so many artists and writers for the Old World. Their disgust with what America had evidently become in the post-war years made Williams curious to discover the sources of the current tide of money grubbing, Red baiting, and neo-puritanism. He felt certain these things did not arise from a vacuum but were products, results. Others were searching for answers as well: Harold Stearns's collection, *Civilization in the United States;* George Santayana's *Character and Opinion in the United States* (1920); and in the decade's final years Vernon L. Parrington's *Main Currents in American Thought* (1927–30). In fiction two of the most striking depictions of what America had become would appear in 1925 along with *In the American Grain*: Dreiser's *An American Tragedy* and Fitzgerald's *The Great Gatsby*.

But the writer who came closest to Williams's specific concerns was D.H. Lawrence in his *Studies in Classic American Literature,* published the same year Williams began his reading. Lawrence started on his study of our classic authors—Hawthorne, Cooper, Melville, Whitman—as early as 1917, but it was not until he had come to America in 1922 that he had actually felt what he called "the Spirit of Place." When you are actually in America, Lawrence writes in his chapter on Cooper's Leatherstocking novels, America hurts because it is "full of grinning, unappeased aboriginal demons" that persecute whites and will do so until the whites give up their absolute whiteness and the "Spirit of Place is atoned for." The American landscape, he claims in the same chapter, "has never been at one with the white man. Never." Americans, he went on in his chapter on Dana's *Two Years Before the Mast,* "have never loved the soil of America as Europeans have loved the soil of Europe." For Americans the real homeland was still the Old World.

These were feelings Williams had had himself for several years, and it is possible that the fortuitous arrival of Lawrence's *Studies* spurred him to make his own discoveries and claims. He too believed in the spirit of place, though he would not want to capitalize it as Lawrence had. His mother had long claimed she could communicate with the spirit world, and her efforts to do so had upset Williams as a child. Later, he had been able to dismiss them, but now he found himself engaged in his own spirit quest, looking through the old chronicles, as he says in prefatory remarks to *In the American Grain,* for "the strange phosphorus of the life, nameless under an old misappellation."

Since it was his conviction that Americans took far too much for granted about their history, his strategy in the book was to go back over the most familiar, beaten ground of American history and look at it as if it were terra incognita, as he believed it really was. So here again are the imposing figures of childhood's textbooks: Eric the Red, Columbus, Cortez, Ponce de Leon, de Soto, the Puritan fathers at Plymouth, Boone, Washington, Franklin, Lincoln. Mixed in are portraits of lesser-known figures— Father Rasles, Sam Houston, Aaron Burr, Edgar Allan Poe—whom Williams felt could throw new light on the major figures.

In each essay Williams is concerned to show that America as it currently exists is the product of the deeds enacted in its past and that the national historical amnesia is a major cause of that cultural thinness of which the fleeing artists and intellectuals were complaining. It is not, he says, only American poets like himself who need to clear the ground by plowing it, it is the nation as a whole. In the Rasles essay he represents himself in conversation with Valéry Larbaud, saying to the Frenchman,

"It is an extraordinary phenomenon that Americans have lost the sense, being made up as we are, that what we are has its origin in what *the nation in the past* has been; that there is a source in AMERICA for everything we think or do. . . ." Americans, Williams tells Larbaud, "recognize no ground as our own. . . ." When he tells Larbaud of the Puritans' vicious persecution of the Quakers, Larbaud asks him why no one in America speaks of these matters. "Because," Williams says, "the fools do not believe that they have sprung from anything: bone, thought and action. They will not see that what they are is growing on these roots. They will not look. They float without question. Their history is to them an enigma."

Once we do begin to plow the soil of the American past, turning it up to the light of close inspection, we begin to see that our past has a character far other than popularly supposed. "History, history!" Williams exclaims at the outset of his essay on Ponce de Leon. "We fools, what do we know or care? History begins for us with murder and enslavement, not with discovery." From Eric the Red, outlaw, through the Spaniards' savage slaughter of the tribes, to the Puritans hanging witches at Salem, America's beginnings as Williams presents them are almost as dark and bloody as the secret sacrificial chambers of the great temple of Tenochtitlán. From the first the invading whites wanted from the New World only what riches they could tear off and take with them back to civilization. In his diary Columbus noted the magnificent greenness of the trees on the islands of his first landfall, yet, writes Williams, within the Admiral's lifetime the flower of the untouched paradise had already shriveled, victim of a mindless rapacity that was to repeat itself with incremental intensity on through the successive waves of exploration and settlement.

When Williams arrives at the Puritans' contributions to American culture—by far the longest, most impassioned section of the book—he is ready to identify the crucial failure of all the first-comers. In their lust and greed they had not truly seen nor touched the new land. For them it was but a featureless wilderness, its only value its exportable riches. The Puritans were not greedy for gold, but in their hatred of the "squallid, horrid American Desart" (Cotton Mather's words) where they lived like exiles a particular nadir was reached. Here, says Williams, is their legacy to us: in opposing the spirit of the land so fiercely with their own spiritual force the Puritans bequeathed a pervasive negation of America that persists to this day. "Our resistance to the wilderness has been too strong," Williams writes. "It has turned us anti-American, anti-literature. As a violent 'puritanism' it breathes still.'' He tells Larbaud that no one who has not lived in America can really understand how profoundly affected the country still is by this Puritan legacy:

> There is a "puritanism"—of which you hear, of course, but you have .never felt it stinking all about you—that has survived to us from the past. It is an atrocious thing, a kind of mermaid with a corpse for a tail. Or it remains, a bad breath in the room. This THING, strange, inhuman, powerful, is like a relic of some died out tribe whose practices were revolting.----

Against all these fierce original negations of which the Puritans were the nadir Williams proposes his heroes, Father Rasles, Boone, Burr, Poe: surely an odd pantheon but figures united and distinguished by their ability to imaginatively and actually experience the New World, to come into creative contact with it. "Contrary to the English," Williams writes, "Rasles recognized the New World. It stands out in all he says. It is a living flame compared to their dead ash.'' Living thirty-four years among the Abnaki tribe, "TOUCHING them every day," the lone Jesuit became absorbed in the land, lost in it, swallowed by it—no horrid fate, as the Puritans had feared, but a singular triumph. Williams represents Boone as a "great voluptuary born to the American settlements against the niggardliness of the damning puritanical tradition." It is not, says Williams, so important that Boone settled Kentucky and blazed a westward path but that he fell in love with the land and locked himself into a life-long embrace with it. Thus his life remains for us "still loaded with power—power to strengthen every form of energy that would be voluptuous, passionate, possessive in that place which he opened."

As for Burr, stigmatized as the murderer of Hamilton and a traitor, Williams contends that his true character lies buried beneath these black marks. Here, he suggests, was a man of remarkable imaginative energy who envisioned an open, liberal America and saw the conservative forces beginning to fence the country off, closing it up in behalf of the few and the privileged. Hamilton is thus seen not as Burr's brilliant victim but as a mean-minded man who like so many others before him meanly conceived of the land as merely a field for economic opportunity. He wished, says Williams, "to harness the whole, young, aspiring genius [of the nation] to a treadmill":

Paterson he wished to make capital of the country because there was waterpower there which to his time and mind seemed colossal. And so he organized a company to hold the land thereabouts, with dams and sluices, the origin today of the vilest swillhole in christendom, the Passaic River; impossible to remove the nuisance so tight had he, Hamilton, sewed up his privileges . . . through his holding company. . . . *His* company. *His* United States: Hamiltonia—the land of the company.

Like Burr, Poe carries about him his stigma, that of a failed and impoverished drunkard. Williams salutes him as a trailblazer, "A new De Soto (sic)": it is "a *new locality* that in Poe is assertive, it is America, the first great burst through to expression of a re-awakened genius of *place.*" Poe in his criticism was the first to "clear the GROUND" and dare to act as if literature in America was a serious matter, not a game of croquet. Turning his back, Boone-like, on all that was derivative, Poe wrestled with the sullen spirit of his place, Williams claims, and so was made to suffer for his originality. Speaking for himself as well as for those artists and writers unhonored in their native land, Williams writes, "Invent that which is new, even if it be made of pine from your own yard, and there's none to know what you have done."

In the American Grain is an offbeat, highly idiosyncratic book, quite as its author meant it to be. How many, for instance, would claim for Poe (who wrote no poetry with an American context and but two short stories with recognizably American settings) a more radical literary Americanness than Hawthorne? But, though he believed—passionately—in the views he expressed in the book, finally Williams was more interested in the cleansing, clearing effect of having voiced them. History, he wrote, must remain open, by which he meant that it must remain in our minds available for new looks into its old phenomena. When a writer looks anew at anything, Williams believed, the strangeness of his perspective challenges readers to become discoverers themselves and look closely at the old ground to see what might be there. Williams had done that for himself with *In the American Grain,* and he had done it for others, too. For despite the fact that many of its judgments are highly debatable, the book stands as a major twentieth-century document of American cultural history.

Now, the ground cleared, the obscurities exposed, the spirit of the land evoked, Williams was ready to move on to the evocation of the spirit of his own special place, Paterson, Hamilton's proposed capital, through which now coursed the "vilest swillhole in christendom, the Passaic River," of which he had, somehow, to sing.

Late in life Williams told an admirer from nearby Fairleigh Dickinson University that he had always known he wanted to write a long poem but that for many years he had not understood how to go about it. The very size of the ambition was both daunting and baffling, and even after the publication of the fifth and final book of *Paterson* in 1958 Williams continued to think of the poem as "impossible." Paul Mariani, who has made the most thorough use of the writer's papers, believes Williams's thinking about his long poem can be traced back at least to 1926 and to an 85-line poem he wrote then and entitled "Paterson." Here, Mariani says, was a kind of trial balloon which Williams sent up to see how it might feel to write of his native place in an extended form. He had at the same time the astonishing example of Joyce's *Ulysses* before him. Sylvia Beach had published the book in 1922, and it could have done nothing but quicken Williams's epic ambition that he had come to know Beach and had dined and drunk with Joyce himself in Paris in '24. In that book Williams saw how Joyce had taken the apparently unpromising cityscape of Dublin and its river, the Liffey, and had transformed it into a mythic geography. It showed, Williams thought, that no place was unavailable for art. The question was, though, how to see your place imaginatively enough and comprehensively enough so that its prosaic defects could become your best creative assets.

By the late 1920s for an artist or writer living anywhere along the upper reaches of America's eastern seaboard those prosaic defects were far more evident than were the ways one might creatively use aging mill towns, dump heaps, city slums, polluted swamps and rivers. If for Thoreau the problem had been, as he said, to learn how to make a literature out of robins and rail fences, now the problem for the American writer in the older settlements was of a different character. It seemed to them greater, too, so that in the 1920s and 1930s some of them could look back at Thoreau and the American Romantics and tell themselves how much easier the literary ancestors had had it.

There had been, to be sure, some singular examples of how this modern American landscape might be used in literature. Stephen Crane (a New Jersey man) had published *Maggie: A Girl of the Streets* in 1893, a short novel about a New York City streetwalker trapped by her bleak environment. Dreiser had used the landscapes and cultures of Chicago and

New York for *Sister Carrie* (1900). More recently Abraham Cahan had drawn on his deep knowledge of the Lower East Side's needle trades, unions, and immigrant cultures in *The Rise of David Levinsky* (1917). But for poetry the urban-dominated landscape of the eastern seaboard remained largely undiscovered territory.

Except, of course, for Whitman, and by the late 1920s Williams had already taught himself that he should not try to write like Whitman. Nobody should. Whitman's example remained powerfully suggestive in many ways, however. He had taken on virtually every aspect of the modern America that was emerging in his time: westward expansion, the Civil War, the huge influx of immigrants, the explosion of the city, industrialism. He had even written a hymn to a locomotive, that seeming antithesis of the pastoral and the poetic. In poems such as "Crossing Brooklyn Ferry," "A Song for Occupations," "A Broadway Pageant," and the first "Manahatta," the reader feels the crush of crowds, the grit of city life, the enormous electric current pulsing through a nation speedily leaving its agrarian beginnings far behind. But Whitman was Whitman, unrepeatable, inimitable, and in that sense though his example was so suggestive, it was after all finite. The thing, Williams had decided, was to learn to see as Whitman had, not to try to write like him.

Besides, Whitman's world had been New York, and that, Williams knew, was emphatically not his home. "New York City was far out of my perspective," he wrote in his autobiography (1951). "I had no wish, nor did I have the opportunity to know New York . . . and I felt no loss in that." What he wanted to write about instead were the people and places close to him, to know them "to the whites of their eyes, to their very smells." This, he continued,

> is the poet's business. Not to talk in vague categories but to write particularly, as a physician works, upon a patient, upon the thing before him, in the particular to discover the universal. John Dewey had said (I discovered it quite by chance), "The local is the only universal, upon that all art builds." Keyserling had said the same in different words.

Feeling so, he faced his unavoidable choice: the immediate, the local as the subject for the long poem he felt compelled to attempt. He thought of several places on the Passaic River since he was from the first determined to have the river running through the poem. But he kept coming back to Paterson with its falls, its rich and variegated colonial history, its suggestive associations with Alexander Hamilton and what Williams thought Hamilton represented. Even its ravaged, worn look had its literary potential, he felt, for this, *this* was what had become of "Nature's Nation," and this would be its future. In earlier books of poems he had already taken on the challenge of his landscape with a Whitmanian courage. In *The Wanderer* (1913) there had been a poem about the big strike of silk weavers in Paterson that same year. In it Williams writes of the bruised, beaten look of the laboring people of the city,

> The flat skulls with the unkempt black or blond hair,
> The ugly legs of the young girls, pistons
> Too powerful for delicacy!
> The women's wrists, the men's arms red
> Used to heat and cold, to toss quartered beeves
> And barrels, and milk-cans, and crates of fruit!

And yet the sight of these people, their faces "knotted up like burls on oaks," does not depress the poet/observer so much as it brings him an odd feeling of peace since in seeing these realities he can come to terms with the truth of where he lives. In "St. James' Grove" from the same collection the poet asks his muse, an old hag, to arrange his marriage to the Passaic, "that filthy river." The muse consents, crying fiercely,

> "Enter, youth, into this bulk!
> Enter, river, into this young man!"
> Then the river began to enter my heart,
> Eddying back cool and limpid
> Into the crystal beginning of its days.
>
> But with the rebound it leaped forward:
> Muddy, then black and shrunken
> Till I felt the utter depth of its rottenness
> The vile breadth of its degradation
> And dropped down knowing this was me now.

In *Al Que Quiere!* (1917) he had written the only kind of pastoral poem left him, writing of his prowls through the city's back streets,

> admiring the houses
> of the very poor:
> roof out of line with sides

the yards cluttered
with old chicken wire, ashes,
furniture gone wrong;
the fences and outhouses
built of barrel-staves
and parts of boxes, all,
if I am fortunate,
smeared a bluish green
that properly weathered
pleases me best
of all colors.

"No one," he concludes, "will believe this / of vast import to the nation." To him it was all-important. It was, as he would say many years hence, the poet's business to speak of these things, to reveal local truths to readers so that, reading a poem about, say, the streets of Paterson the reader could find his or her own particular landscape there.

Then in his bellicose modernist manifesto, *Spring and All* (1923), there had been an untitled poem in which the poet defiantly claims as his own the trashy, fouled landscape where any newness must struggle up through old death and decay. By the road to the contagious hospital the poet witnesses the first, stark beginnings of spring, the first thrusting, tentative plants coming up in the "waste of broad, muddy fields / brown with dried weeds" and gripping down to the poor soil, their only reality.

By the mid-1930s Williams knew himself to be in strenuous training, like a marathoner, for his long poem. He still did not know what form it would take, but he did know he wanted it to be written in the jagged, racy idioms of local speech, and he wanted it to arise out of the ashes of his appointed place like the stiff, spindly trees he'd written of in the poem about first spring: "Spirit of place rise from these ashes," he writes in "Morning." More and more purposefully he haunted Paterson in what odd moments he could steal from the daily demands of his practice, writing notes to himself of what he had seen there, perhaps while he sat in his car at a railroad crossing, waiting for a slow freight to haul itself through. He went to look at the falls in the very midst of which he began to discern a great hydrocephalic head of stone that jutted against the ceaseless roar of water, as if a grotesque giant lay there on his right side. Or he might drive in summer to the Third Street Bridge over the river at the border of Passaic and Wallington to watch kids swimming in the river or simply to contemplate the river, his river, running wearily between its refuse-burdened banks. Below the foundation of a huge factory a pipe spewed its used water out into the turbid stream. As in his youth, he tramped Garrett Mountain and from its stony summit looked down on the city he would imaginatively possess. Always he jotted down scraps of his patients' talk on his prescription pads, trying to catch those rhythms, those odd, characteristic turns of phrase he hoped to get into the poem. In *Life Along the Passaic River* (1938), a collection of short stories, he composed most of the stories out of simple exchanges in dialogue between a doctor and his patients, trying in the genre of prose fiction to get down those rhythms.

As in the early '20s when others about him were running off to Europe and wondering why Bill Williams persisted in being as local as a mud turtle, so now he seemed to fairly wallow in the seamiest circumstances of his place, tasting it, running its talk over his tongue, reading in its history. When the great expatriate presence, T.S. Eliot, in a lecture at Harvard, sneeringly wrote Williams off as a poet "of some local interest, perhaps," Williams was incandescent with rage. Yet it was true that he was more and more surely restricting the scope of his imaginative energies, still certain that the truest work was that founded securely in a specific locality. That was the trouble with Eliot and with his old friend Pound, too, Williams thought, despite what he admired of Pound's epic effort, *The Cantos*: their work evaded American realities; it wasn't grounded in anything. "Of only one thing," he was to write during these years of training, "relative to a work of art, can we be sure. It was bred of a place. It comes from an application of the senses to that place, a music, and that place can be the middle of the African jungle, the Mexican plateau, a Parisian whorehouse, a room where Oxford chippies sip tea together or a down-hill street in a Pennsylvania small town. It is the particularization of the universal that is important."

Yet as the forties came on and with them the war years, Williams knew that if earlier he had been in training, he was now verging on the overtrained. He had accumulated an enormous amount of material, enough, he judged, for at least three books of the poem. He had developed the poem's governing imaginative equation, that a man is himself a city. And yet he could not get fully launched into the thing. He made little stabs at it, then shied away, as he confessed in a letter of late summer 1943 to his old friend Bob McAlmon. "I want to work at it," he wrote, "but I shy off whenever I sit down to work. It's maddening but I have the hardest time to make myself stick to it. In spite of which I have done a hundred pages or so. . . ."

The more he struggled to lash his wildly heterogeneous materials into a coherent, comprehensible form, the more he despaired of ever doing so, and that failure, he feared, would be his finish as a poet. So he told the critic Horace Gregory in a New Year's Day letter, 1945. But by March he had done what he thought at this point he could do and had sent a sixty-page manuscript off to his friend and publisher, James Laughlin. When in August Laughlin sent back the galleys Williams was appalled at how the thing looked and read. The long training, the years of painstaking gathering of materials, the search for controlling images, his work on speech rhythms—it had come to this mess: the poem was "tight, pebbly, cracked, rubbly—full of dried shit," as he described it to a friend. To the critic and old friend Kenneth Rexroth, Williams wrote that when he had seen the galleys he had "almost vomited. I was heartbroken, the mess looked so foul. I slashed right and left, reedited bits here and there, pulled things together as best I was able—and the Lord only knows what the final result will be." He told Rexroth he knew that most readers would be baffled by the thing and would wonder whether in fact it deserved to be called a poem. But his hope, a slender one now, considering what his hopes for "Paterson" had once been, was that here and there a reader would "realize what I have been up against in taking the crude mass which I determined to attack and make it into SOMETHING, anything."

What Williams had been up against was the epic impulse versus the muddy, chaotic flux of a modern democratic republic. It was from the outset an unequal contest, as the poet himself had long known: that was why with a telling consistency he referred to his attempt as "impossible." And the longer the poem became—*Paterson 2* (1948), *Paterson 3* (1949), *Paterson 4* (1951), *Paterson 5* (1958)—the more evidently unequal and oddly heroic that contest became. It was as if a good welterweight boxer were attempting to go toe-to-toe with a good heavyweight. Yet Williams persisted in the contest, trying in the successive extensions of the poem to dominate it, to give it a manageable shape. Among the papers found after his death in 1963, and after multiple strokes had so numbed and fuzzed his brain that he could hardly type a line, were fragmentary notes toward a *Paterson 6*. Like Whitman who worked at *Leaves of Grass* right up to the turning of what he called the "mortal knob," Williams labored at his epic right up to the end, stubbornly convinced that in the very rush and tumble of what he'd brought together—the falls of the poem—there was an order. To discover that order was not only the task of the poet, though it was this, too. It was also a task of great cultural importance, since the poet's task was parallel to that of his readers, as they read his poem, as they lived their lives. For they, too, lived amidst a roar, a welter, a landscape dominated by the city. Like the others who had tried to write American epics—Joel Barlow, Whitman, Pound, and Hart Crane—Williams had been driven by the need to create a poem that would include the history of his country and himself, so to instruct his readers in the nature of their nationality and their own individualities. Where he differed was in his intensely local approach.

The form Williams finally settled for might best be thought of, Henry Sayre suggests, as collage. The analogy has its limitations, especially considering how very experimental and provisional *Paterson* is and was meant to be; and considering, too, the fact that Williams's ideas about the poem and its form underwent some evolution during the more than fifteen years that intervened between the composition of *Paterson 1* and *Paterson 5*. That said, it is helpful to think of the poem as a collage, a visual art form that seeks to establish a kind of compositional harmony between disparate elements. Crusading modernist that he was, Williams was temperamentally and philosophically drawn to the sorts of visual experiments he first saw at New York's famous Armory Show in 1913. Later, in Paris, he was moved by what he saw there and what he sensed Braque, Picasso, Brancusi, and Juan Gris were striving after. For at least two decades Gris was his favorite artist because he could see that Gris had made the effort to bring his art into direct contact with the things, the objects of daily living. The introduction of "real" objects (newspapers, cloth, string) into the field of the painter's composition affirmed, Williams believed, the vital and vitalizing connection between art and life. Further, it brought art to viewers with an immediacy and punch that made the art, for all its modernity, newly available. Here again was the new and the living thrusting boldly up through the past, through decay, detritus, and death. And the boldness of Gris and his contemporaries had the salutary effect of denying viewers the refuge of illusion: looking for the familiar landscapes, portraits, narratives, you were instead stuck with the demands the artist's imagination made on you, forced to enter yourself into the act of composition. As ever, Williams admired such a confrontation rather than escape.

The collage of *Paterson* includes a literally stunning variety and amount of "real" objects. There are, as with *In the American Grain,* numerous extracts copied out of history texts, "for the taste of it." Here all of them have to do with Paterson and its environs: an account of a mid-nineteenth-century find of mussel pearls; a description of a hydrocephalic dwarf who lived near the falls in the eighteenth century; a woman lost in the falls in 1812; Hamilton's vision of Paterson as the nation's capital; the Paterson silk strike of 1913; tightrope walkers across the falls; an early nineteenth-century case of witchcraft; diagnosis of a case of salmonella poisoning; an infanticide discovered on Garrett Mountain; a hired man's murder of his employers and his public execution on the mountain.

There are letters Williams received through the years, letters from unidentified correspondents as well as from the prominent—Edward Dahlberg, Pound, the young Allen Ginsberg. There are letters Williams was given access to: those from a young Paterson nurse to her alcoholic father; letters to a black maid working for Williams's friend and typist, Kitty Hoagland. By far the most extraordinary letters are those of an unknown poet, Marcia Nardi, whom Williams met in the spring of 1942, and with whom he had a stormy and tangled relationship.

Also included here are a weather report, a political advertisement, a portion of a radio interview with Williams on the subject of the incomprehensibility of modern poetry; a snippet from a book on Greek and Latin poetry and one from George Santayana's novel, *The Last Puritan;* ethnographic reports form Africa and North America; a report of soil samples taken from an artesian well at a Paterson mill.

Dozens of historical figures stalk through these pages, some mentioned only fleetingly, others like Hamilton and the daredevil diver, Sam Patch, given fuller treatment: Washington, David Lilienthal, Governor Altgeld of Illinois, John Reed, Big Bill Haywood, Lautrec, Billy Sunday, Madame Curie, Columbus, Carrie Nation, Allen Tate, Walter Reuther, the jazz artists Mezz Mezzrow and Leon Rappolo, Gertrude Stein, Ben Shahn, Phideas, Nathaniel West, Audubon, Sappho, Picasso, Klee, Brueghel. . . .

The major figure is the poet himself in his several personae but chiefly as a man called Paterson. He walks through the poem a witnessing, judging, commenting presence, sometimes sagacious, sometimes a figure of mockery, by turns angry, saddened, occasionally joyous. Above all, he is the controlling imagination who attempts to make a composition of this world. He does not uniformly succeed and admits it, once even counseling himself to give it all up, to quit writing the impossible poem. But he does not.

Paterson 1 introduces the reader to what Williams calls the "elemental character of the place." Here are the falls and within them the giant lying on his side, the spent waters outlining his back. This is one of the poet's personae. Here, too, is the mountain above the falls and the city, imagined as a woman, the giant's mate. And yet, though they are mates, they are not mated, separated by form, substance, and geography, a condition the poet uses to stress one of the poem's major themes: divorce. "Divorce is/ the sign of knowledge in our time,/" Williams writes, "divorce! divorce!" Men and women, we are to understand, are divorced from a sympathetic understanding of one another, as Marcia Nardi's letters to Williams so poignantly reveal. But this is not all, though surely it would be enough since it points to a fundamental human problem. The city's residents are also divorced from one another, from the place where they live, from their history, even from themselves: automatons, they sleep-walk through existence, rarely in touch with anything. The poet's task, again, is to reattach the things of this world, to make a whole out of the separated parts. But in the figure of Sam Patch, another of the poet's personae, Williams seems to admit the impossibility of this.

The historical Patch was a cotton-spinner from Pawtucket who achieved national prominence when as a stunt he jumped into Niagara Falls in 1829. Earlier he had done the same thing at Paterson. His career ended the same year it began when he failed to survive a leap into the Genesee Falls. Subsequently Patch became a folk hero of the stridently confident Jacksonian era. Clerks called themselves "Patch" as they vaulted counters; so did country boys leaping fences. Plays were written about him: one had him coming up out of the falls in China. His mottos—"Some things may be done as well as others" and "There's no mistake in Sam Patch"—suited the mood of his time. But in the poem Patch, for all his daring, is a failure. In the very midst of his leap at the Genesee Falls he appears to want to say something, and his body twists into a fatal sideways slant. Here, Williams says, is a man (the poet himself) doomed by the very audacity of his ambition. He leaps into the roar of the waters, which are also language and history, and so is lost. Still the effort had to be made, whatever the cost, and thus the books of the poem that follow.

In *Paterson 2* the poet walks among his people on Garrett Mountain to give a more particular view of local life. His feet touch common ground, "the picturesque summit, where/ the blue-stone (rust-red where exposed)/ has been faulted at various levels/ (ferns rife among the stones)", the mountain's surface scarred by boot nails more than it was by a glacier. It is a Sunday in the park, and in contrast to, say, Seurat's famous Neo-Impressionist painting, "Sunday Afternoon on the Island of La Grand Jatte," with its brilliant order, Williams gives us this vision of the working class trying for some sort of ease:

> . . . the white girl, her head
> upon an arm, a butt between her fingers
> lies under the bush   .   .
>
> Semi-naked, facing her, a sunshade
> over his eyes,
> he talks to her
>
> —the jalopy half hid
> behind them in the trees—
> I bought a new bathing suit, just
>
> pants and a brassier   :
> the breasts and
> the pudenda covered—beneath
>
> the sun in frank vulgarity.
> Minds beaten thin
> by waste—among
>
> the working classes SOME sort
> of breakdown
> has occurred. Semi-roused
>
> they lie upon their blanket
> face to face,
> mottled by the shadows of the leaves. . . .

Here are Adam and Eve at rest in the paradisal garden of the New World.

Against this tableau Williams places the scarcely attended rantings of a Pentecostal preacher who stands in the hot sun with his little retinue: four musicians, three of them middle-aged men "with iron smiles," and a woman with a portable organ. The man's sweating sermon, the new Sermon on the Mount, takes the form of a discourse on wealth versus salvation. But pathetic as old Klaus's sermon may be, in the context of the poem it assumes a charged relevance, for America, this garden, was itself potentially salvation until men with the vision of Hamilton could see its falls and mountains as reservoirs of wealth. Surveying the Sunday scene, the lolling lovers, the preacher, the general lack of privacy, the whole messy and pathetic effort these people make at play, the poet is moved to ask, what beauty is left here? All he can find of it is a "fresh budding tree," as yet undamaged by the heedless crowd. "Be reconciled, poet," he admonishes himself, "with your world," for it is "the only truth!"

The last word in *Paterson 2* is Marcia Nardi's: six-and-a-half pages of small-print invective directed at the poet's (Williams's) divorce from life, from the reality of human relationships. It is a brilliant, impassioned description of the inability of a man to understand a woman, an inability Nardi traces to the poet's tendency to view anything—people, their lives, even their deepest anguish—solely as potential literary material. Maybe it is true, Williams suggests in placing the letter here. And yet, how else is Paterson, the poet, to begin to make an order out of all this? Must he not compose out of the welter of his given world? As with the figure of Sam Patch in *Paterson 1,* Williams leaves the reader to ponder the question of cost.

The theme of Nardi's letter is continued in *Paterson 3* where the poet turns away from the foulness of his city's hot summer streets, seeking escape from their reality in the library where he reads of the tornado, fire, and flood that reduced Paterson to mangled, charred ashes in 1902. But the more he reads, the less successful his attempted escape becomes until the library air is more stifling than the streets outside and the poet is forced back out and into an embrace of what they are: "Embrace the/ foulness." So he must contemplate anew the significances of divorce, not the least of which, surely, might be the divorce of books, literature, from life itself. And perhaps here the ultimate divorce would be that of this very poem, *Paterson,* from those for whom it is ostensibly written, Williams writing in Section II as if he had shown a portion of it to a patient:

> . . . Geeze, Doc, I guess it's all right
> but what the hell does it mean?

At the end of *Paterson 3* the poet assents to the conditions of the world he must compose out of: an uncomprehending, beaten audience living in a blasted landscape from which they are divorced. The very stones along the Passaic emit an ineradicable "granular stench" when the poet takes a few of them home for "garden uses." If he would make his little garden beautiful, he must make it so out of what materials he has at hand: "—of this, make it of *this,* this/ this, this, this, this   "

Originally Williams planned *Paterson 4* to be the concluding book of his epic effort as in it he follows the Passaic down into New York City and its end in the sea. The river's end was not, however a true end, he asserted (nor was it to prove the end of the poem), since as rain the same waters would return to replenish the sources of the Passaic in the eternal hydrological cycle. Moreover, among the flotsam the river carries to the sea are seeds, some few of which would be carried ashore where they might take root and bloom. Thus the inclusion here of a letter from the young, unknown Paterson poet, Allen Ginsberg, who had been inspired to write of the city by what he had read of Williams's work. Here in truth was the poet's seed taking root. I know, Ginsberg's letter reads, "you will be pleased to realize that at least one actual citizen of your community has inherited your experience in his struggle to love and know his own world-city, through your work, which is an accomplishment you almost cannot have hoped to achieve."

If in a sense this was the organic conclusion to *Paterson,* it is testament to the strength of Williams's commitment to his epic and to his staying power that the last book does not read like an afterthought. It is instead a rebellious, old-age affirmation of the immortal power of the human imagination. Williams's mind was clearly much on his great Camden predecessor, Whitman, in these pages as, Whitman-like, he looks back on his earlier efforts and finds them justified. Climbing Garrett Mountain once more, the poet looks out on the field of his effort, "to witness / What has happened / since Soupault gave him the novel / the Dadaist novel / to translate—/ *The Last Nights of Paris.* / 'What has happened to Paris / since that time? / and to myself'?" Then comes the poet's ringing response:

A WORLD OF ART

THAT THROUGH THE YEARS HAS

SURVIVED!

If in some ways he had failed in *Paterson,* as he had predicted in the figure of Sam Patch in the first book (and as he had confessed in a letter to Marianne Moore having finished *Paterson 4*), still he had not utterly failed. For he had indeed imagined a whole, teeming, boiling world in his poem. So, too, had those anonymous medieval weavers who had stitched together the famed Unicorn tapestries that now hung a few miles away in The Cloisters in New York: that had been a world which, save for the efforts of the weavers, might have vanished beyond our recall. Instead, there it vibrantly, magically was, preserved by the power of the anonymous imagination. The hunters in the medieval legend thought they had slain the Unicorn, but they had not. It rose from the dead, a triumph. And so was *Paterson* a triumph, whatever its defects. It would survive, a seed, a flower, an inspiration in the mind of a mad poet haunting Paterson's River Street. Those who thought humans were trapped in mortality as if in a bag were wrong:

> It is the imagination
> which cannot be fathomed.
> It is through this hole
> we escape

> So through art alone, male and female, a field of
> flowers, a tapestry, spring flowers unequaled
> in loveliness.

> Through this hole
> at the bottom of the cavern
> of death, the imagination
> escapes intact.

II

YEARS AGO I had taught with Paul Mariani, and we had remained friends despite the diverging paths of our lives. Now I had called him asking him to meet me in Rutherford to guide me around that territory he had come to know so well while writing about Williams. Could he give me a couple of days? He could.

Sitting in a Saddle Brook motel/restaurant where we had met for a bite before going over to Paterson, I said to him that after rereading his massive Williams biography, *A New World Naked,* the subject appeared an inevitable one for him. "I was struck again and again," I said, "by how much there is of you in Williams, how much of you there is in your book." I was thinking, as I explained then, of his immigrant background, lived so close to here, of his own real accomplishments in poetry, and even of his struggles to remain faithful to the true spirit of his marriage.

"Yeah," he said, somewhat shyly. "You know, it's really true, Fred. There's a lot of the autobiographical mixed up in my work on Williams. I guess there must be in any biography, don't you think? Anyway, I know there was in mine. And even after all the years I spent on the guy, all the research, all the writing and so forth, the guy still fascinates me.

"I can still remember the very first time I saw a reference to the book, *Paterson*. I thought, 'Who in hell would write a book about *that* place?!' See, I knew Paterson. I used to play there when I was a kid. We had a cousin that lived there, Turk Lacherza, and we thought coming over from the city [New York] to Paterson was a trip to the country because Turk had a lawn in back. Can you imagine that?!" Surprise and delight were mingled in his voice as he hunched his heavy shoulders over the table, closing the gap between us. "A book about Paterson! Not that I despised it. It wasn't that at all. It was simply that it was local: it was a place *I* knew all about, and somehow I'd gotten the idea that poetry wasn't written about local places that had street names and so forth that you knew, about factory towns where you grew up. You know? Anyway, that's what started me thinking all those years ago, though not in any conscious way, not as if I'd said to myself I'd write about Williams.

"At the time I was in graduate school at Hunter. I'd never had more than fifty minutes of Williams [as an undergraduate] at Manhattan College. And now here was this mention of his book, *Paterson*. I'd been reading Hopkins and Eliot and Pound, and so forth, and all that was fine: I was really excited about those guys. But who was speaking for *my* people? Who was writing poetry about the immigrant experience in these tough towns? Nobody.

"Then later I read the book, and it was tough going—you know that yourself. I put it aside. Then, still later, I came back to it. You know how a thing will hang around in the back of your brain. Well, then I wanted to try to teach it—and I found there was no criticism to speak of on the thing or on Williams, really. Think of that! Hardly a thing worth reading. And so then I said to myself, 'That's it: that's your job.'"

Talk turned to the marvelous variety of sources Williams had used to make the poem, of the collage effect, and here we found ourselves talking of the Nardi letters that play so significant a part in the poem. Mariani said some feminist critics had attacked Williams for his use of the letters and for his use of Nardi as a person. And some of them had attacked Mariani as well for being too easy on Williams's treatment of women generally and of Nardi in particular, claiming that Mariani ought to have made more of Williams's flagrant womanizing. I said I recalled a conversation we'd had on the subject years ago at which time Mariani had said all his efforts to find Marcia Nardi had come to nothing.

"Oh, I found her, all right," he said now with a small smile. "It was after I'd published the book, though. Then one day I get this phone call, and it's her, Marcia Nardi, and she's calling me from Cambridge [Massachusetts]. 'I've read your book, and young man, I'm going to sue you,' she said. Then I went to Cambridge to meet her, and there she was in this tiny garret. But dressed to the nines. She was wearing a suit, her hair was well done, and she was properly posed. She had her feet pointed out like this [sliding his feet out from under the table and holding them together, primly pointed].

"She wanted me to see right from the start that she was a lady of great culture. She could talk of Rimbaud and Mallarmé, and so forth. She knew her French better than Williams ever did, and she made sure I got that. 'I don't need Williams,' she told me. 'I don't need that son of a bitch.' She wanted to emphasize that he wasn't as cultured as she was. Who was he, anyway, to take her under his wing?"

As it developed, the Nardi/Mariani relationship was an eerie reprise of hers with Williams, Mariani told me. "It became clear to me that I was the Williams substitute. But what could I do about it? She used to say to me the things she wished she'd been able to say to him: these long monologues that in every respect recapitulated that relationship. She even wanted me to get her poems published like she'd once wanted Williams to, and I told her I would try. And they were good poems, Fred, many of them. She let me read them in her garret. But when I asked her to send them to me, she never would. She never would. She'd read them to me over the phone, but that was all."

Still talking of Nardi and of Williams's relations with women, we drove over to the falls of the Passaic, going through the handsomely refurbished mill district on the river and then out to the falls. There, overlooking them, with sheaf of papers (plans, no doubt) in hand and pensive look, was a statue of Hamilton, Williams's *bête noire*. Leaning on an iron bridge railing over the waters, Mariani spoke of visiting his relatives in nearby Clinton and how on those occasions he'd come over here with his brother Walter to play about the edges of the precipice. They had seen it in all seasons—encased in ice, in muggy summer, and, as now, in an end-of-autumn solemnity. He pointed out the profile of the great stone head, the giant, who had become Williams's secret image of himself before he announced it at the opening of *Paterson 1*. At the edges of the still pond above the falls we encountered the messy, trashed landscape Williams had had to use for his poetry: floating plastic containers, condoms, papers, Styrofoam cups half sunk. Nor was this a recent development, for Mariani said that when the silk mills had operated here "the falls would run purple or blue or red, depending on what the color of the dye was that day." His grandmother had worked in those mills. "She was Paterson to me," he said, gazing steadily down. "You could see the streets of the city marked right there in the wrinkles of her face."

As we walked the streets leading up and away from the falls the heavy, steady roar of the waters gradually was replaced by the street sounds of cars, buses, school kids just sprung for the afternoon. At a football stadium a few blocks up small units of high-school athletes trooped by us carrying gym bags, heading for practice. Most of them were black. Up on Totowa Street we stood amidst the flow of student walkers, talking about why Williams had chosen to live here when he might easily have gone across the river to New York City where there was all the artistic stimulation and cultural ferment anyone could ask for. "It was a temptation to him, for sure," Mariani said. "It was like a seduction. Right there across the river and the meadows. It was like a beautiful flower. Sometimes he talked about it like it was a woman he wanted to fuck. There was tremendous energy there—its life—that he found tremendously attractive. But," he paused a moment, "I think he knew somehow he was better off with it over there, available, if you know what I mean. For one thing, when he was thinking about writing a poem about a city, the city as a poem, he knew he couldn't do New York. It was simply too big. No one could do New York anymore. Whitman couldn't do it now. When he [Williams] thought of the ideal size for a city, he thought of Dante's Florence or Shakespeare's London, not Dickens's London and not twentieth-century New York."

As we stood there on the corner of Totowa and Redwood, surrounded by schoolyards, schools, an Italian grocery Mariani remembered from his youth, a tightly bound, bouncy little man wearing a Greek sailor's cap came up to us. Pointing at the camera I held in one hand, he asked if we were "shooting the city." He was Alfonso Giella, back for a visit to the city of his birth after forty-five years in Los Angeles and now Tucson. Born in Paterson in 1912, he told us he couldn't get used to what the city had become since he left it. Even the street names were different: what was "Rosa Parks Boulevard," anyway, he wondered? "It used to be Graham Avenue. I used to walk Graham Avenue all the time. Where is it?"

"My father," he told us, his bright eyes flashing behind his spectacles, "was the first Italian pharmacist licensed in the state of New Jersey. But now I look around here, and I don't see anything familiar. Where are all the people like the ones I grew up with? The place's falling apart! It's like it's been invaded by a foreign army! I think it'll take a hundred years to assimilate all this [nodding to the river of black school kids streaming past us]! I feel like a stranger walking around where I used to live. Or a ghost. The culture, don't you know, looks strange. And it isn't just a matter of skin." I was getting a trifle nervous here for Giella was making no concessions to the situation, and his voice had risen. I saw a look out of Mariani's eyes that said, "Well, let's just see." "It's the language," Giella went on, "and its cadences. I walk around here, and I hear Haitian, I hear the islands. It's the way they walk. It's their dress."

He knew Williams's work and had read it with some attention, as it appeared when we told him how we happened to be standing at the intersection of Totowa and Redwood with a camera. But he claimed that Williams, too, had been corrupted by the influence of the Old World just as Williams had said Pound and Eliot had been. "You know," Giella lectured, "even in the old days—and I left here in '37—Williams didn't really have this place. Not really. He didn't really get into it in a gutsy way. You see, by the time he came to write *Paterson* he already had an international perspective. He'd already been to Europe—Paris and all that. When he's writing with his eye on Eliot and Ezra Pound, he's not really looking at this specific place."

I disagreed, saying I thought Williams had made a heroic effort to reproduce that "granular stench" of this place, the look of it, the idioms he heard. Mariani introduced another angle when he said to Giella, "*Paterson*'s not just about this place. He's writing a kind of autobiography there, too. But Charles Olson agrees with you. He said Williams wasn't really getting into his city. You gotta get down into it, name the streets, and so forth. And that's one of the impulses behind what he [Olson] was trying to do in Gloucester."

Back at the Hamilton statue Mariani and I continued with the subject Alfonso had raised. It would indeed be ironic to think, I said to Mariani, that Williams couldn't have written about his chosen locality had he not sampled Europe and the expatriate experience. If that were true, it would seem an oblique confirmation of what Pound had always told Williams was the special virtue of being abroad. Could Williams, I wondered, have written *Paterson* without Europe?

"I can't see it," he said finally. "At least not as we have it. He had to go there to learn a certain kind of esthetic distance—how to look at things from that sort of remove, that perspective. Going to Paris, he wouldn't have been so much interested in the Louvre, the Eiffel Tower. What he wanted to find, what he wanted to see was some evidence of a Europe in its infancy—something like a beautifully executed fifteenth-century piece of stonework. That would be a clue to the history of that place before it had become the home of Modernism. What he was after was the older order that underlay all this contemporary sophistication, this urban landscape. The thing was to find an equivalent, sort of, for what had happened in America. If he could find that equivalent, make that bridge, then he could confront Modernism and begin in his own voice. Go back, go back: re-see: that's what Europe gave him."

By this time we were driving back and forth trying to get on the east side of a freeway so that Mariani could show me the river at the Third Street Bridge separating Passiac and Wallington. Finally we managed access, crossing the whizzing expanse of the freeway, then dodging up some curling, narrow avenues and onto the bridge. At the other end of it was a factory, monstrous, Blakean in proportion and menace, and beside it the self-same pipe Williams had described in the opening paragraph of the title story of *Life Along the Passaic River*. It was still belching its used water out into the river. There was low water here again as at the falls, the banks a beaten slough of mud in which were embedded oddments of motor parts sticking up from the goo like last glimpses of drowning creatures. Late autumn light fell on the scene where it was not blotted by the factory, mellowing a harsh tableau. From the east end of the bridge a pride of teenage boys strode toward us bouncing basketballs. The scene— pipe, river, factory, boys—clearly was as evocative for Mariani, seasoned observer of all this, as it was for one discovering Williams's landscape. He turned to me in the middle of the bridge, talking now with his whole large torso. "This guy is Whitman's literary descendant, Fred. He was the one who said, '*This* is what we have to sing. You don't like this America? Well, this is what we've made.'"

Nine Ridge Road in Rutherford, long home to Williams and Floss, is now home to their son, William Eric Williams, who still practices medicine in the same office his father had. Mariani had come to know the doctor well in the course of his long research, and now on this soft October evening we were pulling into the driveway behind the house, the bells of the Presbyterian church across the street heavily striking off "God Bless America." The house's outdoor lights thrust Williams's little backyard garden in deep shadow, but Mariani knew the way.

Doctor Williams (thereafter simply "Doc") greeted us warmly at the door: a middling-sized, vigorous, upstanding man, bald, and with skin the color of parchment. He took us into the well-lighted kitchen where he hurriedly finished up his bachelor's supper, his second wife, Mimi, being just then away on a trip. He seemed glad of the company, a night with the boys over the bottle of wine we'd brought along. The impression was strengthened later when he insisted we come by the next day for lunch. He wanted, he said, to take us over to the edge of the Meadowlands where his father used to take him when he was a boy. The Meadowlands there, he said, had once been a "famous white cedar swamp. It still was when my father was a boy. But, of course, that's all gone now. Giants Stadium's there now. The ships used to anchor in the Hudson and send their wagons over the Palisades to get that swamp water. You could haul it around the world, and it would never spoil. Just full of tannic acid. It was a little on the tea-colored side, but it wouldn't spoil."

Here we were again, I thought, right in the middle of talk about that fouled, polluted landscape Williams had claimed as the very ground of his epic poem, and so I asked Doc about the recent reports of the suspiciously numerous cases of cancer that had turned up on the Giants football team in the last few years. "I have to believe there's something to it," he returned. "What precisely I can't say. But I've seen too much of it in my own practice. Too many cases here: leukemia, Hodgkins. It may turn out, though, to be something so simple as the fact that we're right in the middle of all these radio waves—the airports in New York, Newark, Teterboro, and so on. This area is just a tangle of them. Then, that area [the stadium] was apparently a mercury dumping ground years ago for people who were manufacturing thermometers and whatnot."

Sitting in the living room under a Ben Shahn painting called "Paterson," and next to it a small oval photo (Williams, Floss, and Shahn with the as yet uncompleted painting before them), I asked Doc about his father's poetic program. He had wanted to be the voice of this place with its ruined meadowlands, its dirty streets, its working-class people, their minds "beaten thin" by work and poverty. How successful, I wondered now, did Doc think he had been in this huge and singular ambition? There was a certain diffident resistance here, as if the son would be wholly gracious as long as talk remained at a general level, with the poet there, all right, but there in the background. Or was it that the son believed his father had failed to be the voice of this place, that he hadn't made his poetry accessible to those for whom and to whom he was writing? What Doc answered now arguably suggested this latter, for he asked, by way of anecdote, how truly accessible the poetry was.

"There was a grade-school teacher here," he said, "who called up the other day to tell me her kids—these are sixth-, seventh-, eighth-graders we're talking about—were putting on a play using Dad's stuff. 'They're all volunteers!' she told me. Come on! Jesus! What could they possibly understand? You could put the words in their mouths, they could ape his lines, but what could they really understand? It's just too difficult.

"You know, I never had any idea my father had any stature whatsoever as a writer until I was in college at Williams in, oh, '32, '34, along in there. And someone asked a question on an exam, and it was about William Carlos Williams. And a kid said later, 'Who the hell is William Carlos Williams,' and I thought, 'Wow! that's my dad!'"

He thought I would like to see the business end of the house, a portion of which he had remodeled since he had set up his own practice in it. In the same small office where his father had practiced before him he opened a file cabinet to show us Williams's files on his patients. "Some of 'em," he said, riffling through the fading, stiffening cards, "are sort of laughable. Dad didn't really keep records that were that good." No, Williams didn't have any practice to speak of over in Paterson. "Most of it was right here [tapping the desktop]. He really didn't go over there that much that I can recall. Except to the falls. There he went a lot. And, of course, he went to the library a lot to research. They have all those records over there, what he took out, and so forth. And he went there to get the image. He had to have that." He sat down in the old chair behind the desk under the somewhat bleak fluorescent overhead and suddenly offered up a comment that was either summation or signal. It might have been both.

"Dad and I never had much to say to each other. He was always either in here or else up there [nodding ceilingward toward the second-story study where Williams in his stolen moments composed], banging away at the typewriter. He tended to go his way, and I tended to go mine. He was a loner, and so was I. Now, with the grandchildren, it was completely different by then. By then he had time for 'em. I guess he thought he'd done pretty much what he had to do. Anyway, he took Paul [Jr.] fishing with him a lot."

We went out then through the waiting room, and in its emptiness and scuffed, used look under the white glare of the overhead I felt that urgent professional necessity that had filtered Williams's days and nights and about which he had at times raged through the years when he felt the need to compose rather than to attend. The scraps of talk, the notes jotted on his prescription pads told a story, and I wished I might have seen some of those. As it was, we said good night at the front door, made plans for the morrow, and Mariani and I stepped out into the fall night, leaving behind us the house with its blazing windows and the writer's son with his rooms, his memories.

Saturday morning dawned bright though misty. A fog, doubtless mixed with some sort of soot, hung about the roofs and chimneys of Paterson as Mariani and I rode through the still streets heading for Garrett Mountain. The woman's huge shoulder, Mariani pointed out, had been shaved off to accommodate the freeway that now whips around its base. Climbing the winding road through the rusty fall foliage, we gradually left the mist below, and when we arrived at Lambert Tower we were in the unalloyed sun of a new day.

Lambert Tower was the remnant of the feudal estate of the one-time Paterson manufacturing lord who regarded his employees as serfs and of whom Williams had written in *Paterson,* quoting him to the effect that he, Lambert, reserved the right to fire any son of a bitch in his factory simply because he didn't like his face. The tower was empty, gutted, and wore that relentlessly assaulted look of older public park buildings in the Eastern states. Its lower walls were bedizened with slogans: "FUCK LA BAMBA. SPEAK ENGLISH OR DIE! GO HOME BANANA BOAT FREAKS! ⌘": evidence that even those who hadn't been away as Alfonso Giella had were bothered by those changes in complexion, cadence, and gait of which he had so feelingly spoken. There was no one else in sight at this early hour, and Mariani seemed sunk in thought, perhaps forgetting for the moment why it was we'd come up here. Then out of his depths he said, without looking at me, "Don't forget, Fred, this was once envisioned as the *capital of the United States!* Think of that when you look at what's down there. What's left behind [interstate] 80 was in Williams's time hundreds and hundreds of slums of Poles and Italians, each of whom was cultivating his little plot of green, his piece of America. You remember how small that patch of lawn looked behind Turk Lacherza's old house yesterday? Jesus! You could roll a mower over it once, and that would be all you needed. And that was country to us. And these gardens were the New World to these people. You know: Fitzgerald's 'fresh, green breast. . . .'" He left the rest of the quotation from *The Great Gatsby* unfinished.

In his company the mountain was humming with Williams's presence. One of the effects of reading Williams's poetry—and surely one he intended—is that under its influence the commonest phenomena become potentially poetic: that was what he meant when he said that his readers could find themselves in his poem about his place. So now on this bright autumn morning on the mountain all we saw seemed utterly appropriate, as if another poem could be written of what we saw: "Saturday in the Park." Up there that morning was a high-school track meet, the parking lots filled with school buses and the trails through the littered woods filled, too, with panting, fresh-faced runners. And there were other kids up there at a sport of their own: hurling rocks from the summit in hopes of hitting cars on the freeway below. There was a tattered gazebo at the edge of a still pond, perhaps the very one at which the glabrous-skulled preacher delivered his sermon. And the pond itself: filled at its margins with half-submerged beer cans, Styrofoam cups, Crunchy Cheez Doodle bags, plastic knives and forks, soft-drink cans, and plastic bags: make it of *this* and this, and this.

As he had made his poetry of what was there for him to use in his new Jersey landscape, so Williams had made his backyard garden at 9 Ridge Road out of native materials. There were trees in it that he had transplanted from all over the state, some of them now of a good size. And there were the stones, too, that had emitted their ineradicable stench when he'd pulled them from the banks of the Passaic to use as borders for flower beds and walkways. Sitting in the garden in the warm noontime sun, Doc Williams, Mariani, and I had Beck's beer and ham-and-cheese sandwiches and talked of the garden and the pleasure the poet had taken in tending it. "That used to be a sort of stage," Doc said, pointing to a small platform of raised stones in the garden's corner. "Dad would have friends in and they'd put on plays or give readings. It seems to me I once saw Isadora Duncan dance . here, but maybe I didn't." Planes droned overhead on their radio waves to touchdowns in Teterboro or Newark or New York. From an adjacent yard kids shouted while they played on their swings. Jays screamed through the mellow air. Leaves drifted slowly down around us, settling on the neat beds and grass. Bees buzzed the sweet tops of our beer bottles. They would be playing football games in towns like Rutherford this afternoon, and after lunch as we drove over to the Meadowlands we passed two fields where bands were practicing their moves and junior-varsity players slammed into one another on the brilliant grasses of late fall.

When Williams had been a member of the local mosquito control board he'd often brought young Bill and his pals out to the Meadowlands. Later, the boys went on their own to shoot rats and feral cats. "The trucks used to bring loads of leaves out here," Doc was saying as we rolled along a narrow road with high reeds on either side. "They'd just dump them along the road here and there. And that's where the rats would go because those piles had a warmth, you know. So we'd come along, kicking the piles, and out they'd jump. *Pow!!*" He pointed out the changed composition of the meadows themselves, how tougher, pollution- and salt-resistant reeds had replaced the original vegetation. And, of course, the white cedar swamp had long ago vanished.

The road ended abruptly in a stretch of grooved granite blocks. They were, Doc told us, ballast that ships of long ago had used. Beyond them lay a smoking heap of rubbish and beyond that the Hackensack. On its eastern banks were rotting wharves, rusted boats, and huddles of small buildings, many painted an aqua green. Then the meadows again and sitting in their midst, seemingly sunk to its mid-section, the huge, ghostly stadium. Walking to the river's edge we passed the rubbish heap, and Doc paused a moment to turn over with his shoe tip some busted books that lay at the pile's edge. "Dad liked to poke around trash heaps," he said, as much to himself as to us. "And it is amazing what you'll come up with sometimes in these places."

On the bank we looked toward the stadium. No, he had never gone to a Giants game there. "Dad loved the [baseball] Giants. Loved to listen to Frankie Frisch broadcast the games. 'Ohhh, those bases on balls,' he [Frisch] used to say. They drove him nuts. One time they were going to walk Jo-Jo Moore intentionally, and he just stuck out his bat like this and flicked that ball into the outfield. Dad loved that. Later, it was Johnny Mize."

When, late that afternoon, we finally said goodbye to him, Doc Williams seemed to me just a trifle wistful, as if it might have been fun to spend the rest of the day with us, tooling about the old neighborhoods that Williams had reclaimed for literature, talking of one thing and another with just enough of those glancing references to the life and works of his father. But we said we had other appointments to keep, and he smiled warmly, waving his hand to us from his drive as we backed out. Though we had not said so, our last appointment here was the cemetery where Williams was buried.

It was a big enough place so that, though Mariani had been to the grave before, he had at last to ask directions from a groundskeeper raking leaves. The stone next to Williams's bears the name "Florence," for Floss. But she elected not to be buried beside the man who was her husband for fifty-one years. I asked Mariani how he read this final statement of hers. "Maybe," he said, looking down at the two, small, modest stones lying close together, "it was her way of saying, 'Bill, you've hurt me once too often.'" Turning away at last, he beckoned me. "Come here, Fred. I want you to see something." We walked through the silent ranks of stone to where a hillock crested, then fell away eastward, and there it was: New York, shimmering through the shiny, dirty air, a purplish mass of skyscrapers. "That," Mariani said, "would have swallowed anyone's imagination or effort. But Paterson was the right size. And it had all that history."

TEN

VOICE OUT OF THE LAND: LESLIE MARMON SILKO'S <u>CEREMONY</u>

I

❧ IN HIS *Studies in Classic American Literature* D. H. Lawrence alleges that Ben Franklin had worked out what Lawrence calls a "specious little equation in providential mathematics: Rum + Savage = O." By this he meant that Franklin saw in the Indians' fatal predilection for alcohol the fine hand of Providence which had chosen this unlikely method of ridding the continent of its original inhabitants. But, Lawrence continues, "Rum plus Savage may equal a dead savage. But is a dead savage nought?" Sitting up there at the ranch above Taos and with an enormous bowl of American earth beneath, he thought it could not be so. Perhaps it was only that he saw so many Indians at the Taos Pueblo and had attended numerous Pueblo ceremonies that he was encouraged to think there was simply too much of an aboriginal presence left to make Franklin's alleged equation an accurate one. But if this was a factor in his thinking, it was only one factor. He believed profoundly in place spirits, and he felt them all around him in the American Southwest. They convinced him that in some curious, real way America was still Indian country.

Williams, too, believed in place spirits and in the ghostly presence of the Indian, writing in *In the American Grain* that present-day Americans were dominated by the spirits of those they thought they had killed off. The Indian's spirit, he writes in his essay on Ponce de Leon, "is master. It enters us, it defeats us, it imposes itself." The Indian, he believed, was natural to these lands, the "right" expression of them, whereas the white conquerers were always crucially divorced from the lands they called theirs. And now that the whole country was "theirs" they could feel for

323

the first time that divorce. "The land!" he writes in his essay on Champlain, "don't you feel it? Doesn't it make you want to go out and lift dead Indians tenderly from their graves, to steal from them—as if it must be clinging even to their corpses—some authenticity, that which—.''

At the time Lawrence and Williams were speculating about the power of aboriginal spirits in America theirs were lonely voices indeed. Indians were popularly supposed to be dead as peoples and cultures (a belief Williams himself may then have shared). They had always been considered dead as sources of imaginative material, except on the broad level of popular culture. Our earliest writers, as we have seen, utterly failed to make effective imaginative use of tribal cultures, and it is a sign of the persistence of that failure that when Lawrence and Williams wrote, Longfellow's tedious, galloping *The Song of Hiawatha* (1855) remained the most significant attempt any white writer had yet made to write about aboriginal life. In Williams's time Indians flitted through the pages of a few of his major contemporaries—Hemingway, Faulkner—and Indians were, of course, the stock of Western pulp magazines and Hollywood films. But really no one was thinking much about American Indians in the twenties, thirties, and forties.

Out on the reservations, where most Indians lived, or tried to, hardly anyone there was thinking about writing literature about Indian life, either. It was hard enough just surviving. Besides, only the tiniest minority of tribal members had sufficiently mastered the tongue of the conquerors, and those who had used that skill to assimilate into the cultural mainstream.

There had, to be sure, been a sparse, thin line of Indian authors dating back to the turn of the twentieth century, notably two Sioux, Charles E. Eastman (educated at a mission school in Nebraska and then at Dartmouth) and Luther Standing Bear. A Potawatomi, Chief Simon Pokagon, had written a novel published in 1899, and in the 1920s and '30's John M. Oskison (Cherokee) published three novels. John J. Mathews's novel about life on the Osage reservation, *Sundown,* was published in 1934, and four years after that came D'Arcy McNickle's *The Surrounded,* a novel about a young man's tragic return to his Flathead reservation in Montana. *The Surrounded* was one of the best American novels to appear in 1938, and it was far and away the best work of fiction yet published by an American Indian. But there could in no way be said to be anything like a Native American literary tradition.

For those who have studied the history of American Indian written literature it is now customary to characterize N. Scott Momaday's *House*

Made of Dawn (1968) as the "breakthrough" book, one that announced to America the advent in the national letters of a tribal consciousness; and one that at the same time announced to school-educated Indians the artistic success of one of their own. There can be no doubt of the importance of *House Made of Dawn* as a cultural event: considered as an "Indian" novel it *is* a breakthrough, while considered simply as a fiction it is a very fine piece of work, maye even a great one. Certainly on a shelf of fifty volumes to represent American culture it would find a place.

Yet it is important to note that the very excellence, modernity, and singularity of Momaday's novel made it less exemplary than this characterization suggests. In a way, it was for nascent Indian writers the equivalent of Whitman's entry into the stream of nineteenth-century American poetry for his contemporaries. Whitman, as Henry Miller rightly said, was in his own time and subsequently a rude hieroglyphic, and so to some extent was Momaday: how the hell did he do it?

Part Kiowa, part Cherokee, part white, Momaday spent his formative years on the Jemez Pueblo reservation in northern New Mexico. He has a Ph.D. in literature from Stanford, and at the time he wrote *House Made of Dawn* he probably had learned more about the literary tradition of Western Civilization than any Indian before him. The fruits of that learning show in the novel he wrote about the return to the reservation of a young man wasted by drink and the empty temptations of white urban civilization. Here are flashbacks, stream-of-consciousness, multiple points of view, jagged fractures in the narrative flow, purposeful ambiguity— i.e., most of the major devices of literary modernism. But if this were all, *House Made of Dawn* would not be the breakthrough book it clearly is; instead, it would be an illustration of successful artistic assimilation. But this is not all there is in the novel. Here for the first time in our letters is an effective, convincing rendering of the spirit of tribal oral traditions. Here for the first time the ancient, violated yet inviolable spirit of the aboriginal land speaks in art. Brilliantly, Momaday suggests—he does not tell—that his story is neither a beginning nor an end but instead a portion of the long story that has been telling itself through nameless narrators since tribal time began.

Still, what kind of example was this, really? How imitable was it, how available? On America's desperate reservations and in its atomized inner cities where those like Momaday's main character went to seek an entrance, how many would even know of Momaday's achievement, to say nothing of how to use it for their own purposes? A great irony is here. For aspiring American Indian authors were now in a situation analogous to that faced by our earliest white writers. *They* were now colonials, surrounded (as McNickle had said in his novel) by an alien culture. It was they, now, who could not understand the language or the literature of the new majority. And they could hardly imagine how to make an entrance into print, books, there to tell something of what was true about them. Simon Ortiz (Acoma Pueblo) recalled that growing up in his western New Mexico pueblo in the 1950s he had harbored a vague desire to be a writer, but he knew no one of his race who had actually done this:

> I didn't know of any Native American writers until 1966 when I enrolled at UNM [University of New Mexico] and actively sought for works by Native American writers. D'Arcy McNickle and Charles Eastman were the only two writers I found . . . , and I was happy to discover there were at least two current ones, N. Scott Momaday, who was just starting out, and Vine Deloria, a socio-historical-legal commentator.

A year or so thereafter, Ortiz continued, he was surprised and delighted to learn that there were two others like himself, young Indians struggling to find a voice, to write out of their tribal backgrounds. One was James Welch, a Blackfoot/Gros Ventre up in Montana. In 1974 Welch would make his own breakthrough with *Winter in the Blood,* a novel about the salvation of a young Blackfoot man through rediscovering his tribal roots. The other was Leslie Marmon Silko, a Laguna Pueblo woman who had grown up on the reservation that was but a few miles from Ortiz's Acoma.

So, as Ortiz suggested, the situation seemed only bleak, not quite hopeless. There were the few literary ancestors—Eastman, McNickle— and there was the current singular and shining achievement of Momaday. And here was a tiny band of young unknowns (there were, of course, others, but in the early 1960s they were unknown to each other), starting out without precedent, without real trail blazes of any sort, and without an audience. In the last century Emerson had employed that arresting figure of a "stone cut out of the ground without hands" to suggest to himself what it might take for the new literary generation to create an indigenous literature. More than a century later Williams would write at the outset of *Paterson 1* that his epic was a "reply to the Greek and Latin with the bare hands," so unsponsored and ill-equipped did he feel for the writing of his major work. For Simon Ortiz, Welch, Leslie Marmon Silko, and the others, "what it came down to," Ortiz said, "was that we, Native American writers, represented less than a handful, and there was no tradition of a Native American literature; in a sense we barely existed [,] and in some critics' and publishers' eyes we didn't exist at all." The situation was not unfamiliar to the aspiring Indian writers: like the rest of the Native American populace they were used to being invisible to white America. Only the specific terms of that invisibility had altered for them.

The 1960s, though, were a different time in America. Encouraged by the Civil Rights Movement to take a more positive view of their experiences and by the Vietnam war to take a more critical view of mainstream culture, minority groups were beginning to express themselves in new and challenging ways. The relationship between the rise of Native American political activism and the emergence of Indian writers is probably a good deal more complex than the equation of the one equaling the other would suggest. Nevertheless, there is a relationship, and many of the young writers felt a new if obscure sense of empowerment, a daring hope that they, too, had something to say and that it might even be heard. As Simon Ortiz put it, Indians might be powerless and traumatized by change, but they did exist, after all. "We had our communities, traditions, a portion of our lands, some of our languages, and we had our stories. There were always the stories, and we believed in them because they were the truth; they verified our existence."

"Leslie was always a listener," her father Lee Marmon was saying. Lucy Hilgendorf and I were with him in his Chevy Blazer as it bounced over the rocks, sand, and yucca of the Laguna reservation where much of Leslie Marmon Silko's *Ceremony* is set. "She wasn't like the other kids in that way. Oh, she played and all that, like kids do, you know. But she was always hanging around adults, listening to their talk. She'd keep still while they talked, and when they told stories, she'd ask questions. Then, when she'd be out on Joey [her childhood horse] you'd see her stopped to talk here and there with old people. She did a lot of that all through her childhood. Plus she had a great memory. She seemed to remember everything anyone ever told her. I told her once she ought to be a writer because a writer doesn't have to go anywhere. 'You could stay at home and do your work,' I told her. 'You could even live out here if you wanted to.'" He mentioned to Lucy Hilgendorf that Aunt Susie, one of Silko's major sources of stories, had recently died at age 110. Hilgendorf said she'd heard a lot about Aunt Susie when she and Silko had been neighbors and close friends on the Navajo reservation at Chinle in the late sixties.

Aunt Susie had been one of those who had always understood the power and value of the stories, and she had communicated this to the little girl who listened so attentively she could ask questions. Like others in the Marmon family, Aunt Susie had been educated in white schools, but unlike many other Indians the experience served to strengthen her conviction that you had to preserve the tribal oral traditions, not discard them as if they were impedimenta in the process of assimilation. Once lost, the stories were gone forever since they were, as Momaday writes in *House Made of Dawn,* "always but one generation from extinction."

Her family had been from Paguate, the village to the north of Old Laguna, right up there near the Place of Emergence. In 1896 she had been sent to the famous Carlisle Indian School and subsequently to Dickinson College in the same Pennsylvania town. Then she returned to Laguna to marry, raise her own family, and teach school there. In a collection of stories, prose poems, and autobiographical pieces she called *Storyteller,* Silko writes that from the time she could remember there had always been Aunt Susie at her kitchen table with her books and papers spread across its oil-cloth surface. "She was already in her mid-sixties," Silko writes,

> when I discovered that she would listen to me
> to all my questions and speculations.
> I was only seven or eight years old then
> but I remember she would put down her fountain pen
> and lift her glasses to wipe her eyes with her handkerchief
> before she spoke.

Aunt Susie was, Silko says, of the last generation at Laguna to pass down an entire culture by word of mouth. Her aunt realized that the old conditions that had maintained Laguna culture through the centuries had been forever altered by the multiple intrusions of European cultural influence,

> principally by the practice of taking the children
> away from Laguna to Indian schools,
> taking the children away from the tellers who had
> in all past generations
> told the children
> an entire culture, an entire identity of a people.

There were others, too, from whom the girl heard the stories. There was Grandma Lillie, Grandpa Hank, who grew up at Paguate and graduated from an Indian school in California. There was Aunt Alice who told the story about an earlier Laguna girl and the *Estrucuyu* (a monster). There was her father, Lee Marmon, himself a gifted storyteller, though he told me he had been too busy as a boy to listen to the old stories: "My brother and I were always off, hiking around in the hills hunting for gold." And there were those less closely related to Silko but no less significant for what they passed along to her, like the woman who told her once,

You must be very quiet and listen respectfully. Otherwise the story-teller might be upset and pout and not say another word all night.

The strength of the Laguna storytelling tradition even so late as Silko's childhood in the 1950s derived from a broader Pueblo tradition. The Pueblo peoples conceived of the world as a continuous story that was still happening and that would continue to happen as long as there were people and a world. In the beginning was the story of creation and then the story of the emergence. This latter narrative is common in the American Southwest, and in it the people are shown ascending through four or five worlds into this present one. During the retellings of the emergence story over four days of the winter solstice, the Pueblo storytellers become voices out of the land, the story itself and their very breath, words, shaped and toned by the lands they tell of. In the old days it was not only the trained and specially gifted storytellers who told of the emergence, of the Gambler's theft of the rain clouds, of a skirmish with the Apaches. Since existence itself was an ongoing story and since the stories were vessels—beautiful bowls—containing the people's understanding of themselves, everyone was expected to contribute something to the retellings. Perhaps it would only be the addition of an incident, even part of an incident, perhaps a name, the color of the inside of the Gambler's lodge where he kept the rain clouds. Perhaps it would be something so substantial as a different version of the story. "I know Aunt Susie and Aunt Alice would tell me stories they had told me before," Silko writes in *Storyteller,* "but with changes in details or descriptions. The story was the important thing and little changes here and there were really part of the story. There were even stories about the different versions of stories and how they imagined these differing versions came to be." In this way the stories were kept living, growing.

The mythic narratives were, of course, powerful vessels of encultura-
tion. But so, in oblique ways, were the stories that appeared to be merely
after-dinner entertainment. Hunting stories were hunting stories first, but,
as Silko says, they also contained

> information of critical importance about behavior and migration pat-
> terns of mule deer. Hunting stories carefully described key landmarks
> and locations of fresh water. Thus a deer-hunt story might serve as
> a "map." Lost travelers, and lost piñon-nut gatherers, have been saved
> by sighting a rock formation they recognize only because they once
> heard a hunting story describing this rock formation.

Since Pueblo peoples commonly understand themselves to have
emerged out of the land, the land itself must figure prominently in the
stories, and not as background or local color but as an active, presiding
force. Thus, as in the hunting stories, the most precise and minute details
of landscape are entered in; these locate the story, but more, they dramatically tell the listeners who they are by reminding them of *where* they are.
They are spiritual and psychological maps just as the hunting stories may
serve as geographical ones. The abiding presence of the land may indeed
be the essential reason why Pueblo peoples conceive of existence as an
ongoing, endless story, for the land is always there. The huge boulder that
is a monster's petrified heart remains a landmark on the state road that
runs south from Paguate to Laguna even though the people now pass it
quickly in their cars and pickups. The same road is the route the Laguna
people traveled from the Place of Emergence to their home site, and so
every spring, rock, and hill exerts a sacred claim on the people. The claims
are made in the emergence myth, but behind the myth is the land, and
its phenomena make their claims every day.

For Leslie Marmon Silko, rapt listener from an early age, the land
was especially alive with its mythic and narrative associations. In an essay
on Pueblo landscapes, interior and exterior, Silko says one of her earliest
and strongest drives was to be out in the land, with its life, its stories she
was then coming to know. Her first ventures were on foot, but then with
the coming of the horse, Joey, she went farther and farther into the northern hills, toward the great mesas, or to the south, across Highway 66 and
up to the Dripping Springs. "I was never afraid," she recalled, because
"I carried with me the feeling I'd acquired from listening to the old stories,

that the land all around me was teeming with creatures that were related to human beings and to me. The stories had also left me with a feeling of familiarity and warmth for the mesas and hills and boulders where the incidents or action in the stories had taken place." What she enjoyed most in those long, solitary rides, she said,

> was standing at the site of an incident recounted in one of the ancient stories Aunt Susie had told. . . . What excited me was listening to old Aunt Susie tell us an old-time story and then for me to realize that I was familiar with a certain mesa or cave that figured as the central location of the story she was telling. That was when the stories worked best.

In high school she began trying to make her own stories, drawing in part on those talks she'd had with her Laguna elders. She stayed in touch with the elders and with the land even though she was now going to school in Albuquerque. "It was either there or Grants [a nearby mining town]," Silko's father said. "We thought the teachers would be better in Albuquerque, and I think they were." So for four years, five days a week the family made the 200-mile round-trip commute. Silko stayed on in Albuquerque to enroll at the University of New Mexico, and it was there in the late 1960s that she encountered Simon Ortiz. If, as he has said, her fledgling literary efforts were heartening news to him as he tried to find his own voice, for Silko Simon Ortiz's example was of immediate and immense benefit. Here was someone of whom she had known for several years, someone from just down the road on the adjoining reservation. And he was writing for publication. "An aunt of his," Silko told me when we talked over the phone from Tucson to Santa Fe, "worked in the [Marmon family] store at Laguna. She told me about Simon, that he wrote stories. Then at U.N.M. I met him at a party. Those of us who were trying to write admired him because of what he was doing, all he had accomplished."

When she herself appeared on the national literary scene it was with Ortiz and other younger Indian writers in Kenneth Rosen's anthology of short fiction, *The Man to Send the Rain Clouds*. The title story was Silko's and there were six other pieces of hers in the collection. "The Man to Send the Rain Clouds" is a fine, moving story of culture conflict on the Laguna reservation, and all of Silko's contributions to the collection demonstrate a remarkable appreciation for the way language and landscape may harmonize in art. Yet none of her stories here prepared readers for her astonishing achievement in *Ceremony* (1977).

Ever since Chief Simon Pokagon's *Queen of the Woods* (1899), Indian writers have been telling the story of a character's return to the reservation from white civilization. This, too, is the situation Silko is concerned with in *Ceremony,* but here she is writing about the survival of a people as well as with that of an individual. The young man, Tayo, returns to Laguna from the jungles of the Pacific theatre of World War II. Tellingly, he is a mixed-blood (as is Silko herself) who now must reassume that peculiar burden on native grounds. He has never known his father, and he recalls his mother only as the wine-soused wreck she had become before her early and shameful death. For a time he had lived with her in cardboard shanties in the river bed outside Gallup, spending long days and nights alone while his mother drank in Gallup's bars, foraging for castoff bones and other garbage in the town's alleyways, hiding barefoot and ragged in the stinking piles of human excrement in the river bed when the police would come through on one of their sporadic sweeps. After his mother's death he had been grudgingly raised by her sister, the fiercely impeccable Auntie, who resents him for the shame he represents. And now in the war's wake he has yet another burden to carry, for he is the one who survived, who came back from Bataan and its Death March, while Auntie's beloved son, Rocky, died in the jungle. And it was Rocky who was to have revived the family pride: Rocky—exemplary student, football star, progressive-minded . . . dead.

Back in Auntie's house, Tayo sleeps in the room he shared with Rocky, his cousin's empty bed accusing him with the ineradicable definitions of the departed body still molding its mattress. His dreams are jungles in themselves, his night sweats the jungle rains that bring rot instead of life. To the blind Grandma who sits all day in the outer room, trying to warm herself at the heater Rocky once bought her, Tayo's dreams are a sign his life is spiritually out of balance. She thinks Tayo needs a traditional healing ceremony. Auntie sniffs in disapproval. She, too, is progressive, and does not want the village to think her family would have recourse to that old-time hocus-pocus.

The one family member who had truly understood Tayo and who might now have been of help to him is his uncle, Josiah. But Josiah like Rocky is now gone in death, and Tayo's memories of his gentleness are more a source of anguish than of comfort. Before the war Josiah and Tayo had tried to get into the cattle-raising business when they had bought a herd of wild-horned roan cattle and turned them out on Mt. Taylor's lower slopes. But with the death of Josiah the cattle strayed, and no one knows where they are now. It is yet another dead end.

Tayo is, however, not alone in his condition. The other returned vets—Harley, Emo, Leroy, Pinkie—feel similarly displaced on the reservation, though they do not have to face the stigma of being mixed-bloods. For a brief time they had felt they belonged in America when they wore the uniform and fought for the flag. Then they had been able to forget that they were after all reservation redskins. But now they are back where they began: on the reservation, without jobs, and without the prospect of ever getting any. And the reservation itself never looked bleaker than it does now that it is in the grip of a prolonged drought. So they drift through the days, riding in their battered pickups and cars up Highway 66 to the string of bars that stretches from the reservation line all the way to Gallup. Sitting in the stinking gloom of these places, drinking themselves numb, they attempt to recapture the illusion of belonging. "We were the best," Emo brags emptily to the group. "U.S. Army. We butchered every Jap we found. No Jap bastard was fit to take prisoner. We had all kinds of ways to get information out of them before they died. Cut off this, cut off these." While he talks, Emo plays with a sack full of teeth he has taken from a Japanese colonel. In a reference to the atomic bomb that ended the war Emo says, "We blew them all to hell. We should've dropped bombs on all the rest and blown them off the face of the earth."

Against the force of such despair, what countervailing strength? This is the question Silko dramatically poses. It is a question that concerns Tayo, the other vets, and the Laguna people as a whole. In a larger sense it is also a question presented to the whole human race now faced with the ultimate despair of nuclear weapons. Silko's answer derives straight from what she perceives to be the heart of her Laguna culture: the old stories. These are the ceremonies of healing that can cure despair. Thus the poem at the outset of the book:

CEREMONY

I will tell you something about stories,
[he said]
They aren't just entertainment.
Don't be fooled.
They are all we have, you see,
all we have to fight off
illness and death.

You don't have anything
if you don't have the stories.

Their evil is mighty
but it can't stand up to our stories.
So they try to destroy the stories
let the stories be confused or forgotten.
They would like that
They would be happy
Because we would be defenseless then.

He rubbed his belly.
I keep them here
[he said]
Here, put your hand on it
See, it is moving.
There is life here
for the people.

And in the belly of this story
the rituals and the ceremony
are still growing.

The speaker in the poem is apparently the Navajo medicine man Tayo at last goes to see at the old man's shack in the yellow stone hills above Gallup. Nothing else has lifted Tayo out of that despair he shares with Harley, Emo, and the rest, and though he hardly believes any more in the old ways, yet he has come at least to believe he has nothing to lose by going to Betonie. The old man's shack is crammed with the castoff junk of white civilization, and Tayo's doubts are hardly assuaged by the sight of piles of soda-pop bottles and railroad calendars. Looking down at Gallup, symbol of the white man's power, of the pervasiveness of his culture, and then at the junk-stuffed shack of the medicine man, Tayo is moved to wonder what possible good "'Indian ceremonies can do against the sickness which comes from their war, their bombs, their lies?'" Betonie has the answer, telling Tayo that the trickery of the witchcraft now loose in the modern world is to get people to focus all their hatred on one group, blaming it for all the evil at work, whereas in fact the sickness is everywhere and in every group. The whites, he tells the young man, did not invent this sickness: "'white people are only tools that the witchery manipulates; and I tell you, we can deal with white people, with their machines and their beliefs.'" He tells Tayo that it was in fact Indian witchcraft that made white people in the first place.

Betonie knows that the only way to deal with the sickness all around and with that within the soul of this troubled young man is to make changes in the old ceremonies, to adapt them to the conditions of contemporary life. "'The people nowadays,'" he tells Tayo,

> have an idea about the ceremonies. They think the ceremonies must be performed exactly as they have always been done, maybe because one slip-up or mistake and the whole ceremony must be stopped and the sand painting destroyed. That much is true. . . . But long ago when the people were given these ceremonies, the changing began . . . if only in the different voices from generation to generation, singing the chants. You see, in many ways, the ceremonies have always been changing.

Thus when Betonie performs his version of the Navajo healing ceremony for Tayo he ends by telling his patient that the ceremony isn't finished yet. It is up to Tayo to finish it, and the old man tells him that there will be others, those infected with the world's sickness, those under the spell of witchery, who will try to prevent Tayo from becoming wholly healthy. "'Don't let them stop you,'" Betonie says. "'Don't let them finish off this world.'"

Tayo's first ceremonial task is to find Josiah's strayed cattle, that is, to recover that which has been lost. Betonie has had a vision in which he sees the cattle, a stellar constellation, a mountain, and a woman. It is for Tayo to put them together into a pattern that can lead him to the strayed herd. Here, as with the healing ceremony in the old man's shack, it is a matter of belief and will: he must believe in the vision enough to follow it out. Subsequently in the foothills of Mt. Taylor he comes across a mysterious woman, and with her help he does indeed find the herd, fenced in by white ranchers on Mt. Taylor's North Top, and brings them home. There remain two more tasks: he must confront the active essence of witchery and prove equal to its strength and seductiveness; and he must bring his new-found knowledge and healthy spirit back into his community.

Living out in the land away from the pueblo, he awaits what he obscurely knows will be the confrontation with witchery, and it is a kinsman who brings him the summons, telling him that his army buddies are saying things about him. Emo has been telling people that Tayo is crazy, that he has relapsed into his wartime sickness and ought to be hospitalized again as he once had been in the army hospital in Los Angeles. They are likely to come for you, his kinsman warns, maybe by themselves, maybe with army doctors. As it happens, it is Tayo's buddies, those still sick with despair, who come for him: Harley, Leroy, Emo, and Pinkie, the old gang who used to drink together along the "line." They come in friendship and in the good old way: drunk, driving a battered pickup, and with a sack full of warm beer. Tayo rides with them and drinks, but he is not fooled, and when Harley and Leroy pass out, Tayo escapes into the hills north of the pueblo. It is there his confrontation takes place as he is compelled to watch from an abandoned mine shaft while Emo and the others torture Harley for allowing Tayo's escape.

It is a uranium mine, part of those vast deposits discovered on the northern portion of the Laguna reservation. This one has been abandoned because of persistent underground flooding, as if the earth sought to prevent this fatal perversion of itself. But there had been other sources of the ore, the bomb had been created and dropped, and the world was forever altered. Now it truly was within the power of the destroyers to finish off this world, as old Betonie had said. Hiding in the shaft, Tayo suddenly knows that he has arrived "at the point of convergence where the fate of all living things, and even the earth, had been laid." Here was witchery's "final ceremonial sand painting." Here on this weirdly powerful spot of ground Tayo cries silently into the night, for he sees the "way all the stories fit together—the old stories, the war stories, their stories— to become the story that was still being told." And that must go on being told, that must not be finished off. He had, he realizes, "only to complete this night, to keep the story out of the reach of the destroyers for a few more hours, and their witchery would turn, upon itself, upon them." This requires that he witness the torture of Harley and refrain from charging down the hill and burying a screwdriver (his only weapon) in Emo's skull. His fingers knot about the weapon's handle as he hears Harley's hopeless screams, and he can visualize the way Emo's skull would give way under the fatal force of that blow he would love to deal him. It had been a close call when he'd been in the truck with Harley and Leroy, but he had gotten away. This, however, is an even more perilous situation, for Tayo's temptation, disguised as the desire to rescue an old friend, is in fact another of the masks witchery wears. The temptation is to kill, to destroy, and this, Tayo remembers, is "their" story, not his. His story and its ceremonial enactment require that he act on the side of Life, not Death, and so Tayo stays in the shaft watching while witchery turns in upon itself. And then he can return to the pueblo with his healthy spirit.

Traditionally, the last step in the story of the hero is his return to the normal world, the community of daily living, to tell the people what he has been through. If he fails to do this, the story is incomplete, and all his heroic adventures are rendered useless. For it is only in the sharing, in the giving back, that the hero truly achieves his special status. In telling the people his story, he makes them party to his achievements and so strengthens and vivifies that larger life of which he is a part. So now with Tayo. Back in the pueblo he enters the kiva and tells his story to the elders and so at once completes his story and his ceremony of healing. It doesn't seem to matter then that Emo, the most deeply sick of any of the vets, gets away with murdering Pinkie at a sheep camp. Accidental homocide, they call it. Emo is banished from the reservation and goes to California. That, Silko is saying, does not show that crime goes unpunished. Instead, it shows that witchery lives on, in some form present always. But for Tayo, his family, and the pueblo it is, as Silko writes in conclusion, "dead for now."

> It is dead for now.
> It is dead for now.
> It is dead for now.
> It is dead for now.

Ceremony is more than a wonderful book. It is a wonderful act, not least because it takes its title word seriously. Who but a tribal person would be able to do this? For we know that it is with "ceremony" as it is with "myth" and "place spirits": all have become seriously degraded in current usage. "Myth," which originally meant a narrative so sacred as to be almost beyond expression—"that which may not be uttered"—now is taken to mean a popular untruth, while "ceremony," which originally referred to a high and communally shared action, now can mean something as empty as a coin toss before a football game. Silko restores the word to the language, for her *Ceremony* is a high and shared action in which the reader must participate in the endless struggle between good and evil, Life and Death. Tayo's temptation at the mine shaft is our own as well. Refraining from killing, Tayo can complete his ceremony. In making Tayo refrain, Silko saves her readers from their own violent fantasies, and the evil of endless retribution can be avoided. The effect of the book is thus purgative and regenerative, as all ancient ceremonies were meant to be. This is a gift from the land, from the voice out of the land, the voice that tells us how much we stand in need of the land and the stories that come out of it:

> I will tell you something about stories,
> [he said]
> They aren't just entertainment.
> Don't be fooled.
> They are all we have, you see,
> all we have to fight off
> illness and death.
>
> You don't have anything
> if you don't have the stories.

II

   "LESLIE WAS ONE TOUGH LITTLE RIDER," her father said as we swung west at Paguate on the road to the St. Anthony Mine. "She was a tomboy, and the herders liked to have her around because she was so good with horses. She rode all through these hills by herself, and we never worried about her." To the north the hills gave way to red rock mesas and beyond, near the Place of Emergence, were the tree-dotted foothills of Mt. Taylor—juniper, piñon pine, and farther up, ponderosa. But the lands all around us now were the dead, poisonous gray of the tailings from the uranium mining: neat cones of the stuff, long, flat ridges of it, deep pits lined with it. Here and there were small bodies of standing water Lucy Hilgendorf referred to as "glow-in-the-dark lakes." Here, as Silko had written, was witchery's "final ceremonial sand painting." Lee Marmon observed that it was "ironic they didn't know all this stuff was here when they were making the bomb up at Los Alamos. And Los Alamos is what, seventy-five miles away? I think they had to send all the way to Africa, or some such place, to get a good supply of the stuff. Well, there was enough right here under their noses for twenty bombs. Now America's put itself out of the uranium business. Personally, I think there's been an over-reaction to the hazards, but maybe it'll come back some time. We know this much: coal isn't going to be the answer." He told us the Laguna tribe had accepted fifty million dollars from the Anaconda corporation to clean up what we were now looking at, though at this point, he said, no one really knew how long the process would take or what it would involve.

Silko writes of the St. Anthony Mine that after it had repeatedly flooded in the early 1940s big vans had come in to the site and hauled away all the machinery. "They left behind," she writes,

> only the barbed-wire fences, the watchman's shack, and the hole in the earth. Cebolleta [land grant] people salvaged lumber and tin from the shack, but they had no use for the barbed wire any more; the last bony cattle wandering the dry canyons had died in choking summer duststorms.

Now everything was gone, except the hole, some wind-loosened fencing and twisted gates, and, on a hill looking down on the road, a toppled water tank. The roadbed itself, laid by the government, was a blinding unnatural white in contrast to the browns, yellows, and reds of the surrounding soil and rocks. Occasional swoops of wind spun up tiny dust devils on the barren slopes, made the fencing hum, and rattled the scorched iron of the gates. I wasn't sure whether Lee Marmon understood—or cared—why I'd wanted to come out to so desolate a place, but having a daughter who was a writer he was probably used to such vagaries. Besides, he was a friendly, obliging fellow who seemed to have arrived at a Zen-like attitude toward life and the obscure drives of his fellow mortals. Here as at other neighboring sites to which he'd taken us he waited patiently while I took in the scene, talking amiably, steadily with Lucy of mutual friends and the shared experiences of the past. Throughout the day of brilliant sun, scudding thunderclouds, showers, and a double rainbow he had very little to say about *Ceremony,* though it was clear enough that he was proud his daughter had written it. What I said to him now as I climbed back into the cab was that I thought it had been a brilliant stroke for his daughter to have made this barren, brooding place the scene of her hero's last, best stand. "What a telling contrast between the forces of creation and destruction!" I said. He nodded and smiled, and we turned back toward Paguate.

Our route to Laguna took us along the mythic migration route the people had followed south from Paguate. Lee Marmon said again he didn't know the significance of the various landmarks along the route, and I wondered whether that were really so or whether he didn't feel like discussing them with this Anglo stranger. At any rate, he did know the location of the dripping, unfailing springs south of the pueblo where in

Ceremony old Josiah had once taken Tayo. The springs, he told us, were actually on Marmon family property, and it would be an easy matter to take us there.

In the novel, the springs are described as coming out of sheer red/ochre sandstone cliffs that drop directly to the narrow floor of a canyon. A small, luxuriant grove of cottonwoods is fed by the small pools the springs make. "'You see,'" Josiah says to Tayo as they stand there listening to the dripping of the cool, delicious water, "'there are some things worth more than money.'" This, he reminds Tayo, "'is where we come from, see. This sand, this stone, these trees, the vines, all the wildflowers. This earth keeps us going.'" Later, when Tayo visits the springs with Harley everything seems changed: Josiah gone, Rocky gone, and even the unfailing springs dried up. Yet when he drinks from a tiny remnant pool there he can still taste "the deep heartrock of the earth" and think that maybe everything isn't changed, that things aren't finished. But Tayo only wonders vaguely about this, for he has not yet arrived at that point where he can see the patterns of all things, of all the stories. That awareness would come later, at the mine we had just left behind.

As we lurched upward toward the springs in the pickup through scattered juniper and piñon, Lee Marmon was reminded of his cousin, Leslie Evans. "Last time I saw him was up here," he said jerking the steering wheel hard left, then hard right. "He looked bad. He was up here with another Indian guy who was driving him around. Les had lost his driver's license and his truck." As he talked of Evans it was clear Silko had drawn upon this kinsman in creating the characters of the cousins, Rocky and Tayo, and the returned and dislocated World War II vets.

"Les was a hothead," Marmon said with slow emphasis. "He was a great football player. Went to U.N.M. on a scholarship. But he got thrown off the team for something—I think he might have slugged a referee. Anyway, when he couldn't play football anymore he said, 'The hell with it,' and joined the army, or maybe it was the National Guard. Then they got sent over, and he was on that Bataan Death March. He was a prisoner the rest of the war. The Japanese beat the hell out of him. I don't know why they didn't kill him because he got in fights with the guards all the time: he'd take them all on. He was one tough fellow who never backed down from anything. Then after the war he came back here and just stayed drunk. He had all that back pay, you see, plus a pension. So he just stayed drunk. Got in fights. Got beat up. His wife shot him once. Wrecked cars, and so forth. I think Leslie talked to him quite a bit. Now? He died. Probably a liver thing."

When Marmon pulled to a stop we were on top of a mesa. In the blue distance we could see other huge mesas and buttes far to the south and to the east, under the flitting cloud shadows, broad flats. Marmon said that was where his cousin ran cattle. We walked over to the mesa's lip, and there beneath us were the springs, spilling out of the sheer walls, staining them a dark green. In the tangle of brush and cottonwoods at the foot of the cliff you could see parts of small pools. A large piñon pine had struggled up out of the canyon floor, welding itself to the seams of the rock and hanging its branches out over the cottonwoods. Lying on your belly at the lip, you could hear the steady splash of the nearest spring from its dark recess a hundred feet beneath. After the silent, desolate menace of the St. Anthony Mine and the landscape around Paguate there was something sane and reassuring about the sound, more so, I thought, than if we had here been in the presence of a thunderous waterfall. The springs were modest, to be sure, but they were fitting, right, in their arid context, and they were doing their slow, steady best, just as they had been for as long as the Laguna people could remember.

We had a light lunch up there above them while Lee Marmon told stories, beginning with one about Gus Rainey, a man, he said, even tougher than Leslie Evans. "He was from over near Grants," Marmon told us, rummaging in his picnic sack for some apples. "First time anybody noticed anything about Gus was when he brought two bodies into town in the back of his truck. They were stiff as boards, and Gus stood 'em up in the truck so he could have his picture taken with 'em. He said he found 'em, and maybe he did. Some people always said he'd killed 'em, but there's no proof.

"But he was known to be a *mean* fellow. Always went armed, and I mean not just with a pistol: I mean when he'd be out in his truck he'd have a couple of pistols, a rifle, a shotgun. After he died, they found something like forty guns in his house.

"Years after the incident with the two bodies that fellow Budd I told you about that owned the bar just off the reservation, well, he went out to Gus's just to bully him, you know. Gus was by then a pretty old man, and Budd thought of himself as a pretty tough guy. Anyway, Budd got to messing around out there, laughing at old Gus, and so on, so pretty soon Gus just strolls into the house and comes out with a rifle and shoots

Budd dead! Killed him right there! Didn't threaten or anything. That's the way he was. Budd had a gun on him, and I think they called that justifiable homicide.

"Then, just a few years ago, there was a couple came out from California to see Gus. They said they wanted to bring him the Word, which was a big mistake right there, because if there was anything Gus purely hated, it was religion. Actually, they had heard some way that Gus had found gold, and they were hoping he'd lead them to it. Next thing you know, one of Gus's neighbors reported to the sheriff that there were two bodies lying out in Gus's front yard. The sheriff knew Gus's reputation, and he wasn't about to go barging onto Gus's property. So he gets way up on a hill away from Gus's with a telescope, and, by God! there were two bodies there.

"Now he didn't want anybody to find out how cowardly he'd been, so he called in the bomb squad, told 'em Gus's neighbors had complained he was making explosives—something like that. So they all go out there: armed, trucks, and so forth. And there's old Gus, feeding the chickens, fixing a door with those two bodies laying there getting blacker and more bloated by the minute. Gus isn't paying any attention to all his company, and so finally the sheriff works up his nerve and says to Gus, 'Say, Gus, what's this about?'

"Well, that time they did get a conviction. But you know what? Old Gus cheated justice after all: he died before they could send him up. I think he was nearly a hundred, but a big, straight-backed fellow with a beard as bushy as this [holding his hands well out from his face], and a fierce face. After he died they discovered he'd been charged with two murders down in Silver City years before, but nobody'd done the paperwork. His wife's name was Sugarfoot, and he named his children Snow, Sleet, and . . . now, I've forgotten the other one's name. Maybe it was Plague?"

On the way back down to Laguna, Marmon talked to Lucy about his recent visit to his daughter at her ranch outside Tucson. He reported on the latest additions to the writer's already sizeable menagerie, especially about the Chinese pot-bellied pig named Tingo that had been a gift from Larry McMurtry, Silko's close friend. He said Tingo had been bitten between the eyes by a baby rattler in the living room of the ranch house and that he had had to "beat the rattler to death with a poker." He said Silko was having a terrific struggle in finishing her huge new novel, which bears the working title *Almanac of the Dead*. "She's just having a time finishing it," he said. "I think she worries too much about every little word. Larry McMurtry told me if he had to work as hard at his writing as Leslie does at hers, he'd quit. Of course, if you can write *The Last Picture Show* in twenty-eight days like he did, I guess things will come to you easier than they will to another."

Back on Route 40 and heading east toward the exit for the pueblo, Marmon pointed out a couple of old bars on the side road that once had been Highway 66 and of which Silko had written in *Ceremony*. Wouldn't he, I asked him now, like to join Lucy and me for a beer at the Dixie? He looked at me as if I were either a child or a fool. Finally he said simply, "I don't like to go in those places. There's always some character who'll try to start something with you."

As we discovered later that evening in Gallup. In *Ceremony* Silko uses Gallup as the antipode to the reservation and its traditions, its mythically infused landscape. Gallup is the local pit of that despair the writer, her hero, and her book wish to defeat, and though Silko is, as ever, very specific about the scenes set there, Gallup is at the same time tragically symbolic of towns throughout the American West where Indians gather to forget that misery which is by now as much (if not more) a part of their birthright as their tribal identities. I had seen Rapid City and Mission, Billings, and Hardin. But I had not experienced Gallup. Lucy Hilgendorf had, back when she and Silko had been neighbors and friends on the Navajo reservation, and I asked her to show it to me. Some things, she advised as we headed west from Laguna into a lowering sun, had changed there. The new highway now cut right through town and had wiped out the old shanty town of which Silko had written. The police, too, were a bit more careful to round up the drunks in cold weather, fearing there might be repercussions if too many of them were found frozen to death. But some things had not changed in Gallup—and maybe never would. If I wanted to look in the face of that despair that is so powerful a presence in *Ceremony*, there would be plenty of it around.

Saturday night. The pawnshops conveniently parked right next to the bars along old Highway 66 were shuttered and gated right down to the very surface of the sidewalk. But the bars were winking seductively: neon arrows blinking down at entrances, beer signs glowing on/off, cocktail glasses tilting up/down. The Talk of the Town tavern was closed, though Lucy had remembered it as a hot spot. But there was no dearth of alternatives: the American Bar, for instance. Inside it was a jungle of large-lettered signs: NO INTOXICATED PERSONS WILL BE SERVED. IF IN OUR OPINION YOU HAVE HAD ENOUGH TO DRINK YOU HAVE TO LEAVE. $500 FINE FOR ANY MINOR WHO BUYS OR ACCEPTS ANY ALCOHOLIC BEVERAGES. BE 21 OR ADIOS. On a wall above some booths another sign, this one in neon: YA TA HEY STROH'S BEER. Lucy said that was, very loosely, "Hello" in Navaho, though a better translation would be, "The sky is blue."

No blue sky in here but plenty of blues: the blue glare of the TV, broadcasting unheeded the Olympic games from Seoul, South Korea; the anemic blue of the fluorescent lights that hung low over the pool tables in the rear of the room; and blues at the bar where among the other clustered drinkers were two young Navajo women consoling one another with mixed drinks in frosted glasses. Marcy told us her sister, Priscilla, was having "big trouble at home." Priscilla looked it, from time to time removing her triple-thick eyeglasses, wiping her eyes, then lowering her face wearily to the bar's formica surface. Marcy said their mother had wanted them to go to a squaw dance that evening but that instead they had come in to Gallup. "We're modern," she explained. "We don't want to go some place way out there and sit in a truck and watch others dance."

As yet there was no dancing in the American Bar, though a man continued to feed the juke box. Above it was a forty-foot mural depicting the slow crawl of a wagon train across the Great Plains. In the mural's foreground were some Indians pointing out to each other this strange spectacle. It seemed a wonder to me that the wagon train had apparently escaped ex post facto attack from the booths that ran along just beneath.

Things got livelier at the Silver Spur up the street. The action at the bar was not especially vigorous yet, but all the pool tables in the big room were in use, and there was a small crowd of interested spectators against the far wall. At the table nearest us a nattily dressed older man in gray slacks, white shirt, and navy sweater was giving some expensive lessons to a Navajo man in his early thirties. The younger man was a pretty fair player himself, yet when he missed a shot the older man would routinely run the table. A big Navajo man at my elbow told me the two were regular adversaries. "He's [the older player] a plumber. Comes in every weekend to play. That guy [the younger] was Special Forces in Nam. You should play him," he said, giving Lucy in her tight jeans a frankly appraising once-over.

A half-hour later I was still telling Ben that I didn't play pool, and he was still telling me I should play Virgil, only by then he was putting it that I *needed* to play Virgil. Also by then Virgil had joined us, and both men were right in my face. Virgil's eyes were utterly flat and dead look-ing, and he seemed to be waiting quietly for whatever would happen, only occasionally glancing at me and Lucy, then looking away into the darkness, as if he might be listening for something. Ben said he thought I was a hustler. Why else, he wondered aloud, would I be in here? "We had business in Gallup," I offered lamely, "and just stopped in here for a drink."

"That's a real pretty lady you got there," Ben said, never moving his eyes from Lucy. Virgil stood looking away, listening, his hand curled lightly around his cue stick. "I wonder whether you know just how pretty she is." He mumbled something then about never understanding the choices women made.

Ten minutes later with both Ben and Virgil still very much there it was plain to me I was going to have to play pool, and I said this to Virgil. He shrugged and headed for the table. What, I wondered, choosing blindly among the racked cue sticks, was my best shot? Dump the game? No. Whatever else Virgil might be—and I didn't want to think further along that line—he was too smart for that. Best just to play my characteristically poor game and let nature take its course. The game was eight-ball-and-call-your-pocket, and as luck would have it I kept getting set up with one easy shot after another until the table was empty except for the eight ball. It was Virgil's shot. He called a corner pocket, stroked the cue ball crisply only to have the eight carom weirdly, then stagger like a drunk down the cushion and plop into the wrong pocket. In the deafening silence that followed I could see Lucy and Ben at the bar, and I could hear Virgil at my elbow asking in his quiet, level voice how much we'd been playing for. "Safe passage to the parking lot," I said to him, scooping Lucy as I passed the bar at a smart pace.

She thought we ought to try one more place, the Shalimar maybe, a motel with lounge and dance floor. "Okay," I said as we pulled out of the Silver Spur lot, "but I can tell you I'm already growing weary of defend-ing your ass in those jeans." She said she could take care of herself, and I had no doubt she could do a better job of it than I.

The Shalimar, because it was so much bigger and had dancing, was more anonymous, safer. It cost $2.00 to enter the hangar-like room with a bar running around three walls, a stage against the other wall, and the dance floor in the middle. An Anglo country band was doing its decibel best on the stage, but the dancers were few, and the bar was five deep with Navajos, most of them men. A sign behind one wing of the bar discouraged temperance, offering iced tea for $1.50 and beer for $.75. When the live band gave up, recordings came on full bore, and immediately the floor filled with black-hatted men in jeans and boots. Their partners seemed to effect no uniform, but all the young women looked as though they had their hands full. "You're looking at the top two percent here," Lucy shouted in my ear. "These would be Navajo yuppies, the ones who have a ride and a few dollars to spend on a Saturday night. Not at all like those Leslie wrote about who lived down by the river."

A young man next to us, his broad-brimmed black hat jammed tight to his ears, said he lived north, at Many Farms, and was surprised when Lucy said she knew it well, then followed up with the names of two families she knew up that way. His beer-swollen eyes widened momentarily before he hit me up for a buck. He worked construction, he said, and when he could he rode rodeo broncs. His friends called him "Cowboy," and when we left some time after one in the morning he seemed on the verge of getting next to a buxom young woman whose tangled black hair hung down her back all the way to her studded belt.

Sunday morning. The bars stay closed all day in McKinley County. Out early into the streets, I encountered the monotypic evidence of the night before: dark figures limping eastward in the lifting light; a great splash of blood on the sidewalk opposite the American Bar; empty six-pack cartons skittering down old 66; a leaning lamppost with a scatter of glass beneath it. As the sun climbed into a high and heedless blue Lucy and I toured the city's back streets north of the freeway, watching the tattered, stained forms wake up out of the weeds and without so much as a preliminary dusting off begin a restless, constant migration that had no apparent end or destination. Tangle-haired, rumpled, some clutching small paper sacks, they walked slowly through salt-cedar lots behind junkyards and out into the Sabbath-emptied streets. Most were men, though here and there was a woman, usually with a female companion. At an intersection a woman gestured violently into a man's numbed face while a small huddle of onlookers watched without emotion. At the Palamino Saloon a group of men stood near the door that wouldn't be opened again for at least twenty-four hours. Troops of men with empty hands walked the roads leading into and out of town, but not a one had his thumb out for a lift. Atop the yellow rocks that overlook the ceremonial grounds, where Silko has old Betonie's shack and where Tayo has his healing ceremony, men in the regulation black hats stood talking, their faces shadowed. The sun glinted on a bottle being passed. A hugely obese young Navajo waddled past us toward the hilltop figures, carrying a big plastic 7-Up bottle filled with water. It was already slopping out behind him, leaving dark gray stains in the dust.

Actually getting to see Leslie Marmon Silko is about as difficult as getting an interview with Brando. She has never been accessible and wasn't in 1978 when I first met her shortly after the success of *Ceremony*. At that time she was in the throes of a bitter custody battle with her second husband and was understandably feeling reclusive. In recent years, something of a literary celebrity, recipient of a MacArthur Fellowship and of a six-figure advance for her new novel, she had been consumed by this project to the point at which almost every other aspect of her life was neglected. In a letter to me from Tucson she called her current condition "novelist's neurosis syndrome," and went on to describe its symptoms:

> First you notice your gut clutches when the telephone rings. Then you leave the bundle of mail unopened, rolled up tight in the Wall Street Journal and the Times. If by chance you open the bundle, you may open only the junk mail which is not personal. If you open a personal [letter] or a letter of professional business, you leave it lying on the pile of "to-be-answered." Each year when I box up the IRS effluent, canceled checks, etc., I also box up the mail.

But she wanted me to understand that it was nothing personal, and in fact if I could make it out to Tucson in mid-August, she would be happy to talk with me there.

August in Tucson is so hot the air seems to make a distinct friction with your skin, but inside the lounge of the posh old Arizona Inn it was cool enough and, in this dead off-season, quiet. There had been enough negotiations for the meeting that it occurred to me Silko might not show up at all, but she arrived right on time, wearing a black pant suit with a low-cut blouse underneath that raised some eyebrows as she strode across the room. She clearly didn't care, dropping down next to me and putting her feet up on an empty chair. While we waited for our drinks to come, she explained that she had her feet up because she'd hurt one of them the other day while jumping from one point of her roof to another. With her rings, bracelets, dramatic clothes and hair she reminded me instantly of Janis Joplin: the Janis Joplin of American letters. The analogy, though limited, was strengthened as we talked that afternoon and, later, that evening up at her ranch, for here was a person, I thought, who like Joplin was out to take her vision of traditional things right up to and maybe through the limits.

Understandably, she was full of her current project. Nothing, to a practicing writer, is as dull and inappropriate a subject as a previous book: the book you're working on is the only book. So I led with a question about *Almanac of the Dead.* Then I had nothing to say for the better part of two hours as Silko rambled, free-associated, and ran variations on the multiple themes of the huge, sprawling weave she had been struggling with for so long. She said she'd been at work on it since January 1984, "but I've been ratting away stuff on it since '81, though at that time I really didn't know what it was I was saving this stuff for. Now I do.

"The basic idea comes from the Mayan concept that the days are alive, that they have spirit, and that they return to us." As for the structure, she said she chose that of the almanac because it was loose and episodic and so provided her with what she called with a laugh "an out: I don't have to make everything cohere, you know, if I don't want to. Just like an almanac doesn't have to knot everything up neatly." She said there was, however, a theme to which everything in the novel—Indian protest movements, U.S. government intrigue, drug running, historic white/Indian relations—could be referred. That theme is the land of the New World, its place spirits, its history, its future. The land, she said, "is the protagonist," and it is what has been done to the land and its native peoples that obviously engages her intelligence, her art, her passion.

"You don't," she said now, her hazel eyes throwing sparks in the sedate lounge, "simply take land away from people and then have them sit there and look at it day after day with less and less to lose, and getting hungrier and hungrier. I don't think things work that way. A nation built on stolen land can't ultimately be a just nation, can't have a real system of justice in it—though we've taken the grand ideals farther than anyone else has: look at me, a mixed-blood, writing—. So America *has* to end in apocalypse. Larry [McMurtry] calls me the prophet of the apocalypse. He's kidding. Sort of. But the resistance of native peoples has never really stopped, you know, and it can't stop—ever. Until things are put right."

Gradually I turned her toward *Ceremony,* not because she didn't see what I was trying to do, but rather, I thought, because for the moment, anyway, she had said enough about *Almanac* and was herself exhausted by that whole effort. Somewhat rhetorically, I asked her how important place had been for her in writing *Ceremony.* "That thing [*Ceremony*] is as accurate as a map," she shot back. "I may have intentionally changed a few details about the location of landmarks, and so forth, but for the most part you could take that thing out there [to Laguna] and use it to get around. Partly, this was because at the time I was writing it I was up in Alaska, and I was obsessive about the specifics of home." She laughed briefly. "I was writing it that way to save myself." Yet the more she talked about being in Alaska (home to her second husband) and about that landscape versus the American Southwest, the more it became obvious that she was deeply sensitive to the smallest nuances of landscape, the way a place looked, the way it affected her. Later, I was reminded of the allegation that Willa Cather could often be either rude to or just oblivious of other people but never to a landscape. Silko seemed not to notice anyone in the cocktail lounge, but when she talked of the land she was there.

"I guess," she said, summing up the importance of place in *Ceremony,* "a lot of that sensitivity must have come out of childhood where I was out in the country a lot, you know. You could take me out now into the country, along some dirt road, and if I'd ever been there before in my life—it doesn't matter when—I could tell you where I was, and where I was in relation to other points. I might have the distances wrong—I might have shortened or lengthened something. But the basic identification would be right." I reminded her that she has a character claim pretty much the same thing in "A Geronimo Story," one of the pieces in *Storyteller.*

We talked then about stories as ceremonies ("A story is an act, just like a ceremony is."); about *Ceremony* as a new story, taking its place with stories retold down the generations ("There aren't any new stories; *Ceremony* couldn't be, couldn't exist, if it weren't for those old stories."); and about the reception of *Ceremony* at the pueblo. Was it, I wondered, regarded as a kind of contribution to the life of that place? "Not really," she said flatly. "The people I'd gone to school with read it and were excited. Mostly it didn't get read. Dad never got through it. My sister Wendy read it. She went through it and identified this and that character. Some read it to make sure I hadn't divulged anything really private about the pueblo—things that shouldn't have been divulged. I was very careful to make the protagonist half-white so that anyone there reading it could simply say, 'Yes, that's just like those Marmons: one of them would write a book!' I was really careful about that. But as far as some sort of recognition, it's just that, 'Hey, someone wrote a book about this place.' And that's fine with me." She went on to observe that no matter how much inspiration a writer might take from a place, it is very hard to stay on that place and be a writer. "Look at Simon [Ortiz]," she said. "He's a full-blood Acoma. He doesn't have any of the mixed-blood problems I do. And Simon lives in Albuquerque. I live in Tucson. It's just very hard to live there and do it [the writing]. And I don't just mean the pressures and distractions. It's that, yes. But you need that distance, too."

Later, at her invitation I drove out of the city's aimless sprawl to her ranch, climbing a dirt road into the naked, jagged mountains, the lower slopes of which were graced by the lordly saguaro. Almost instantly the illusion of Tucson vanished from the mind, and I felt in the presence of the unimproved desert. At the entrance to the ranch there was a bristle of signs warning you off: "Stop Stupid." "Wild Rattlesnake Range: Danger." "Federal Land Ahead. No Access." Farther in, where only the foolish or the invited would dare, there was a rambling white structure. A few cars were parked in front, including a Mercedes, and a pack of semi-feral desert dogs came howling down to meet the intruder in the dust. Above their clamor I could hear Silko telling me it was okay to come ahead. The dogs' feed pans and water buckets bordered the entrance to the house, and in the evening breeze swarms of flies swayed back and forth above them. The dogs remained unconvinced and restless while Silko showed me some of her flora, including several saguaros and a barrel cactus she had rescued from the paths of ignorant construction crews and telephone linemen. Here they were safely replanted and thriving, and clearly, she regarded them as friends. Inside the main door a bolt-action rifle with mounted sight lay poised on the arms of a chair, its barrel pointing down at the road. While numbers of cats prowled and rubbed against my legs Silko showed me other friends. There were Gila monsters in cages, and Silko greeted them affectionately, leaning down to say, throatily, "Bad lizards!" There were boas as well as other snakes I didn't recognize, and there was a roomful of breeding mice, food for the snakes.

Still talking to me and to the members of the menagerie, Silko fetched a bottle of Mumm's Extra Dry from the refrigerator, then led the way out along a wooden walkway pocked with coyote scat to an abandoned swimming pool that now served as a catch basin. As we sat there sipping champagne, the sun dipped behind the mountains, and bats began to skim the water caught in the pool's deep end. The dogs came panting by, too, and trotted down to the water to lap.

I sensed early on that Silko was tired of writing and of talk about writing, and it was pleasant, anyway, to sit there in the last of the desert day and listen to her talk about whatever was on her mind. What was on her mind was how much she loved it up here, and how much she detested what was happening to all the land adjacent to Tucson. "I'm pretty isolated up here now," she said, glancing around her wild domain, "but I know it's just a matter of time until there's a development built up around here." She regarded this as inevitable. But "what I can't take is noise, and up here I don't have that. Probably won't, because over there," pointing her chin westward, "is the beginning of the Saguaro National Monument, so there won't be any building that way." She smiled. She went on about sounds, how she had learned to distinguish them in the years since she had lived here. That one, she told me, was the yap of a coyote. And that, the evening call of a curved-bill thrasher.

A gigantic cloud formation like a cyclotron, its innards lit up moment to moment by blasts of lightning, moved ponderously toward us. Silko gave it an unconcerned glance and spoke of other distinctive cloud shapes she'd come to recognize up here. She spoke of the fifty-year rain they'd had just a few years back, "and that slope over there was just covered with blossoms. You can't imagine how beautiful it was!" The cyclotron moved closer, its accompanying winds beginning to swirl around the pool, blowing Silko's hair straight out behind her. Then she reminded me of the mysterious woman she'd written about in *Ceremony*. I said it was nice to get that cool breeze before the rain. "Yes," she said, "but what's really nice is when it gets to be about 110 or so, and it's dry. Then you sweat, and you're instantly cooled." She threw her shoulders back and arched her back. "That's when you're in perfect balance, inward and outward. Everything's perfect. *That's* the magic of this place."

Bibliography

A list of secondary works and collections consulted

ONE. HOW DIVINE THESE STUDIES: THE CHALLENGE OF THE UNSUNG LAND

WENDELL BERRY. *Standing by Words: Essays* (San Francisco: North Point Press, 1983).

JOHN CONRON (ed.). *The American Landscape: A Critical Anthology of Prose and Poetry* (New York, London, Toronto: Oxford University Press, 1973).

PATRICIA HILLS. *The Painters' America: Rural and Urban Life, 1810–1910* (New York, Washington: Praeger, 1974).

J(OHN) B(RINKERHOFF) JACKSON. *American Space* (New York: Norton, 1972).

_____. *Landscapes: Selected Writings of J.B. Jackson.* Ed., Ervin H. Zube (Amherst: University of Massachusetts Press, 1970).

_____. *The Necessity for Ruins and Other Topics* (Amherst: University of Massachusetts Press, 1980).

HOWARD MUMFORD JONES. *The Theory of American Literature* (Ithaca: Cornell University Press, 1948, 1965).

RUSSELL LYNES. *The Art-Makers: An Informal History of Painting, Sculpture, and Architecture in Nineteenth-Century America* (New York: Atheneum, 1970).

CHARLES L. SANFORD (ed). *Quest for America, 1810–1824* (New York: New York University Press, 1964).

JEAN SEZNEC. *The Survival of the Pagan Gods: The Mythological Tradition and Its Place in Renaissance Humanism and Art.* Trans., Barbara F. Sessions (Princeton: Princeton University Press, 1972).

PAUL SHEPARD. *Man in the Landscape: A Historic View of the Esthetics of Nature* (New York: Knopf, 1967, 1972).

BENJAMIN T. SPENCER. *The Quest For Nationality: An American Literary Campaign* (Syracuse: Syracuse University Press, 1957).

MOSES COIT TYLER. *A History of American Literature, 1607–1765.* 1878. Reprint. (New York: Collier, 1962).

LARZER ZIFF. *Literary Democracy: The Declaration of Cultural Independence in America* (New York: Viking, 1981).

TWO. NEARER HERE:
HENRY DAVID THOREAU'S *WALDEN* AND *THE MAINE WOODS*

SHIRLEY BLANCKE and BARBARA ROBINSON. *From Musketaquid to Concord: The Native and European Influence* (Concord: Concord Antiquarian Museum 1985).

CARL BODE (ed.). *The Portable Thoreau* (New York: Viking, 1947, 1981).

JOHN ALDRICH CHRISTIE. *Thoreau as World Traveler* (New York: Columbia University Press, 1965).

JAMES M. COX. "Autobiography and America." J. Hillis Miller (ed.), *Aspects of Narrative* (New York: Columbia University Press, 1971).

RICHARD F. FLECK. (ed.). *The Indians of Thoreau: Selections from the Indian Notebooks* (Albuquerque: Hummingbird Press, 1974).

WALTER HARDING. *The Days of Henry Thoreau* (New York: Knopf, 1965).

WILLIAM HOWARTH. *The Book of Concord: Thoreau's Life as a Writer* (New York: Penguin, 1983).

J. PARKER HUBER. *The Wildest Country: A Guide to Thoreau's Maine* (Boston: Appalachian Mountain Club, 1982).

RICHARD LEBEAUX. *Thoreau's Seasons* (Amherst: University of Massachusetts Press, 1984).

LEO MARX. *The Machine in the Garden: Technology and the Pastoral Idea in America* (New York: Oxford University Press, 1964, 1977).

SHERMAN PAUL. *The Shores of America: Thoreau's Inward Exploration.* 1958. Reprint. (Urbana: University of Illinois Press, 1972).

F. B. SANBORN. *The Life of Henry David Thoreau* (Boston, New York: Houghton Mifflin, 1917).

ROBERT F. SAYRE. *Thoreau and the American Indians* (Princeton: Princeton University Press, 1977).

GLENN STARBIRD. "Joseph Attean (1829–1870)." Unpublished paper, n.d.

———. Untitled paper on Penobscot tribal affairs, n.d.

ROBERT F. STOWELL. *A Thoreau Gazetteer.* (Princeton: Princeton University Press, 1970).

EDWARD WAGENKNECHT. *Henry David Thoreau: What Manner of Man?* (Amherst: University of Massachusetts Press, 1981).

MATTHEW WALD. "Town's Future Shrinks Along With Paper Mills." *New York Times* (March 31, 1986).

GORDON G. WHITNEY AND WILLIAM C. DAVIS. "From Primitive Woods to Cultivated Woodlots: Thoreau and the Forest History of Concord, Massachusetts." *Journal of Forest History* (April 1986).

GEORGE WUERTHNER. "Northeast Kingdoms." *Wilderness* (Summer 1986).

THREE. A PILOT'S MEMORY:
MARK TWAIN'S *LIFE ON THE MISSISSIPPI*
AND *HUCKLEBERRY FINN*

HERBERT ASBURY. *The French Quarter: An Informal History of the New Orleans Underworld* (New York: Knopf, 1936, 1974).

WALTER BLAIR. *Native American Humor* (San Francisco: Chandler, 1937, 1960).

ROBERT M. COATES. *The Outlaw Years: The History of the Land Pirates of the Natchez Trace* (New York: Macaulay, 1930).

JAMES M. COX. *Mark Twain: The Fate of Humor* (Princeton: Princeton University Press, 1966).

JAMES R. CURTIS. "The Most Famous Fence in the World: Fact and Fiction in Mark Twain's Hannibal." *Landscape* (Vol. 28, No. 3, 1985).

BERNARD DE VOTO. *Mark Twain's America* (Boston: Houghton Mifflin, 1932, 1967).

_____. *Mark Twain at Work* (Boston: Houghton Mifflin, 1942, 1967).

DELANCEY FERGUSON. "Huck Finn Aborning," in Sculley Bradley, Richard Croom Beatty, E. Hudson Long, and Thomas Cooley (eds.), *Adventures of Huckleberry Finn: An Authoritative Text, Backgrounds, Sources, and Criticism* (New York: W.W. Norton, 1961, 1977).

DANIEL G. HOFFMAN. "Black Magic—and White— in *Huckleberry Finn,*" in Bradley, Beatty, Long, and Cooley (eds.), *Adventures of Huckleberry Finn. . . .*

WILLIAM DEAN HOWELLS. *My Mark Twain: Reminiscences and Criticisms* (Baton Rouge: Louisiana State University Press, 1967).

JUSTIN KAPLAN. *Mr. Clemens and Mark Twain: A Biography* (New York: Simon and Schuster, 1966).

KENNETH S. LYNN. *Mark Twain and Southwestern Humor* (Boston: Little, Brown, 1960).

CHARLES NEIDER (ed.). *The Autobiography of Mark Twain.* 1917. Reprint. (New York: Harper & Row, 1959).

_____. (ed.). *The Selected Letters of Mark Twain* (New York: Harper & Row, 1982).

RON POWERS. *White Town Drowsing: Journeys to Hannibal* (New York: Penguin, 1987).

JONATHAN RABAN. *Old Glory: An American Voyage* (New York: Simon and Schuster, 1981).

CONSTANCE ROURKE. *American Humor: A Study of the National Character* (New York: Harcourt, Brace, 1931).

JOHN SEELYE. *Mark Twain in the Movies: A Meditation with Pictures* (New York: Viking, 1977).

HENRY NASH SMITH. *Mark Twain: The Development of a Writer* (New York: Atheneum, 1962, 1967).

JOHN S. TUCKEY (ed.). *The Devil's Race-Track: Mark Twain's Great Dark Writings* (Berkeley: University of California Press, 1980).

DIXON WECTER. *Sam Clemens of Hannibal* (Boston: Houghton Mifflin, 1952).

LARZER ZIFF. *Literary Democracy* (q.v.)

FOUR. COMMUNITY OF SIN: GEORGE WASHINGTON CABLE'S *OLD CREOLE DAYS* AND *THE GRANDISSIMES*

STANLEY CLISBY ARTHUR. *Old New Orleans* (New Orleans: Harmanson, 1946).

HERBERT ASBURY. *The French Quarter* (q.v.).

SIDNEY BECHET. *Treat it Gentle* (New York: Hill and Wang, 1960).

JOHN W. BLASSINGAME. *Black New Orleans, 1860–1880* (Chicago: University of Chicago Press, 1973).

GEORGE W. CABLE. "The Dance in the Place Congo." *Century Magazine* (February, April 1886).

———. Cable Papers, Rare Books and Manuscripts Division, Tulane University Library.

———. Cable Papers, Historic New Orleans Collection, New Orleans.

MARY CABLE. *Lost New Orleans.* (Boston: Houghton Mifflin, 1980).

HENRY C. CASTELLANOS. *New Orleans as It Was* (New Orleans: L. Graham & Sons, 1895).

MARCUS CHRISTIAN. *Negro Ironworkers of Louisiana, 1718–1900* (Gretna, La: Pelican, 1972).

RODOLPHE LUCIEN DESDUNES. *Our People and Our History.* Trans. and ed., Sister Olga McCants (Baton Rouge: Louisiana State University Press, 1973).

FEDERAL WRITERS' PROJECT. *The WPA Guide to New Orleans* (New York: Vintage, 1983).

LAFCADIO HEARN. "The Scenes of Cable's Romances." *Century Magazine* (November 1883).

WILLIAM DEAN HOWELLS. *My Mark Twain* (q.v.).

ROBERT UNDERWOOD JOHNSON. *Remembered Yesterdays* (Boston: Little, Brown, 1923).

JUSTIN KAPLAN. *Mr. Clemens and Mark Twain* (q.v.).

THOMAS J. RICHARDSON (ed.). *The Grandissimes: Centennial Essays* (Jackson: University Press of Mississippi, 1981).

(ADRIAN ROUQUETTE). *Critical Dialogue Between Aboo and Caboo. . . .* (New Orleans: Mingo City Great Publishing House 1880).

LOUIS D. RUBIN, JR. *George W. Cable: The Life and Times of a Southern Heretic* (New York: Pegasus, 1969).

LYLE SAXON (ed.). *Gumbo Ya-Ya* (Boston: Houghton Mifflin, 1945).

ARLIN TURNER. *George W. Cable: A Biography* (Durham, NC: Duke University Press, 1956).

FIVE. THE BISHOP'S FACE:
WILLA CATHER'S *DEATH COMES FOR THE ARCHBISHOP*

E. K. BROWN. *Willa Cather: A Critical Biography.* Completed by Leon Edel (New York: Knopf, 1953).

WILLA CATHER. *Five Stories* (New York: Vintage, 1956).

FRAY ANGELICO CHAVEZ. *The Santa Fe Cathedral* (Santa Fe: Schifani Bros. Printing Co., 1947, 1968, 1978).

PAUL HORGAN. *Lamy of Santa Fe* (New York: Farrar, Straus and Giroux, 1975).

SHEILA MORAND. *Santa Fe Then and Now* (Santa Fe: Sunstone Press, 1984).

SHARON O'BRIEN. *Willa Cather: The Emerging Voice* (New York: Oxford University Press, 1987).

PHYLLIS C. ROBINSON. *Willa: The Life of Willa Cather* (Garden City: Doubleday, 1983).

ELIZABETH SHEPLEY SERGEANT. *Willa Cather, A Memoir* (Lincoln: University of Nebraska Press, 1963).

JOHN SHERMAN. *Santa Fe, A Pictorial History* (Santa Fe: Gannon, 1984).

RALPH H. VIGIL. "Willa Cather and Historical Reality." *New Mexico Historical Review* (Winter 1975).

JAMES WOODRESS. *Willa Cather: A Literary Life* (Lincoln: University of Nebraska Press, 1987).

SIX. PLAINSONG:
MARI SANDOZ'S *OLD JULES*

BRADLEY BALTENSPERGER. *Nebraska, A Geography* (Boulder, London: Westview Press, 1985).

JOSEPH CAMPBELL. *The Way of the Animal Powers: Historical Atlas of World Mythology,* vol. 1. Alfred van der Marck Editions, San Francisco: Harper & Row, 1983.

BRUCE CHATWIN. *The Songlines* (New York: Viking, 1987).

EVERETT DICK. *Conquering the Great American Desert: Nebraska* (Lincoln: Nebraska State Historical Society Publications, Vol. XXVII, 1975).

MARGRETTA STEWART DIETRICH. "Nebraska Recollections." (Privately printed, 1957).

EDWARD EGGLESTON. *The Hoosier Schoolmaster* (New York: Orange Judd and Company, 1871).

VIRGINIA FAULKNER (ed.). *Roundup: A Nebraska Reader* (Lincoln: University of Nebraska Press, 1957).

E. W. HOWE. *The Story of a Country Town* (Atchison, KS: Howe & Co., 1883).

FRANKLIN C. JACKSON. *Echoes from the Sandhills* (Lincoln, NB: Word Services, 1977).

EDGAR LEE MASTERS. *Spoon River Anthology* (New York: Macmillan, 1915).

CHARLEY O'KIEFFE. *Western Story: The Recollections of Charley O'Kieffe, 1884–1898* (Lincoln: University of Nebraska Press, 1960).

DARLENE MAE RITTER. *The Letters of Louise Ritter from 1893–1925: A Swiss-German Immigrant Woman to Antelope County, Nebraska* (Ann Arbor, London: University Microfilms 1980).

MARI SANDOZ. *Love Song to the Plains* (New York: Harper, 1961).

———. *Sandhill Sundays and Other Recollections* (Lincoln: University of Nebraska Press, 1970).

———. "Stay Home, Young Writer." *The Quill* (June 1937).

MARK SCHORER. *Sinclair Lewis: An American Life* (New York: McGraw-Hill, 1961).

ADDISON EDWIN SHELDON. *History and Stories of Nebraska* (Chicago, Lincoln: The University Publishing Company, 1913).

HELEN WINTER STAUFFER. *Mari Sandoz, Story Catcher of the Plains* (Lincoln: University of Nebraska Press, 1982).

WALTER PRESCOTT WEBB. *The Great Plains* (Boston: Ginn and Company, 1931).

FEDERAL WRITERS' PROJECT. *Nebraska: A Guide to the Cornhusker State* (New York: Viking, 1939).

WALT WHITMAN. *Leaves of Grass and Selected Prose.* Ed., John A. Kouwenhoven (New York: Modern Library, 1950).

SEVEN. PLACE SPIRITS:
WILLIAM FAULKNER'S
ABSALOM, ABSALOM! AND "THE BEAR"

HAMILTON BASSO. "William Faulkner, Man and Writer." *Saturday Review* (July 28, 1962).

JOSEPH BLOTNER. *Faulkner: A Biography.* 2 vols. (New York: Random House, 1974).

RONALD BLYTHE. *Characters and their Landscapes* (New York, San Diego, London: Harcourt Brace Jovanovich, 1982).

CLEANTH BROOKS. *William Faulkner: The Yoknapatawpha Country* (New Haven: Yale University Press, 1963).

MALCOLM COWLEY (ed.). *The Portable Faulkner* (New York: Viking, 1946, 1961).

JOHN B. CULLEN (in collaboration with Floyd C. Watkins). *Old Times in the Faulkner Country* (Baton Rouge: Louisiana State University Press, 1961, 1975).

MARTIN J. DAIN. *Faulkner's Country* (New York: Random House, 1964).

STEPHEN B. OATES. *William Faulkner, the Man and the Artist* (New York: Harper & Row, 1987).

PEGGY WHITMAN PRENSHAW (ed.). *Conversations with Eudora Welty* (New York: Washington Square Press, 1985).

FRANCIS LEE UTLEY, LYNN Z. BLOOM, ARTHUR F. KINNEY (eds.). *Bear, Man, and God: Seven Approaches to William Faulkner's "The Bear"*. (New York: Random House, 1964).

JEAN STEIN VANDEN HEUVEL. "William Faulkner," in Malcolm Cowley (ed.), *Writers at Work*, 1st series (New York: Viking, 1958).

ROBERT PENN WARREN (ed.). *Faulkner: A Collection of Critical Essays* (Englewood Cliffs: Prentice-Hall, 1966).

———. "William Faulkner and His South," unpublished essay (1st Peters Rushton Seminar in Contemporary Prose and Poetry, University of Virginia, March 13, 1951).

EUDORA WELTY. *The Eye of the Story* (New York: Random House, 1978).

JACK CASE WILSON. *Faulkners, Fortunes and Flames* (Nashville: Annandale Press, 1984).

EIGHT. THE VALLEY OF THE WORLD:
JOHN STEINBECK'S *EAST OF EDEN*

JACKSON L. BENSON. *The True Adventures of John Steinbeck, Writer: A Biography* (New York: Viking, 1984).

PASCAL COVICI, JR. (ed.). *The Portable Steinbeck* (New York: Penguin, 1976).

STEVE CROUCH. *Steinbeck Country* (Palo Alto: American West Publishing Co., 1973).

ALFRED KAZIN. *On Native Grounds: An Interpretation of Modern American Prose Literature* (New York: Reynal & Hitchcock, 1942).

PETER LISCA. *John Steinbeck, Nature and Myth* (New York: Crowell, 1978).

V. S. NAIPAUL. "Steinbeck in Monterey," in *The Overcrowded Barracoon: Essays* (New York: Random House, 1984).

PAULINE PEARSON. *Guide to Steinbeck Country,* pamphlet (Salinas, 1984).

RANDALL REINSTEDT. *Where Have All the Sardines Gone? A Pictorial History of Steinbeck's Cannery Row* (Carmel, CA: Ghost Town Publications, 1978).

SALINAS CHAMBER OF COMMERCE. *Steinbeck Country Starts in Salinas,* pamphlet (Salinas, 1975).

BRIAN ST. PIERRE. *John Steinbeck: The California Years* (San Francisco: Chronicle Books 1984).

JOHN STEINBECK. *Journal of a Novel: The* East of Eden *Letters* (New York: Viking, 1969).

ERNEST W. TEDLOCK and C. V. WICKER (eds.). *Steinbeck and His Critics: A Record of Twenty-Five Years* (Albuquerque: University of New Mexico Press, 1957).

NINE. OF LOCAL INTEREST:
WILLIAM CARLOS WILLIAMS'S
IN THE AMERICAN GRAIN AND *PATERSON*

JAMES E. B.BRESLIN. *William Carlos Williams, An American Artist* (New York: Oxford University Press, 1970).

MALCOLM COWLEY. *Exile's Return: A Literary Odyssey of the 1920s* (New York: W.W. Norton, 1934).

D. H. LAWRENCE. *Studies in Classic American Literature* (New York: Seltzer, 1923).

PAUL MARIANI. *William Carlos Williams: A New World Naked* (New York: McGraw-Hill, 1981).

ROY HARVEY PEARCE. *The Continuity of American Poetry* (Princeton: Princeton University Press, 1961).

CONSTANCE ROURKE. *American Humor: A Study of the National Character* (New York: Harcourt, Brace, 1931).

HENRY M. SAYRE. *The Visual Text of William Carlos Williams* (Urbana: University of Illinois Press, 1983).

WILLIAM CARLOS WILLIAMS. *The Farmers' Daughters: The Collected Stories of William Carlos Williams* (Norfolk, CT: New Directions, 1961).

_____. *The Selected Letters of William Carlos Williams.* Ed., John C. Thirlwall (New York: McDowell, Oblensky, 1957).

_____. *Selected Poems.* Ed., Charles Tomlinson (New York: New Directions, 1985).

TEN. VOICE OUT OF THE LAND:
LESLIE MARMON SILKO'S *CEREMONY*

WALTER HOLDEN CAPPS (ed.). *Seeing with a Native Eye: Essays on Native American Religion* (New York: Harper & Row, 1976).

RICHARD ERDOES and ALPHONSO ORTIZ (eds.). *American Indian Myths and Legends* (New York: Pantheon, 1987).

SUSAN FELDMANN (ed.). *The Story-Telling Stone: Myths and Tales of the American Indians* (New York: Dell, 1965).

CHARLES R. LARSON. *American Indian Fiction* (Albuquerque: University of New Mexico Press, 1978).

D'ARCY McNICKLE. *The Surrounded.* 1936. Reprint. (Albuquerque: University of New Mexico Press, 1978).

N. SCOTT MOMADAY. *The Way to Rainy Mountain* (Albuquerque: University of New Mexico Press, 1969).

_____. *House Made of Dawn* (New York: Harper & Row, 1968).

_____. *The Names: A Memoir* (New York: Harper & Row, 1977).

SIMON J. ORTIZ. "Always the Stories: A Brief History and Thoughts on My Writing." Unpublished speech at Institute of American Indian Arts, Santa Fe, NM (January 11, 1984).

_____. *Going for the Rain* (New York: Harper & Row, 1976).

KENNETH ROSEN (ed.). *The Man to Send Rain Clouds: Contemporary Stories by American Indians* (New York: Viking, 1974).

———. (ed.). *Voices of the Rainbow: Contemporary Poetry by American Indians* (New York: Seaver Books, 1980).

LESLIE MARMON SILKO. "Interior and Exterior Landscapes and the Pueblo Migration Stories." Unpublished essay.

———. *Storyteller* (New York: Seaver, 1981).

BRIAN SWANN AND ARNOLD KRUPAT (eds.). *Recovering the Word: Essays on Native American Literature* (Berkeley: University of California Press, 1987).

FREDERICK W. TURNER, III (ed.). *The Portable North American Indian Reader* (New York: Viking, 1974).

JAMES WELCH. *The Death of Jim Loney* (New York: Harper & Row, 1979).

———. *Winter in the Blood* (New York: Harper & Row, 1974).

Index